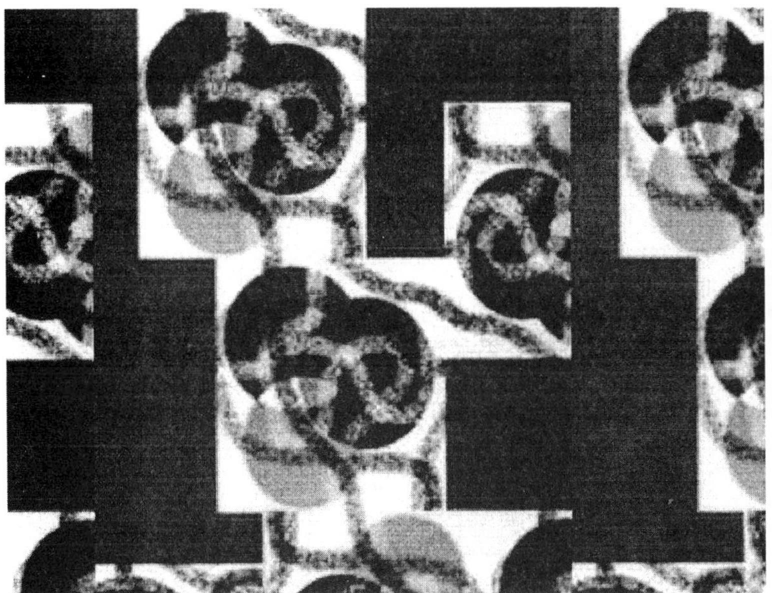

Preface and table of cont

About this workbook . . .

This workbook is designed to present a contemporary introduction to information technology. We cover a brief history of computers, hardware, software, databases, and network concepts. Building on this foundation, we explore the Internet and the World Wide Web, and how it supports modern e-business. Our concluding chapters cover Enterprise Resource Planning (ERP) systems, system development methodologies including the traditional SDLC "waterfall" model and several newer methodologies, business process reengineering, software maintenance, and business continuity planning.

Our goal in creating this workbook was to create a modern resource that could be used freestanding or with any standard text introducing information technology concepts. A workbook and web site combination is ideally suited to meeting this goal. In this arrangement, each component contributes what it is best suited to.

This workbook contains podcast scripts we composed to form brief "mini-lectures" on over 50 relevant topics, note-taking copies of our highly graphic lecture slides, study questions, tear-out "modular" homework assignments, and a sample final exam. Our web site, www.ambriana.com, is organized by "chapters" matching those in this workbook. The web site provides links to required readings, downloadable podcasts produced from the text here, and our PowerPoint lecture slides.

Our web site: an integral part of this workbook!

Visit **www.ambriana.com** for links to all of the readings associated with each chapter of this workbook, over 50 "mini-lecture" podcasts produced from the scripts contained in this workbook, and downloadable lecture slides. Our web site and podcasts are freely available to everyone! Your comments and suggestions are welcomed.

Jim Janossy and Laura McFall
DePaul University CTI, Chicago

For instructors

This workbook is organized into **10 chapters**. Each chapter is designed to be covered in about 7 to 10 calendar days, providing flexibility for various class meeting patterns ranging from 10-week quarters to 16-week semesters. Like this workbook itself, each chapter begins with a list of learning objectives that clearly indicates what it is intended that the student gain from the chapter. We find this helps students see the point of the effort to be expended on the chapter.

Each chapter includes original writing in the form of carefully prepared **podcast scripts** summarizing the web readings and highlighting important concepts. Each chapter also contains note-taking prints of the lecture slides. The slides themselves are available for free download at www.ambriana.com.

We designed this workbook to provide flexibility in your use of weekly or periodic homework. At the end of each chapter you will find **ten single-page exercise sheets**. Each of these sheets can be removed from the workbook by the student and completed. Exercise questions are based on assigned web readings, slide content, and podcast scripts. Questions are scaled from minimally challenging to the more challenging. Exercise sheets 9 and 10 in each chapter generally deal with questions of ethics, legality, or data privacy, issues that inherently arise in the modern use of information technology.

You can assign as many or as few of these exercise sheets as homework for a chapter as suits your instructional purposes, or combine them with assignments of your own creation.

Appendix A provides a sample final exam that draws content from the assigned readings, slides, and podcast scripts. Whether or not you choose to compose a final exam based on the format of the sample final, it's handy to point out the sample final to students at the beginning of the term for use as a study guide. The questions on the sample final indicate the type of knowledge each student is expected to gain during the course. Students should approach the course with the goal of developing a sound answer to each of the questions on the sample final.

How our podcasts were created

We often are asked how the podcasts for this workbook, freely available at www.ambriana.com, were created. As academics we are most concerned with accurate content, well formulated and clearly expressed. We are not trained announcers. We composed each scripts as a formal, carefully edited document, then used TextAloud software with a separately-licensed Scansoft voice to "read" each script to produce each of the .mp3 audio files available as podcasts. Each podcast is designed to provide concise coverage of one topic in a five to ten minute "mini-lecture." We feel this approach utilizes the advantages of downloadable sound files as a modern pedagogical tool to enhance the learning process. (Once you hear one of these podcasts, it makes sense to state that while both Laura or Jim watch a lot of British comedy, neither speaks with the inflections of a BBC-trained announcer...)

Table of contents

Chapter	Topic	Page

Learning objectives

This workbook, in combination with the web readings, web links, and podcasts available at www.ambriana.com, is focused on providing the knowledge to achieve these learning objectives:

- Learn how information is used by organizations to conduct business and solve problems
- Understand how information systems form an integral part of modern organizations
- Become knowledgeable of the technological aspects of information systems and the Internet, including a basic history of digital information storage, computing machinery, and database management
- Develop an understanding of basic information systems principles and concepts, the system development life cycle, and modern methodologies for system development
- Learn the basic aspects of e-commerce and how it functions in the modern business and consumer environment
- Gain familiarity with organizational processes relating to information system specification, acquisition, and support
- Develop an awareness of information technology security and issues of privacy, intellectual property rights, and the ethical uses of data.

Specific learning objectives for each chapter of this workbook are provided as a part of each chapter introduction.

Data, Information, and Systems

Electronic digital computers were commercialized more than 55 years ago, with the debut of the Univac computer in 1950. The first machines of this type were named "computers" because they were envisioned by their designers as fast "automatic" calculators. But since their invention and eventual mass-production their use has expanded far beyond performing calculations with numbers.

A better name for computers would now be **information appliances**. In the modern world computers are usually linked together to capture, access, and share information. The information, represented digitally, can consist of text, numbers, sounds, pictures, or combinations of these.

A business computer system consists of the machinery itself (**hardware**) and programs that direct the action of the machinery (**software**). The same computing machinery can do many different types of tasks as governed by software. The combination of hardware and software is called a **system**. Systems are usually named by the function they are designed to perform, such as a payroll system, an air traffic control system, or a manufacturing control system.

Three major classes of computer systems exist:

- personal productivity/office automation systems
- transaction processing systems
- reporting systems.

Each of these types of systems can be further broken down into various categories according to intended use. The main goal of this chapter is to help you understand these different types of computer systems and their purposes.

Learning objectives

Chapter 1
Data, Information, Systems

This chapter, in combination with the web readings, web links, and podcasts available at www.ambriana.com is focused on providing the knowledge to achieve these learning objectives:

- Understand the origin of electronic information technology leading to the development of the personal computer and the Internet, and some of the consequences of this development on computer users
- Understand the nature of transaction processing and batch processing
- Understand how transaction processing is a subset of interactive processing
- Recognize and be able to describe various types of office automation
- Understand the nature of data warehousing, the types of reporting capabilities it supports, and the reasons for its implementation
- Be able to distinguish between computer hardware and software
- Understand the basic levels of an organization and the differing time perspectives and information reporting requirements of personnel at each level
- Develop a basic understanding of some of the ethical issues involved in the application of information technology and the basic principles of ethics useful in addressing them, including the role of professional codes of ethics.

> ***Podcast Scripts – Chapter 1***
>
> These readings of the **Information Technology Workbook** are also accessible in audio form as .mp3 files for free download. Visit the workbook web site, www.ambriana.com. You can listen to this material on your own Ipod or .mp3 player or computer, read it, or both listen and read.

Podcast 1
TRANSACTION PROCESSING

We interact with computers in two ways. These ways are named transaction processing, and batch processing. Transaction processing requires interaction with a user, while batch processing can take place without a user being present. We'll consider batch processing in podcast 2. Here, we focus on transaction processing.

Transaction processing is **interactive**. It is used for immediate work. The computer responds to our requests as soon as we enter them, and each request is considered to be a transaction. For example, a patient is admitted to a hospital, and the event is immediately recorded. A customer order for an mp3 player is fulfilled, and inventory records are immediately diminished by one. A car with a radio frequency identification device, such as the Illinois Tollway I-Pass, drives through a toll point. The passage is instantly logged in the tollway database, and the amount of the toll is deducted from the driver's credit account. In all of these examples, transaction processing keeps business records current, to the moment transactions are transmitted to the computer system.

Automated teller machines are another example of transaction processing. Let's consider the operation of an ATM in detail to learn more about transaction processing. You insert your ATM card, enter your pin number, and the amount of money you desire, and the machine provides you with paper currency in this amount. But what appears to you to be one transaction, namely, withdrawing money, actually consists of several separate actions.

Here, at least five **actions** are involved in this transaction.

- One action involves a computer consulting your account balance, to determine if you have sufficient funds to withdraw the amount you have requested.
- Assuming you have sufficient funds, the next action is to check if the ATM has enough cash in it to provide the amount of money you have requested.
- A third action is the deduction of the amount you have requested, from your account balance.
- A fourth action is the deduction of a like amount, from the computer-stored information, about the amount of cash now in the ATM.
- The fifth action is the computer's command to the ATM to dispense the amount of cash you have requested.

Because transactions almost always involve multiple actions, transaction processing is formally defined in the following way:

Transaction processing is an unambiguous and independent execution of a set of actions on data contained in a database.

Since several actions must be completed successfully in processing a transaction, the actions related to a change in a database must be treated as a **single event**. If any of the steps fail, the effects of the entire transaction must be undone. In such a case, anything already affected by the transaction must be undone. This action is called a **roll-back**.

The requirement to be able to roll back all of the related data to a prior condition, is usually provided by a **relational database management system**. In theory, you could arrange transaction processing without a relational database, but it would be very difficult to achieve.

Relational database management systems are constructed with elaborate mechanisms to provide the type of roll-back capability that transaction processing demands.

Now, we must consider an additional formal definition. Data base systems providing the roll-back transaction processing capability, must pass what is known as the **ACID** test. Not the word, acid, but the several letters individually, which form an acronym.

"A" stands for atomicity. Atomicity means transactions are atomic. That is, all of the actions invoked by a transaction must either happen completely, or none of them are allowed to stand.

"C" stands for consistency. Consistency means that the transaction processing system must always be consistent with its own rules. No transaction can take place if an error is encountered in any action. For example, if one of several things in a database that must be updated cannot be accessed, the whole transaction must fail.

"I" stands for isolation. Isolating transactions means that other processes, that might be active in the computer, are not allowed to see the effect of only some of the actions caused by a transaction. Other processes either get to see the affected data before any changes caused by transaction processing, or they get to see the data after all changes have taken place. For example, anyone querying an airline reservation system for seating will see seats not reserved at that specific moment. But if two people simultaneously try booking the last seat on a flight, only one can succeed.

"D" sands for durability. The effect of completed transactions must be durable. This means that once the last seat on an airline flight is reserved, and the customer receives notification of the booking, that transaction is permanently recorded. Even if the computer system suffers a failure after the transaction was completed, a transaction processing systems must be able to retrieve the result of the transaction.

Let's review that acronym, ACID:

Atomicity, Consistency, Isolation, Durability

These are the principles of transaction processing.

Not so long ago, transaction processing systems were the exclusive domain of large, expensive, mainframe computers. Airline reservation systems, and banking systems, are examples of such transaction processing systems. In the mainframe era, transaction processing systems were generally unknown in the world of personal computers. Now, with personal computer system prices so low, and capabilities so great, many smaller organizations are discovering computer applications that benefit from transaction processing. Extending transaction processing capabilities to the **Internet** frequently appears as easy as building a **graphical user interface** and defining business logic for an application. And **e-commerce** needs effective transaction processing mechanisms, since they form the basis for e-commerce.

But meeting the ACID requirements of transaction processing on the web, is especially complex. It involves special security arrangements including encryption, since the web, and transmissions made using it, are accessible to a very large audience.

Three challenges must be addressed when any large scale transaction processing system is developed, on any type of computer, using the web or local communication arrangements. These challenges are:

1. Handling hundreds, or even thousands, of users at the same time.
2. Allowing many users to work on the same set of data **concurrently**, and
3. Handling errors in a safe and consistent manner.

We'll consider how hardware and software solutions are achieved, to overcome these challenges, in the following podcasts of the Information Technology Workbook. Make sure you read the material at the web links provided at the workbook web site, at www.ambriana.com.

Podcast 2
BATCH PROCESSING

We interact with computers in two ways, which are named transaction processing, and batch processing. Whereas transaction processing requires interaction with a user, batch processing takes place without a user being present. Here we focus on batch processing. Batch processing is particularly useful for operations that require the computer or a peripheral device for an extended period of time. Batch jobs can be stored up during working hours and then executed overnight, or whenever the computer is not heavily loaded.

Once a **batch job** begins, it continues until it is done, or until a serious error occurs. An example of batch processing is the way that credit card companies process their monthly billings. Customers don't receive a bill for each separate credit card purchase, but rather one monthly bill, for all of that month's purchases. The bill is created through batch processing. All of the data about each purchase is collected and stored in a database. The bills for many thousands of customers are prepared in a batch, at the end of each billing cycle. The program that works its way through the data for each customer, then goes on to the next customer, may run for several minutes or even hours, without ever having any interaction with a human being. Many jobs can be more economically processed in this way:

- Month end reports are generally produced by batch processing.
- Payroll checks produced at the end of a pay period, are created with batch processing.
- The electronic transactions documenting telephone calls are stored in a computer until month end, when they are matched against customer data for the production of monthly bills.

Let's consider the frequency with which a given batch process is initiated. The **batch cycle** is the period of time between one running of a batch process, and its next running. Sometimes the batch cycle is as frequent as daily. A transaction processing system may be used to gather data and store it, and the stored data then used as input to a batch program at the end of the day. For example, point of sale transactions, generated by the laser scanners at grocery check out counters, can be stored throughout the day. Uploaded at night to a central computer, they serve as input to a batch program, which uses them to update a database, and compose and transmit purchase orders for the replenishment of stock. These automatically generated orders, will cause the replacement of inventory, sold that day.

Let's consider an environment you are probably very familiar with, the world of personal computers, and word processing. Direct interaction between a user and document is actually a form of **event driven programming**, which, like transaction processing, is a subset of interactive processing. But batch processes do exist in the personal computer environment.

When you compose a document using a word processor, and want a hard copy of the document, you invoke a print function. You don't separately request each page of a document to be printed. Instead, the print function you invoke prints all pages, without further manual involvement. Such a print function is actually a batch process. Once initiated by a user, printing would run on its own, to completion. If the document was a lengthy one, the printing process might take several minutes, but it would proceed without your intervention.

In the next several podcasts of the Information Technology Workbook, we'll consider office automation. We'll look at how common office tasks are supported by information technology. We'll also examine how some reporting is handled in the modern environment, with the techniques of data warehousing. And finally, we'll discuss how concepts of ethics have been addressed, in the application of information technology, through the development and publishing of codes of ethics.

Podcast 3
OFFICE AUTOMATION

The term **office automation** refers to the use of computer systems to support a variety of office operations, such as word processing, accounting, and e-mail. Office automation almost always implies the use of a network of computers, ranging from as little as two computers, linked together, to a larger **local area network** or **wide area network**. Personal computer based office automation software has become an indispensable part of office management, all over the world:

- **Word processing programs** have replaced typewriters.
- **Spreadsheet programs** have replaced handwritten accounting ledgers.
- **Database programs** have replaced written or typed data lists, such as inventories, customer, and personnel lists.
- **Calendar and scheduling programs** have replaced paper appointment books.

Beginning with crude word processing and spreadsheet programs in the 1970s, office automation programs have become much more sophisticated. These programs empower office workers by enabling them to complete tasks at their desks, that once had to be sent offsite for specialists to perform.

For example, **desktop publishing programs** allow novices to produce professional quality publications, where once even the simplest of typesetting tasks had to be sent to professional printers. Database and spreadsheet programs, running on high-powered personal computers, make it possible for users to input, store, and output data in flexible ways. Such data management tasks once would have been possible only on large **mainframe computers**, operated by specialist programmers.

A typical suite of office automation software may include the following kinds of programs:

- a word processor, such as Word, or WordPerfect
- a spreadsheet, such as Excel
- a database, such as Microsoft Access
- a presentation aid, such as PowerPoint
- e-mail, such as Outlook, or Novell Groupwise
- an internet browser, such as Internet Explorer, or FireFox

Office automation software also includes other programs less widely used, but ofthe personal productivity variety:

- a project manager, such as Microsoft Project
- a desktop publisher
- a file manager
- an internet publishing tool, such as Microsoft FrontPage, or DreamWeaver
- a personal organizer
- contact management software, which assists people who call on and market to potential customers.

Office automation software is often provided with the appropriate hardware, as part of a computer purchase agreement. Many magazines and web sites can help people choose office automation software.

Modern office automation software is very powerful. Skilled users can develop very sophisticated products using it. For example, books, brochures, forms and newsletters can now be produced in the office. Composing, editing and printing all correspondence can be done in a standard, corporate style, including pictorial and graphic elements. Employee records can be stored and accessed. Inventories of products can be maintained. Corporate websites can be created and updated. Documents can be scanned into electronic form and shared by many people.

Once installed, office automation software is often upgraded. Upgrades are typically released every year or two by the vendors of the most popular products. **Upgrades** generally provide more powerful versions of existing features, new features, and correct known problems. Upgrading office automation software can be expensive and time consuming, especially if the organization is large, and maintains hundreds, or thousands, of personal computers. The management of an organization must judge, if and when office automation software should be upgraded. If the current office automation software is performing acceptably, there may be no compelling reason to upgrade it as soon as the software vendor releases a new version.

One reason for delaying an upgrade is to wait for other users to identify problems with the new software. Problems in new software, detected prior to the release of a new version, are corrected by the installation of adjustments known as **patches**. Usually, patches become available to adjust the operation of new versions of software, within weeks of the time new software becomes available.

Pressure usually exists on an organization to upgrade its office automation software, as clients or customers move to an upgraded version. Upgraded software is often **"backwards compatible."** This means that the new version can read documents and files created under the old version. But old software versions often can't read files created using the new software version. When this becomes a significant problem, it is time to upgrade.

Office automation software provides the capability to locally perform many of the regular day-to-day information reporting tasks of an organization. Office automation software may also make it possible, to meet unanticipated, one time data extraction and reporting requirements.

In the next podcast of the Information Technology Workbook, we'll examine how some of the more demanding reporting requirements of an organization, especially those of a historical or trend nature, are fulfilled using the techniques of data warehousing.

Podcast 4
DATA WAREHOUSING

A **data warehouse** is a collection of information gathered from one or more transaction processing databases. A data warehouse is used to create information outputs that support business analysis, and decision making. In the modern environment, this analysis and decision making reporting is sometimes called **business intelligence**.

Why do data warehouses exist, when transaction processing systems already capture and store data? Couldn't we just gather data from a transaction processing database, and use it to produce reports? The answers to these questions reveal why data warehousing was invented, and is a growing phenomenon.

By 1990, large organizations using computer systems began to need more information about their business operations than simply day to day reporting, to improve productivity and competitiveness. These organizations found that traditional transaction processing systems could not readily provide data for analytical and reporting purposes of the type they needed. Completing reporting requests could take days or weeks, using antiquated reporting tools. And often, the data needed for historical analytical or trend reporting was not retained, or very difficult to access, in transaction processing databases. The data structures in transaction processing databases are designed to optimize the type of updating of individual data items that are the substance of transaction processing. Such a database organization is not conducive to the retrieval of data for historical and trend reporting.

The solution to the problem of data access for reporting was to copy the data in a transaction processing database to a second database, designed not for efficient updating, but for efficient data analysis and reporting. Thus the concept of data warehousing was born.

Data warehousing is not usually meant to provide live current data. A data warehouse is periodically loaded from live data. Its contents then remain static until the next load of live data is received. The live data that has changed or been added to the transaction processing database, since the last load of the warehouse, is added to the warehouse. As part of the loading process, live data coming into the warehouse is validated and standardized in format.

Unlike a transaction processing database, the data warehouse keeps a **history of changes**, so that the data at any point in time in the past can be retrieved.

The data warehouse is a place where data needed for **strategic management reporting** can be stored in a form optimized for rapid retrieval. The key word here is the word strategic. Most executives are less concerned with day-to-day business operations, than with an overall look at the functioning of the business over a period of time. That is, executives are generally interested in **trends**. In order to map trends, historical as well as current data is needed.

Data warehouses solve additional reporting problems as well. The quantities and types of databases in many organizations has increased dramatically in the last two decades, especially as mergers and business acquisitions have combined what were previously separate firms. Many large businesses now find themselves with different formats of data, scattered across multiple different types of computer systems, supported by several different kinds of technology. It is often difficult to acquire and use data from multiple sources.

A central idea within data warehousing, is to take data from multiple sources, and place it in a **common database**, and to provide a **common reporting tool** to access it. This way, transaction processing databases can operate on whatever computer system is most efficient for the purpose. But data needed for reporting is copied to a common database, and accessed using a common language.

Modern data warehouses take this a step further, by standardizing the nomenclature and format of data from various contributing systems, and establishing definitions for each data item. This allows decision support information extraction to be accomplished rapidly and flexibly, without any adverse effect on transaction processing response time.

Data warehouses frequently hold large amounts of information. In order to make reporting by separate business units more convenient, a data warehouse is sometimes subdivided into smaller units called **data marts**. A data mart allows easier reporting for a specific business area, by keeping relevant data together in one location. Like data warehouses, data marts contain a snapshot of operational data, that helps executives formulate strategy based on the analysis of trends.

Specialized software tools exist to create and report from data warehouses. A tool known as an **Extract, Transform, and Load tool**, is often used to obtain data from various sources, and load it to a data warehouse.

An Extract, Transform, and Load tool is also known by the acronym, **ETL**. An ETL speeds the process of extracting data from one or more sources, transforming it to fit business analysis needs, and periodically loading it into the warehouse. Some data warehouses deal with huge amounts of activity. An efficient ETL is vital in these situations, to handle millions of rows of data in every night's loading process.

Specialized reporting tools also exist in the data warehouse environment. These reporting tools provide rapid means to access the data structures of the data warehouse, and to list, tabulate, and depict counts, averages, and trends in multiple graphic forms.

Two designs for data warehouses exist. One type of data warehouse is a **relational database**, similar to a transaction processing database, but with data structures and indexes designed to facilitate reporting. This type of data warehouse is associated with the writings and designs of **William Inmon**.

The other type of data warehouse follows what is known as the **dimensional model**. Dimensional data warehouse models have been popularized by **Ralph Kimball**. In a dimensional data warehouse, descriptive data such as date, the zip code in which a given customer lives, the gender of the customer, and the customer's age, are housed in tables called **dimensions**. (Zip code is often used to categorize people as to socio-economic status and income level, using publicly available census data.) Descriptive data in dimensions is also sometimes called **attributes**.

On the other hand, in a dimensional data warehouse, data about events such as purchases, the usage of a resource, or any other item about which reporting is desired, is loaded to a fact table. **Fact tables** contain data known as **metrics**, which are the measure of purchase or resource usage. A metric is a property that can be summed or averaged, such as the quantity of a given item purchased. Metrics are typically received in the transactions that bring new activity into the warehouse, such as purchases recorded at a grocery store checkout counter.

Business intelligence reports are generated from a dimensional data warehouse, by accessing descriptive items such as date, and product type. By specifying date and product type as a selection criteria, the appropriate rows in a dimension and a fact table are quickly accessed, and counts, sub totals, and totals generated. By specifying multiple dates, trend reporting can rapidly be accomplished, for example, to determine how sales of a given product vary over a time period.

Both relational and dimensional data warehouse models have a following, and each are appropriate to a different form of business intelligence reporting. Both data warehouse models are used by organizations large and small.

The largest data warehousing applications contain huge volumes of data, up to hundreds of billions of rows of information. It would be difficult, or impossible, to produce meaningful reporting from such large repositories of data, without data warehousing tools and techniques.

You should now have an overall conception, of the ways in which modern organizations employ information technology, in productive ways.

Information technology, as any tool, can be applied in ways that serve purposes mutually beneficial to organizational productivity and to the customers and clients of an organization. Because this technology often deals with sensitive and private data, it is important to use information technology in the appropriate ways. Inappropriate ways to use Information technology need to be identified, and avoided.

As a guide to information technology professionals, many organizations compose and publish **codes of ethics**. In the next pod cast in the Information Technology Workbook, we'll consider what the field of ethics has contributed, to an understanding of ethical uses and implementation of information technology.

Podcast 5
CODES OF ETHICS

What are the ethical implications of information technology? Is **privacy** being lost to massive databases and electronic spying? Is **intellectual property** in jeopardy, through the quick and easy sharing of digitized products? What do we do about the **digital divide**, the dividing line that modern technology is forcing, between those able to interact with computer technology, and those unable to do so? This has the effect of disenfranchising whole segments of the population, due to economic, geographic, or other similar circumstances, often beyond an individual's immediate control. These issues are all related to the general question of ethics.

Ethics deals with principles that can guide individuals in their behavior in civilized human society. Let's begin by examining three widely accepted **ethical principles**.

A primary ethical principle, independently originated in many civilizations, since ancient times, is the **Golden Rule**. The Golden Rule states that you should do unto others, as you would have them, do unto you. Stated in the reverse, the golden rule admonishes you not to take an action towards any other person, that you would not appreciate someone taking towards you.

A second important guiding ethical principle is termed **universalism** If an action is not right for all people, it is not right for any person. This homes in on the issue of consistency, fairness, and the elimination of discriminatory actions.

A third important ethical principle is called the **perfect information rule**. This rule provides a test of a proposed action, to help determine if the action is ethical. Would you like the action to be reported on the front page of the newspaper? Perfect information about the action, means everyone else will know that you did it. Actions that you feel must be hidden from everyone else, are likely to be unethical.

Expectations of ethical behavior have increased over the past several years. More and more, customers, clients, and employees seek to have their business partners, clearly define the ethical ground rules, that govern their behavior. Many business organizations have developed and published **codes of ethics**, to make their position on professional conduct clear and concise.

Codes of ethics are developed and published for the following five reasons:

1. To identify acceptable behaviors.
2. To promote high standards of practice.
3. To provide a benchmark for members to use for self evaluation.
4. As a vehicle for occupational identity.
5. As a mark of occupational maturity.

Codes of ethics are created by organizations based on the general principles of ethics. These codes identify specific actions that are considered ethical, and those actions that are considered unethical. The codes also specify sanctions for violations of the ethical code.

Codes of ethics are mutually beneficial. They benefit an organization's members, and the people with whom they deal. Let's look at two organizations active in the computer and electronics industry, each of which have established a code of ethics.

The **Institute of Electrical and Electronic Engineers**, the IEEE, is a voluntary, cooperative organization of electrical and electronics engineers, and computer professionals. The IEEE has established and published its code of ethics, by which all of its members must abide.

The **Association for Computing Machinery** is the oldest organization of computer professionals. The ACM has established a code of ethics, and requires its members to know it, and to act in accordance with it.

Members of these organizations must conform to their ethical rules, which guide professionals in their conduct. Each member has a stake in maintaining compliance with the code of ethics. Codes of ethics also enhance the sense of community among members of an organization or business, defining their common values and mission.

The exercise of developing a code of ethics is, in itself, worthwhile. It causes a large number of people to think about their mission, and the obligations they have with respect to others.

You have now completed all of the podcasts for Chapter 1 of the Information Technology Workbook. Make sure you review the material at the web links for Chapter 1 provided at the workbook web site, at www.ambriana.com.

Lecture Slides 1

Introduction to Information Technology

Data, Information, and Systems

(C) 2006 Jim Janossy and Laura McFall, Information Technology Workbook

Slide 1- 1

Course resources

PowerPoint presentations, podcasts, and web links for readings are available at **www.ambriana.com** > IT Workbook

Print slides at 6 slides per page

Homework, quizzes and final exam are based on slides, lectures, readings and podcasts

Slide 1- 2

Main topics

- Brief history of information technology
- Nature of information systems
- Transaction processing, batch processing, and data warehousing
- Organizational structure and information system integration
- Ethical implications

Slide 1- 3

Brief history of information technology

Slide 1- 4

Beginnings of information technology

- Electronic computers invented 1939-46
- Commercialized 1950 by Univac, IBM
- Early computers were **mainframes** built with vacuum tubes to do computations
- Very expensive, very large, not very powerful

Slide 1- 5

Information technology's early era: 1950 - 1960

- Acquired by large business and government organizations
- Dedicated to accounting and records - keeping **batch** applications
- Nothing interactive about them: users fed data in with punched cards, got paper **reports** back

Slide 1- 6

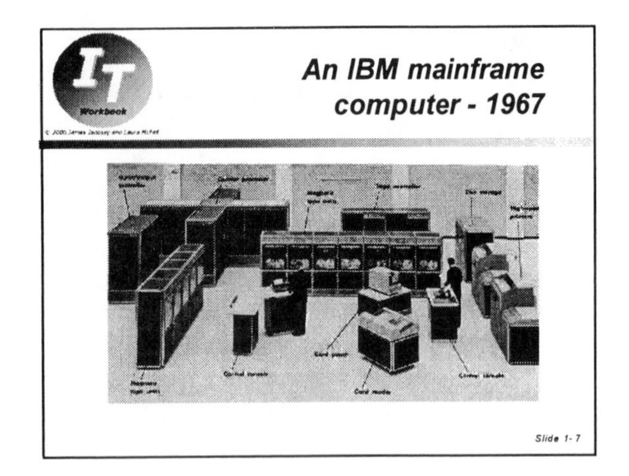

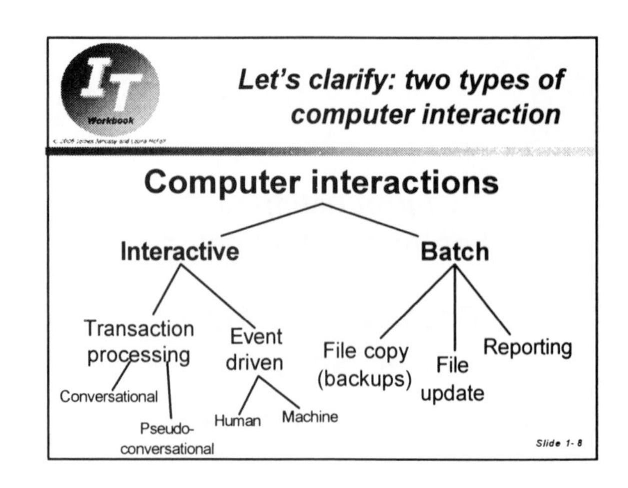

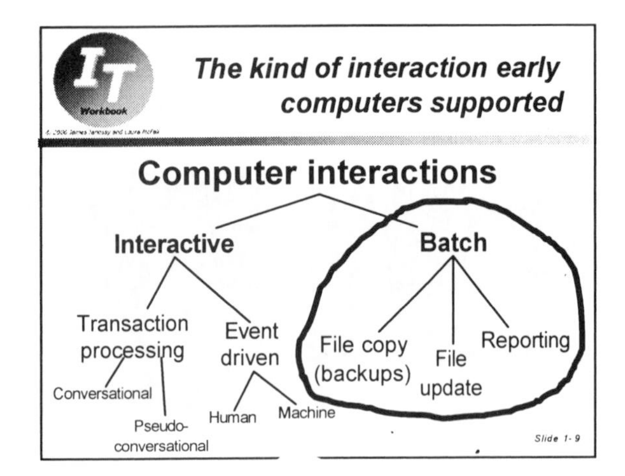

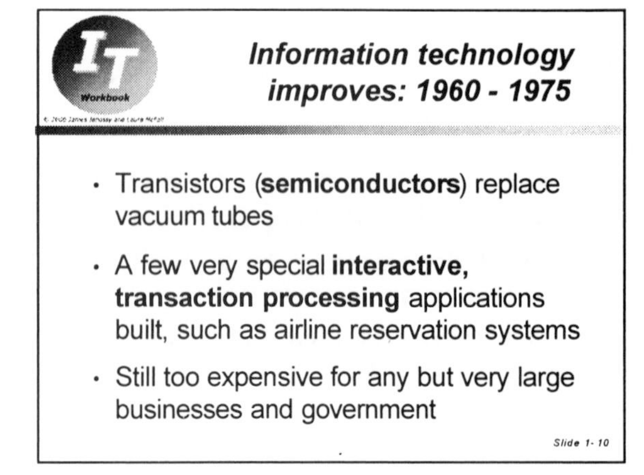

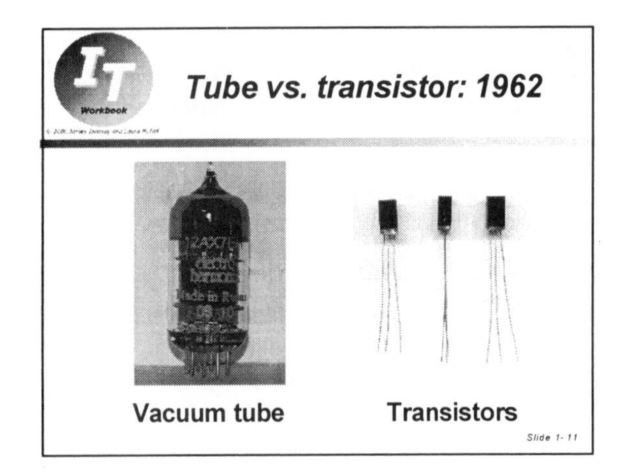

IT

Workbook

Interactive applications begin to grow: 1975 - 1990

- **Minicomputers** smaller than mainframes; cheaper but still $100,000 and up!
- Highly interactive
- **Transaction processing** environment (**C**ustomer **I**nformation **C**ontrol **S**ystem) grafted onto batch mainframes
- **PC** developed 1976, "legitimized" by the IBM PC in 1981; "smart terminals"

Slide 1- 12

Information technology before the Internet: 1990s

- **PC** and **Mac** become common in many organizations
- **Office automation** applications abound
- Computers in **local networks**
- Collaborative **client/server** interactive applications split processing between central computer and desktop

Slide 1- 14

Information technology after the Internet: 1995 and later

- **Internet** invented 1970-80s, but it's a pipeline without a graphic interface!
- **WWW** invented 1989, develops through 1994, then bursts onto the scene; HTML
- Computers (large, minicomputer, and desktops) now connected and accessed across the world by WWW and Internet

Slide 1- 16

Information technology now

- **Globalization:** trade barriers drop, markets are international
- WWW facilitates **outsourcing**; "brain work" as well as manufacturing can now be shipped "offshore" (overseas)
- Documents, music, pictures, movies become **digital products**

Slide 1- 17

Consequences of information technology

- Business is speeding up
- New markets are now accessible
- Consumers have more information so competition increases and drives prices down
- Ethical implications...

Slide 1- 18

Ethical implications of information technology

Ethical implications of information technology

- Is **privacy** being lost to massive databases and electronic spying?
- **Intellectual property** in jeopardy through quick/easy sharing of digitized products?
- **Disenfranchisement:** lack of access
- **Digital divide**
- What is the basis for ideas of ethics?

Slide 1- 20

General ethical principles

- **Golden rule:** Do unto others as you would have them do unto you.
- **Universalism:** If an action is not right for all situations, it is not right for any.
- **Perfect information rule:** Would you like the action to be reported on the front page of the newspaper?

Slide 1- 21

Organizations' codes of ethics

- Created by organizations based on the general principles of ethics
- Identifies **specific actions** of members of the organization that are considered ethical, and those that are not
- Specifies **sanctions for violations** of the ethical code

Slide 1- 22

Information systems

Slide 1- 23

What is an information system?

- A set of **interrelated components** that work together to **collect**, **store**, **transform**, and **distribute** information
- Information systems are found at all levels of an organization
- Consist of **hardware**, **software**, **databases** and **network** arrangements

Slide 1- 24

Two basic modes of operation

- **Transaction processing:** interactively processing a purchase, deposit, withdrawal, registration, reservation
- **Batch processing:** retrieve, summarize, select, calculate new information from stored data, or apply bulk updates; "reporting"

Slide 1- 25

Data warehousing

- Eliminates batch reporting from the same database as transaction processing systems use
- Covers **longer time periods** for trend analysis than data is maintained for in transaction processing systems
- Makes data **easier**to access for reporting

Slide 1- 26

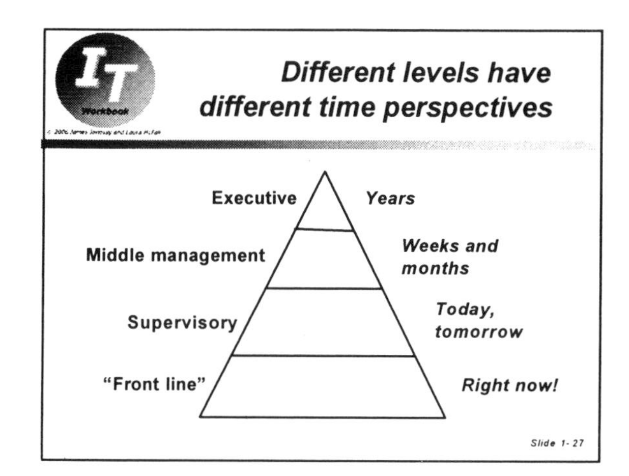

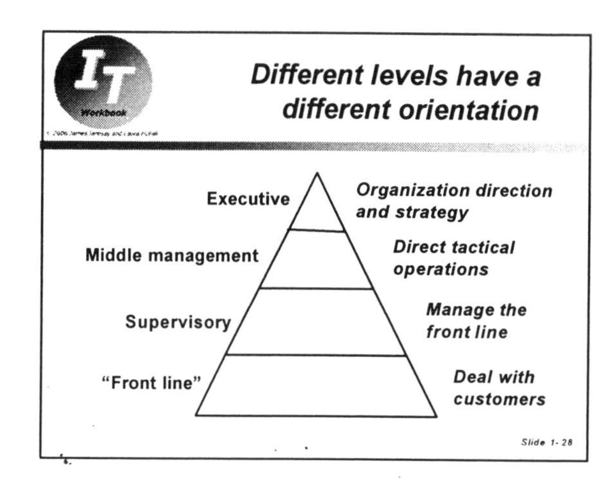

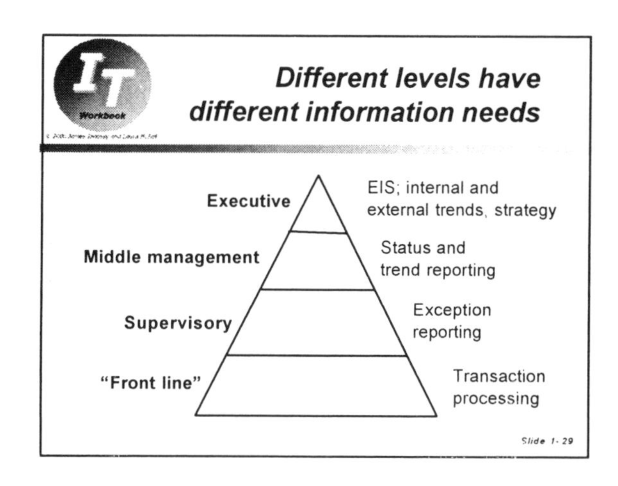

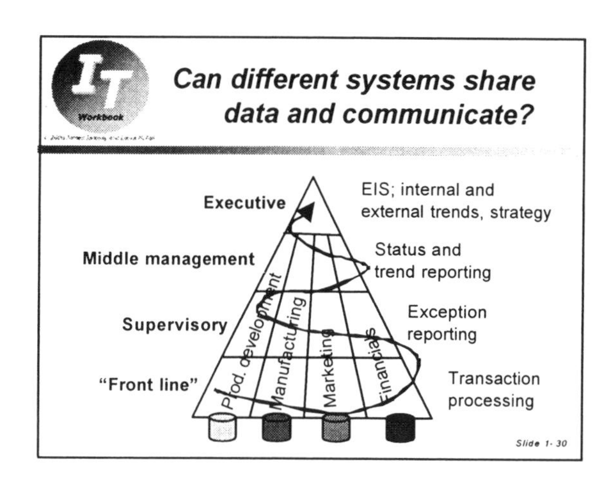

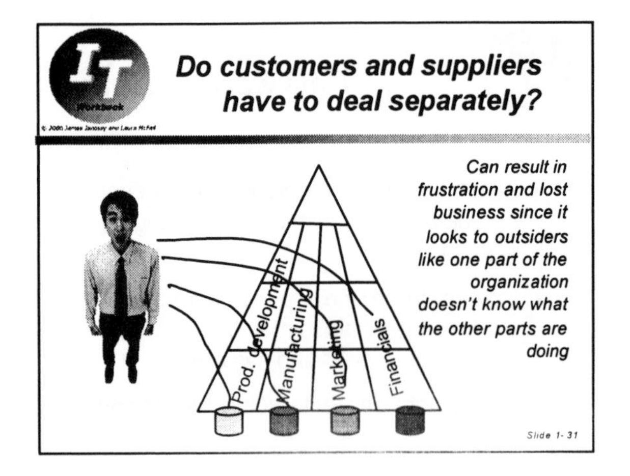

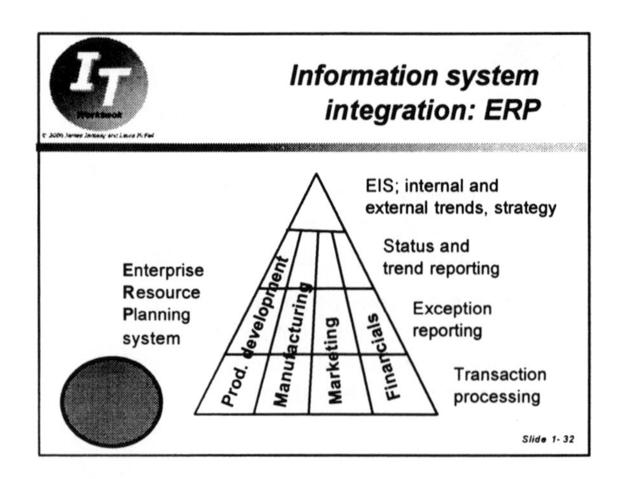

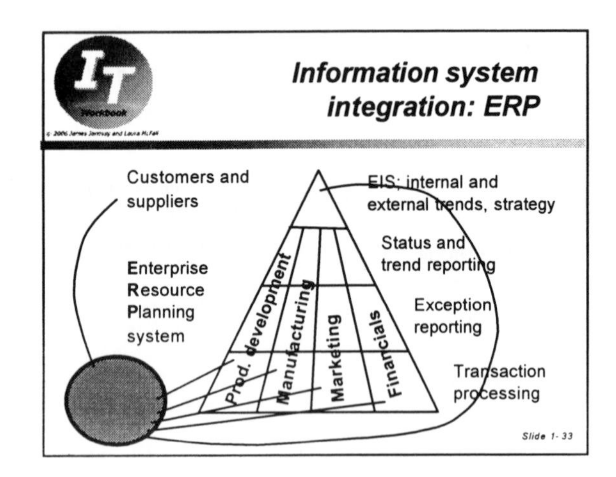

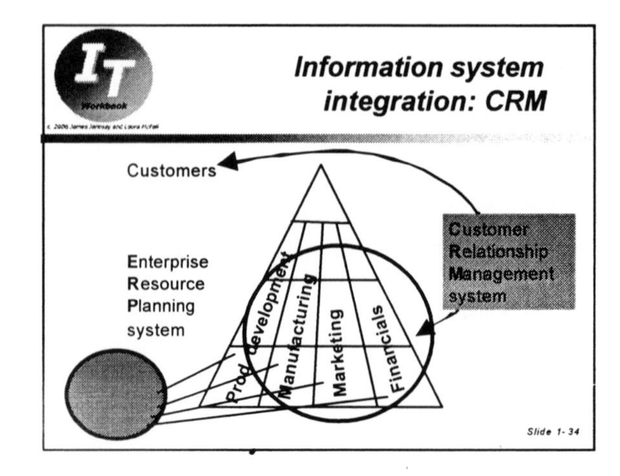

ERP systems

- Ambitious, complex, expensive
- Implementation has had both successes and failures
- SAP, PeopleSoft, Oracle, and others
- We'll look ERP systems more closely in Chapter 8

Slide 1-35

IT Workbook Assignment 1.1

Student name:

Write your answers to the questions below within the box. In each case, please choose your words carefully to answer the specific questions. Avoid simply copying passages from your readings to these answers.

Describe what **transaction processing** is:

1)

Describe what a **database** is:

2)

Describe the **relationship** that exists between transaction processing and a database in modern business organizations:

3)

IT Workbook Assignment 1.2

Student name:

Write your answers to the questions below within the box. In each case, please choose your words carefully to answer the specific questions. Avoid simply copying passages from your readings to these answers.

Describe what **batch processing** is:

1)

Describe some of the **roles** that batch processing plays in modern business organizations, giving two specific examples of its use:

2) Describe some roles:

3) one example:

4) another example:

IT Workbook Assignment 1.3

Student name:

Write your answers to the questions below <u>within the box</u>. In each case, please choose your words carefully to answer the specific questions. Avoid simply copying passages from your readings to these answers.

Describe the nature of **transaction processing** and how it relates to **interactive processing**:

1) interactive processing:

2) how transaction processing relates to interactive processing:

Provide a definition that answers the question **"what is a computer program?"**

3) A computer program is…

IT Workbook Assignment 1.4

Student name:

Write your answers to the questions below within the box. In each case, please choose your words carefully to answer the specific questions. Avoid simply copying passages from your readings to these answers.

Describe what **office automation** is and name some **common tasks** that are handled by office automation programs:

1) Office automation is…

2) Some common tasks office automation handles:

Identify and describe **what type of processing** is involved when you use a spreadsheet to keep track of expenditures as you make them, each time seeing what the spreadsheet tells you about how much money you have spent since the beginning of the current month:

3)

IT Workbook Assignment 1.5

Student name:

Write your answers to the questions below *<u>within the box</u>*. *In each case, please choose your words carefully to answer the specific questions. Avoid simply copying passages from your readings to these answers.*

In connection with information technology, define what an **"end user"** is:

1)

What is an application? Provide a definition for it relevant to information technology, and discuss whether an application can involve interactive processing, transaction processing, batch processing, or a combination of these.

2)

IT Workbook Assignment 1.6

Student name:

Write your answers to the questions below <u>within the box</u>. In each case, please choose your words carefully to answer the specific questions. Avoid simply copying passages from your readings to these answers.

Identify and discuss **what type of processing** is involved when a bank prints 40,000 monthly credit card statements and bills between midnight and 6:00 AM the next day for mailing to credit card holders, and why this type of processing is used for this process:

1) Identify and discuss what type of processing is involved:

2) why this type of processing is used in this instance:

Both **computer memory** and **mass storage devices** house information in digital form. Compare and contrast these forms of information storage, noting their similarities and differences:

3) computer memory ca be described as:

4) mass storage devices are described as:

5) how computer memory and mass storage devices are similar:

6) how computer memory and mass storage devices are different:

IT Workbook Assignment 1.7

Student name:

Write your answers to the questions below within the box. In each case, please choose your words carefully to answer the specific questions. Avoid simply copying passages from your readings to these answers.

Describe what the word **"atomic"** means in connection with transaction processing systems:

1)

Reporting is usually a batch process that selects, formats, and presents information to an end user from a database that accumulates information from transaction processing. Describe how using a **data warehouse** differs from reporting directly from a database accessed by a transaction processing system, and (b) list three reasons why a data warehouse is often used.

2) how data warehousing differs from reporting directly from a transaction database:

3) three reasons why a data warehouse may be implemented:

IT Workbook Assignment 1.8

Student name:

Write your answers to the questions below within the box. In each case, please choose your words carefully to answer the specific questions. Avoid simply copying passages from your readings to these answers.

Describe in general how a dimensional data warehouse is organized:

Identify the different levels of a typical business organization, the time perspective of personnel at each level, and the type of reporting that best serves each level:

a) levels:

b) time perspective at each level:

c) type of reporting suited to each level:

IT Workbook Assignment 1.9

Student name:

Write your answers to the questions below within the box. In each case, please choose your words carefully to answer the specific questions. Avoid simply copying passages from your readings to these answers.

www.microsoft.com/smallbusiness/resources/technology/security/danger_danger_5_tips_for_using_a_public_pc.mspx is a web page provided by Microsoft. In this article at this page, Kim Komando discusses the **dangers of using a public computer**, such as a computer in a cyber cafe, for sensitive types of transactions. Read this article and then answer the following questions:

1) what does a **keystoke logger** program do?

What are the five things that Komando suggests you do or don't do when using a public computer?

2)

3)

4)

5)

6)

IT Workbook Assignment 1.10

Student name:

Write your answers to the questions below within the box. In each case, please choose your words carefully to answer the specific questions. Avoid simply copying passages from your readings to these answers.

Describe the simple basis for almost all systems of **ethics**, and **three questions** that help identify if a given action is ethical:

1) the basis for almost all systems of ethics:

2) one question that helps identify if a given action is ethical:

3) a second question that helps identify if a given action is ethical:

4) a third question that helps identify if a given action is ethical:

Identify what a **code of ethics** is, and why organizations create codes of ethics:

5) what a code of ethics is:

6) why an organization creates a code of ethics:

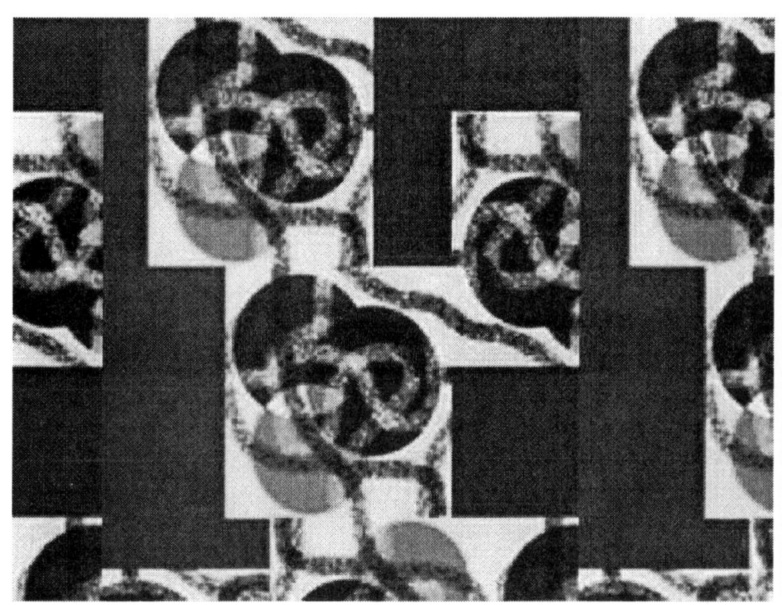

Computer Hardware

The machinery of a computer is divided into several units. While computers vary greatly in physical size, information manipulation capabilities and speed, all digital computers share these generic parts.

The computer must have some form of **input/output device** where information can enter and leave it in a recognizable form. A keyboard, CD/ROM drive, and internet connection are the most common input devices. A display screen and printer are common output devices.

The computer must also have some form of local information storage called **random access memory** (RAM) and usually a **hard disk**. These form the primary and secondary information storage resources of the computer. And the computer must have a "thinking circuit" known as a **central processing unit** (CPU). The CPU is where the circuits that actually perform arithmetic and logical comparisons reside.

Different "classes" of computers exist. **Desktop and laptop computers** are the most common, and often contain input, storage, and output devices all in one enclosure. A personal digital assistant (PDA) contains rudimentary versions of these basic components. Larger computer systems such as **workstations**, **minicomputers**, **mainframe computers**, and **supercomputers** also exist. These classes of computing machinery are faster and have greater storage and processing capacity ("throughput") than desktop computers, and they support multiple concurrent users. Computers represent text information as series of on-off “bits” in a standard encoding system known as ASCII, which we explore here.

This chapter is designed to help you understand the parts common to every computer and the characteristics of the different classes of computers.

Learning objectives

Chapter 2
Computer Hardware

This chapter, in combination with the web readings, web links, and podcasts available at www.ambriana.com is focused on providing the knowledge to achieve these learning objectives:

- Be able to distinguish the various level of computers, their purposes, and differing capabilities
- Be able to identify and describe the four eras of computing equipment that have existed since ancient times, and the basic operation and impact of the stored program computer design
- Recognize and describe the purpose of the major components of a modern desktop computer, and how different types of computers displays and printers operate, and the tasks to which each type of printer is best suited
- Understand the various ways that digital information has been and is stored, within and external to the computer, and the advantages of modern storage techniques
- Understand terminology commonly associated with modern computers and data storage, such as kilo, mega, giga, and tera, and how these are applied to memory, tape, disk, CD/ROM and DVD media
- Understand what digital bits and bytes are, and how text is encoded digitally in the ASCII code set
- Understand the vulnerabilities of sensitive data and the consequences of unauthorized data disclosure, including the potential for identity theft and the way in which encryption can be used to minimize this potential.

Podcast Scripts – Chapter 2

These readings of the **Information Technology Workbook** are also accessible in audio form as .mp3 files for free download. Visit the workbook web site, www.ambriana.com. You can listen to this material on your own Ipod or .mp3 player or computer, read it, or both listen and read.

Podcast 6
THE HISTORY OF COMPUTERS

Although many historians trace the roots of computing to Charles Babbage, in the mid-nineteenth century, four periods of development can be mapped for information technology, dating back to thousands of years before present.

The earliest **pre-mechanical age** was marked by the beginning of writing and alphabets. Paper and pens were the input technologies, and books and libraries were the permanent information storage devices of the day.

The major technological advance that moved the world into the second period of computer development was the invention of the printing press. From 1450 to 1840, the **mechanical age** saw the development of the first computers, which were actually people working with or computing numbers.

Toward the end of the mechanical age, a Frenchman named Joseph Jacquard invented a "automatic" weaving loom based on punched cards that stored weaving patterns. When the punched cards were run through the weaving machinery, even an unskilled operator could create complex patterns, over and over again, because the pattern of holes in the cards controlled the weaving process. You might say that the looms used for weaving were "programmed" using the cards.

Charles Babbage undoubtedly noticed Jacquard's technology during the early 1800s and conceived of a mechanical computing device called the **Analytical Engine**, a general-purpose calculating machine. Punch cards held the machine's instructions. Unfortunately, the complex mechanism required for the Analytical Engine was beyond the manufacturing capabilities of the time.

The third, or **electromechanical** age of computer development extended from 1840 to 1940, and saw the creation and development of the electric telegraph and Morse code, the telephone, and the electric light. The most primitive of these electrical technologies were put to use by the U.S. Census Bureau in 1890. By that time it was taking seven years to hand tabulate the results of the eighteen eighty census. The bureau estimated that it would take more than ten years to tabulate the eighteen ninety census, clearly an impractical phenomenon since the census is made every ten years. So the bureau sponsored a competition to find a faster approach.

Herman Hollerith, a patent office worker, won the contest. His machine used punch cards on which the result of a person's response to census questions was coded as a series of punched holes. When the reading machine examined a card, the holes sent electrical signals to mechanical counters and tabulated the entire 1890 census in only six weeks. Hollerith later founded the Tabulating Machine Company, which became International Business Machines, IBM, in 1924. Punched cards and electrical tabulating equipment were widely used by large business organizations from the early 1900's through the 1970's.

The incandescent light bulb, which is a glass globe containing a vacuum and a glowing wire, was the precursor of the electronic **vacuum tube**. The vacuum tube was the first electronic component, and was invented in 1904. It could be used as an amplifier of electrical signals, or an oscillator, producing radio waves. It was later discovered that the vacuum tube could also function as a switch, which made it possible to use it to represent and manipulate bits, the coding scheme of digital computers.

The fourth and current period of computer development is known as the **electronic age**. During World War II, from 1939 to 1945, the first electronic digital computers were developed, using a combination of electromechanical and electronic parts. In 1943, British code breakers developed a computer named Colossus. Its job was to help decipher what the Nazis considered to be their unbreakable code, Enigma. Colossus was able to decrypt Enigma encoded messages at the rate of around five thousand characters per second. After the war, the only ten Colossus machines that were ever built were destroyed as a security measure, being broken into pieces no larger than a man's hand, by order of British Prime Minister Winston Churchill.

At the same time, the U.S. military was developing new artillery as quickly as possible. Extensive mathematical aiming tables were needed for each new type of cannon, showing settings for a variety of elevations, distances, and weather conditions. A group of women, who were called "computers," developed these tables at the University of Pennsylvania. In an effort to speed the development of these highly complex tables, the United States Army gave money to the University to develop and build an electronic calculating machine. Their efforts led to the Electronic Numerical Integrator and Computer, or ENIAC, the world's first completely digital computer.

The ENIAC computer cost $500,000 to build. It was about half a block long and contained approximately 18,000 vacuum tubes. By modern standards ENIAC was fairly unreliable. One or more of its vacuum tubes would frequently fail in the middle of a calculation, and ENIAC could not store its programs. Still, inventors considered it a success. It was able to calculate an artillery trajectory in about thirty seconds, in comparison to the twenty hours a human took to do the same task.

The first electronic computers were used exclusively by the military and universities. The first general purpose computer for commercial use was the Universal Automatic Computer, or Univac. The first Univac was delivered to the Census Bureau in 1951.

During the 1950's and 1960's computer scientists made a number of improvements to computers, replacing vacuum tubes with **transistors**, and shifting from punched cards to **magnetic tape** and **disks** for external data storage, and moving from programming in **machine language** to **high level programming languages** such as FORTRAN and COBOL. Individual transistors were later replaced by **integrated circuits**. Magnetic tape and disks completely replaced punch cards as external storage devices, and internal memory technology moved from delay lines, which were tubes filled with mercury, cathode ray tubes that stored bits as glowing dots, to **magnetic cores** that stored bits in small ferrite cores, and finally to high density **semiconductor memory**.

In the early 1970's the Intel Corporation developed the world's first **microprocessor**, which contains all of the circuitry required to run a computer on a single postage stamp-sized chip. In 1976, Stephen Wozniak and Steve Jobs developed the Apple I computer using microprocessor chips. The success of the Apple I convinced Jobs that there was a market for an easy-to-use personal computer. Apple introduced the Apple II in 1977, and the **Macintosh** in 1984. The Macintosh came with a **mouse**, and the user controlled the machine by clicking on pictures, or icons, which gave it an interactive interface. Microsoft eventually used this type of interface in developing its **Windows** operating system for personal computers.

In 1981, IBM released its first personal computer. The IBM PC came with a monochrome cathode ray tube monitor and used a cassette tape for storage. It also came packaged with an operating system name PC-DOS, supplied by Microsoft. Due to the popularity of the IBM PC and the standardizing effect it had on the architecture of desktop computers, Microsoft became a dominant force in personal computers by the 1990's, rivaling IBM itself in terms of revenue.

Since 1990, computers have become smaller and more powerful, dropping in price and accessible to many people. The development of large scale and very large scale integrated circuits, and microprocessors that contained memory, logic, and control circuits on a single chip, has led to an explosion in the use of personal computers. By 2006 more than 675,000,000 personal computers are in use worldwide, compared with only 148,000,000 in 1992. And microprocessor chips are used in millions of electronic devices such as cell phones, appliances, and cars, which many people do not realize are actually controlled by computer chips and software.

In our next pod cast, we'll examine how information is represented in digital form, and how the storage capacity of digital computers has expanded due to technological innovation. We'll consider how a modern computer uses several different types of data storage mechanisms, many of which were invented only recently.

Podcast 7
INFORMATION REPRESENTATION

A **computer** is a programmable machine. This means that it is designed to take actions according to instructions instead of being wired to perform only a certain computation. A modern computer abides by two principal characteristics:

1. It responds to a specific set of **instructions** in a well defined manner.
2. Instructions are coded in the same way that data is encoded, so they can be stored in and read from the computer's memory.

Modern computers are **electronic** and **digital**. The actual machinery, wires, transistors, and circuits, are called **hardware**, and the instructions and data are called **software**. All general purpose computers require **memory**, a **data storage device**, such as a disk drive, an **input device** such as a keyboard, an **output device** such as a printer, and a **central processing unit**, or CPU. The CPU is the heart of the computer. It is the hardware component that actually interprets and processes instructions. The CPU **executes** instructions.

An instruction is a programming command . A computer's **instruction set** is the list of all the commands forming the computer's machine language. These commands are developed by the electrical engineers who design the CPU. They consist of instructions to copy bit patterns from one location in memory to another, to input bit patterns from the outside world, to compare quantities, to perform arithmetic operations, and myriad other low level data manipulations and movements. All programs are constructed using combinations of these very basic data manipulations.

To encode information in binary form, we use **ASCII**, which stands for **American Standard Code for Information Interchange**. ASCII codes represent text in computers and other devices that work with text. Every character such as the lower case letter "a", the upper case letter "A", the other letters of the alphabet, Arabic numerals, and punctuation symbols has its own unique ASCII code, each of which is eight bits in length. A bit is just a switch, that can be set to either "on" or "off."

ASCII was invented as a standard in 1967. In addition to printable characters, ASCII defines bit pattern codes for 33 non-printing control characters, many of which dictate how text is spaced when printed or displayed. Of curiousity to most newcomers is the fact that the space character, that is, what you produce when you type the space bar, is actually an ASCII character, with its own eight bit pattern.

Since ASCII specifies a correspondence between digital bit patterns and the symbols of a written language, such as English, it provides the means for digital devices such as computers to communicate with humans and each other, and to process, store, and communicate character oriented information that humans can easily understand. ASCII character encoding is used on nearly all computers. When you view a document on your computer screen you are seeing a translation from ASCII to the characters you know and recognize.

ASCII developed from the telegraphic codes invented in the mid-nineteenth century, when electricity was harnessed to carry messages over wires. ASCII was originally published as a seven bit code. Since 2^7 is 128, this meant that 128 different characters could be represented. Since computer memory evolved using eight bit bytes, ASCII is usually expressed as eight bits to the character, or byte. Since 2^8 is 256, ASCII can now encode up to 256 characters.

ASCII only defines the relationship between bit patterns and characters. It doesn't have any inherent mechanism for describing the appearance of text within a document. Different word processors implement their own coding scheme to represent things such as different type faces or fonts, different text sizes, margins, and line spacing. Hypertext mark up language (HTML) uses ASCII characters to build "tags" that tell a browser how to format the text in a document.

As computer technology spread throughout the world in the latter part of the twentieth century, different groups developed variations of ASCII so that they could express non-English languages and symbols. Many of these were incompatible with one another. In order to overcome these incompatibilities, **Unicode**, a superset of ASCII, was developed. Unicode is expressed in 16 bits to the character in order to have an expanded encoding capacity. ASCII is the lowest 128 characters of Unicode. Text expressed in Unicode is therefore compatible with ASCII. The more than 65,000 bit patterns available in Unicode can be used to express symbols and characters not present in English.

One other digital information coding scheme was developed and used only on large IBM computers. IBM's engineers developed **Extended Binary Coded Decimal Interchange Code**, EBCDIC, pronounced "eb-see-dic", about the time that the rest of the digital world adopted ASCII. This code is similar to ASCII but maps English characters and Arabic numerals to different bit patterns than ASCII. But even the data on IBM personal computers is coded in ASCII. When data is communicated from personal computers to IBM mainframes, the IBM operating system translates the ASCII to EBCDIC and vice versa.

ASCII and Unicode have become the world standard for digital information encoding. The reference chart at the end of chapter 2 in the Information Technology Workbook shows all of the ASCII bit patterns, the character each represents, and the way that each ASCII bit pattern appears in the zeros and ones of binary code. That reference also shows you how the bit pattern in any byte can be expressed in abbreviated form in **hexadecimal notation**, a common way that programmers refer to the contents of computer memory.

In the next several pod casts of the Information techno logy workbook, we'll look at how digital computers were developed to input, process, and output information encoded using bits. We'll also consider how computer memory, and data storage, has been technologically evolved, to store greater and greater amounts of information, encoded in this way.

Podcast 8
ELECTRONIC INFORMATION STORAGE

A modern computer system usually contains several kinds of **data storage**, each with a specific purpose. **Primary storage** can be directly accessed by the central processing unit of the computer. Primary storage is usually what we refer to as the memory of the computer. It is measured and expressed in terms of bytes.

A typical modern personal computer will have 512 megabytes or more, that is, 512,000,000 bytes or more, of memory. This is used as a working location for program instructions and data. This memory is implemented as semiconductor memory chips and is called **random access memory**, or RAM. We say that this form of data storage is volatile, because the bit patterns in it, representing program instructions or data, are lost when power is removed.

A data storage device provides a way to capture the bit patterns in primary storage and preserve them. The term data storage device thus refers to non-volatile, that is, permanent, data storage. The data will remain stored on such a device even when power is removed from it. Data recording on a modern data storage device can be done mechanically, magnetically, or optically.

We rate data storage devices by capacity and information storage density. **Capacity** is the total amount of information that a data storage device or medium can hold. It is expressed as a quantity of bits or bytes. For example, a modern CD/ROM can store about 600 megabytes of data. Since each byte is represented by eight bits, this amounts to about five billion bits of information. A DVD holds much more information. It can store about 4.7 billion bytes, which is about 53 billion bits of data!

Density refers to the compactness of stored information. It is the storage capacity of a medium divided by a unit of length, area or volume. Typically this measure is applied only to magnetic tape. Early magnetic tape storage density was typically 800 bytes per inch. As technology improved throughout the last half of the twentieth century, tape storage density increased to 6,250 bytes per inch, and then to much greater densities.

A list of storage media, from old to new, includes punched cards and punched paper tape as some of the earliest data storage devices. Paper cards and tape were electromechanical devices rather than electronic. Information was recorded by punching holes into the paper or cardboard medium, and was read by sensing whether a particular location on the medium did or didn't contain a hole. Each possible hole location represents one bit of storage. This was the principle used in the Hollerith cards of the 1890's.

Magnetic tape, floppy disks, hard disks, optical disks, magnetic bubble memory, thumb drives, smart media, memory sticks, and flash memory and memory cards are all more modern data storage devices. These devices have a higher information storage density than earlier forms of stoirage; they store more information in a smaller space. They also provide much faster data storage and retrieval speeds.

In a modern computer, volatile storage takes the form of random access memory, to temporarily store the computer programs currently running and the data being processed. A hard disk is used to more permanently store programs and data not in current use. Removable storage media, such as diskettes and CD/ROMs, are used to convey purchased programs or data and to house programs or documents no longer needed on hard disk.

One way to categorize a modern data storage device is its capability to change recorded information. **Read/write storage**, sometimes called **mutable storage**, allows information to be overwritten at any time.

Magnetic disks are a type of mutable storage. They are one of the fastest types of mutable storage for writing out programs or data and reading programs or data into a computer's memory. Some forms of mutable storage are slow to write, and faster to read. Slow write, fast read storage is mutable storage that allows information to be overwritten multiple times, but the write operation is usually much slower than the read operation. An example of this type of storage is rewritable CD/ROMs.

Read-only storage retains information stored at the time of manufacture. A purchased CD/ROM is an example of read only storage. It is an example of **immutable storage**.

Write-once storage allows information to be written only once at some point after manufacture. An ordinary blank CD/ROM is another example of immutable storage. You can generally store data on it once, after which it becomes read-only.

Another way to categorize storage mechanisms identifies where they reside in relationship to the CPU. **Primary storage** refers to the computer's memory since it is directly accessible by the CPU. It is the fastest type of general purpose memory. The CPU itself has a small amount of memory even closer to it, in the form of **registers**, which is where data or instructions currently being acted upon reside.

Secondary storage refers to a hard disk accessible to the CPU. Secondary storage is typically of higher capacity than primary storage, but it is usually much slower. For example, a desktop computer with 512 megabytes of primary storage may have a hard disk providing several hundred billion bytes (gigabytes) of secondary storage.

Offline storage is data storage in which the recording medium can be easily removed from the storage device and thus become inaccessible. Offline storage is used for data transfer and archival purposes. Floppy disks and CD/ROMs are often used for offline storage. The medium has to be manually mounted on a drive in order for the CPU to access the data stored on it.

Network storage is any type of computer storage that involves accessing information over a computer network. Network storage allows the centralization of information management in an organization and can reduce the duplication of information. It is hard disk storage that is even slower to access than hard disk storage attached directly to the computer.

Another way to characterize various types of data storage is its capability to access non-contiguous information. **Random access** means that any location in storage can quickly be accessed. Primary storage (computer memory) is of this type. It has no moving or mechanical parts so access is accomplished at very fast electronic switching speeds.

Sequential access means that reaching a piece of information will take a varying amount of time, depending on how much information has to be read before the desired information is reached. Some types of data storage devices, such as magnetic tape and CD/ROMs, are inherently sequential readings devices. Data on disks can be stored in either a random access or sequential manner.

Secondary access devices, even hard disks which offer random access capabilities, are often mechanical in nature, that is, they have moving parts. The device may need to position the read write head to the appropriate location. This may take a fraction of a second for a hard disk, or several seconds with a CD/ROM, or even several minutes with magnetic tape.

During the experimental era of electronic computer development in the 1940's and the early days of computer commercialization in the 1950's, memory technology was very primitive. The earliest digital computers used delay lines, which were channels several feet long filled with mercury, to store a few thousand bits as ripples which were recirculated through the channel by electronic circuitry. A later development shifted memory storage to cathode ray tubes that stored bits as glowing dots on a screen, with the persistence of the phosphor glow actually accomplishing data storage. **Magnetic cores** later replaced these devices, storing individual bits in small ferrite beads, the magnetic charge of which could be individually set, read back, and reset. Some early computers used magnetic drums for primary storage, making them very slow by comparison to modern day computers. Eventually, during the 1970's, semiconductor memory replaced these earlier primary storage mediums. Semiconductor memory has been greatly improved in storage capacity, information density, and read write speed since that time. Modern memory is of this type.

Semiconductor memory uses silicon-based integrated circuits to store information. A semiconductor memory chip may contain millions of tiny transistors or capacitors. Both volatile and non-volatile forms of semiconductor memory exist. Since the late 1990's, a type of non-volatile semiconductor memory, known as **flash memory**, has gained popularity for offline data storage. This is the type of memory contained in digital camera memory cards and the thumb drives that you can plug in to a personal computer USB port. These devices do not contain any power source. Their semiconductor circuit elements remain set to either zero or one indefinitely, once set by the computer or digital camera, and they can be reset again and again.

In the early days of computers, magnetic tape was often used for secondary storage, much as magnetic disk is used today. Due to delays in sequential reading with tapes, this slowed computer processing considerably. IBM greatly contributed to basic hard disk engineering in the 1950's, developing and commercializing a hard disk of 5 megabytes capacity in 1955. This disk unit, named RAMAC, was a tremendous innovation of that era. The RAMAC unit was as large as a modern refrigerator and weighed several hundred pounds and used more than a thousand watts of electrical energy to operate. Today's hard disks, which are devices small enough to be contained in an Ipod, possess more than 10,000 times the capacity of this first commercial hard disk, but weigh essentially nothing by comparison, and operate on less than ¼ watt of electrical energy.

CD/ROMs and **DVDs** are a form of **optical disc storage**. This type of offline memory relies on tiny pits, etched into the silvered surface of a circular plastic disk, to store information bits. The etched pits are created and read back by a small laser. Optical disc storage is non-volatile and provides sequential access. In general, reading speed is faster than writing speed with this medium.

Today's options and capabilities for storing and retrieving information in digital form are tremendous. This has opened up possibilities for creating and storing digitized textual, audible, and visual types of information undreamed of even a few years ago. In subsequent chapters of the Information Technology Workbook, we'll explore how modern computers access, process and can transform digital information. We'll also be examining ways that modern computer networks, including the internet, provide the means to transmit and share digital information.

Podcast 9
THE STORED PROGRAM CONCEPT

Computer memory refers to the parts of a computer that retain data for some interval of time, either short or long term. In the earliest computer designs, computer memory was not regarded as a critical component, because the first computers were designed to perform calculations rather than to remember data.

Punched cards, which were first used successfully with tabulation equipment made by Herman Hollerith in 1890, were a major step forward in the development of computer memory. Hollerith developed mechanical devices that could automatically read information encoded on punched cards, without human help. Because of this, stacks of punched cards could be used as easily accessible memory. The data for different problems could be stored on different stacks of cards, and accessed whenever needed. From the 1920's through the 1950's IBM refined card punching and tabulating systems that read data from cards and tabulated it, printed reports, and punched computed values on fresh cards. While they were an improvement on many manual tabulation processes, punched card tabulation equipment was very slow compared to today's computers. At the time, however, punched cards were a huge step forward, providing an automated means of input and output, and data storage on a large scale.

For more than fifty years after their first use, punched card machines did most of the world's business computing. Unfortunately, evidence has come to light that the information storage and retrieval capabilities of punched cards were also put to evil uses. It is now known that in the 1930's, IBM's German subsidiary constructed a large installation for the the Nazi government, which used it to record the results of a census conducted in 1933 on punched cards. Punched card tabulation equipment was used by the Nazis in subsequent years to speed the process of identifying many of the 6,000,000 Jews sent to concentration camps during the Holocaust.

World War II, which lasted from 1939 through 1945, created a great need for computer capacity for Allied military forces. To help meet that need, scientists at the University of Pennsylvania built a high speed electronic computer, known as the Electrical Numerical Integrator and Calculator, ENIAC. ENIAC employed over 18,000 vacuum tubes in logic circuits and as devices to remember digital bits. Prior to this time, researchers had experimented with electrical relays as circuit components to form logic circuits and to store bits. Operating without mechanical moving parts, ENIAC operated about a thousand times faster than experimental relay computers, ENIAC is commonly accepted as the first successful high-speed electronic digital computer and was used from 1946 to 1955.

Fascinated by the success of ENIAC, mathematician John Von Neumann contributed a new awareness of how practical, fast computers could be organized and built. His ideas, usually referred to as the **stored program** technique, were universally adopted and became the essential design principle for future generations of high-speed digital computers. The stored program technique involves many features of computer design and function. In combination, these features make very high-speed computer operation attainable.

In Von Neumann's technique, arrangements must be made for parts of a computer program to be used repeatedly. These parts of a program are called **subprograms**, or **subroutines**. In addition, the instructions making up a computer program can be changed by the computer itself, to make a computation proceed differently. Von Neumann's design provided these two capabilities in a very clever way.

In Von Neumann's design, all instructions are stored together with data in the same memory unit, so that, when needed, instructions can be arithmetically changed in the same way as data. In addition, Von Neumann's design called for a special type of machine instruction, named a conditional control transfer instruction, or **logic branch**. A logic branch allows a computer to make a comparison between two values and take a different course of action depending on the outcome of the decision.

As a result of these design innovations, computing and programming became much faster, and more flexible and efficient. Regularly used logic elements could be kept in program libraries and red into memory from secondary storage when needed. Thus, much of a given program could be assembled from the subroutine library.

Computer memory became the assembly place in which all parts of a long computation were kept, worked on piece by piece, and put together to form the final results. At this point, computer memory became a critical component, because more and more of it was necessary to support the new computer design paradigm.

The first generation of modern programmed electronic computers to take advantage of these design improvements was built in 1947. This group included computers using random access memory which is a memory designed to give almost instant access to any particular piece of information. This group of computers included UNIVAC, the first commercially available computer.

Podcast 10
EVOLUTION OF COMPUTERS

Early stored-program computers needed a lot of maintenance, due to the low reliability of memory components based on vacuum tubes. These computers reached about 80% reliability of operation. Digital computers of the 1950's era were very expensive to purchase or rent. They were particularly expensive to operate because of the cost of programming, which required personnel highly skilled in the workings of each individual design of machine. Computers during this era were mostly used in large computer centers operated by industry, government, and private laboratories.

During the 1950's two important engineering advances changed the image of the electronic computer field from one of fast but unreliable hardware to high reliability and even more capability. These engineering advances were **magnetic core memory** and the **semiconductor transistor**.

With magnetic core memory, RAM capacities increased from 8,000 to 64,000 words in commercially available machines by the 1960s, with access times of 2 to 3 milliseconds. The term word means a group of bytes, often two or four bytes.

The semiconductor transistor was a replacement for the vacuum tube. Whereas a vacuum tube required an electric current to heat a filament in order to make it produce a stream of electrons, the transistor needed no such heating. Like a vacuum tube, a transistor could be arranged to function as an amplifier or a switch. As an amplifier, a small electric potential could control a much larger one. As a switch, a transistor could be use to represent a bit. Both applications of the transistor required much less electricity, generated much less heat, and could be made much smaller than a vacuum tube. And, transistors, unlike vacuum tubes, did not wear out, they were inherently much more reliable.

In the 1960's, efforts continued to design and develop the fastest possible computer with the greatest possible capacity. The major computer manufacturers were IBM and a group referred to as the BUNCH, based on their initials: Burroughs, Univac, NCR, Control Data Corporation, and Honeywell. All of these began to offer computers with a range of capabilities and prices, as well as accessories such as consoles, card feeders, page printers, cathode ray tube displays, and graphing devices.

IBM gained a significant advantage in 1964 when it began manufacturing a family of computers named the IBM System 360. The System 360 line consisted of several computers, ranging from a fairly weak, but affordable machine, up through mammoth mainframes costing millions of dollars. The unique thing about the IBM System 360 was that all machines in the family accepted the same programming languages, so that programs written for one level of machine, did not have to be scrapped and rewritten if the organization moved up to a higher capacity machine.

During the 1960's, use of the third generation of programming languages, FORTRAN and COBOL, became widespread. These languages made it possible to write programs using logic statements not dependent on the machine language instructions of a computer. Unlike machine language instructions, which differ from one type of computer to another, FORTRAN and COBOL statements are the same and do the same things in all environments. Software running on each computer, translates instructions written in FORTRAN or COBOL, into the machine language of a specific machine.

Electronic circuit manufacturing techniques improved during the 1960's, and the photo etching of circuit boards gained prominence to eliminate many hand wiring steps in the construction of computer circuitry. A revolution in computer hardware was under way, involving shrinking computer logic circuitry and components by what are called large-scale integration techniques.

In the 1970's the manufacture of transistors in bulk became the norm. Entire assemblies of circuits to add, manipulate data, and perform logic operations could be manufactured at once, on tiny semiconductor chips, as **integrated circuits**. With these developments in circuit miniaturization, and streamlined manufacturing techniques, it became possible to move away from sole reliance on very powerful, single purpose computers, towards a larger range of applications for cheaper computer systems. Many companies, some new to the computer field, such a Digital Equipment Computer Corporation (DEC) introduced programmable minicomputers using the new circuit technology. These minicomputers often were sold with software packages designed to perform office automation functions.

This "shrinking" trend continued with the introduction of personal computers. Apple and Radio Shack introduced very successful PCs in the 1970's. IBM joined the PC market in 1981 when it released its first PC. These machines offered only about 4,000 to 16,000 bytes of memory, and their CPUs operated at less than 5 million cycles per second. But by the mid 1980s, personal computer speeds had increased to 25 million cycles per second, and memory sizes of PCs of 8 to 16 million bytes were common. By the year 2000, CPU chips were running at speeds 40 times faster, at up to one billion or more cycles per second, and individual memory chips commonly offered capacities of 4 million bits or more. By 2006, personal desktop computers priced at less than $600 offered speeds of more than 3 billion cycles per second, and memory capacities have leaped to a half a billion bytes.

Memory devices continue to develop and advance, with miniaturization leading the way to cost effective ways to store, more and more data, more and more quickly. This in turn has made it possible to support high speed graphic applications, and the graphical user interface we all now take for granted, which has done so much toward making the computer an information appliance for everyone.

In the next pod cast of the Information Technology Workbook, we'll take a closer look at how John Von Neumann's innovative design for digital computers actually operates.

Podcast 11
THE CPU AND STORED PROGRAMS

The **central processing unit** (CPU) is microscopic circuitry that serves as the main information processor in a computer. It is made from a wafer of silicon with millions of electrical components in its many layers. These components are logic gates, which respond in precise ways to combinations of signal inputs. The logic gates are arranged in units that interpret and implement software instructions, perform calculations and comparisons, make decisions, temporarily store information, keep track of the current step in the execution of the program, and allow the CPU to communicate with the rest of the computer.

A CPU is similar to a calculator, only much more powerful. The main function of the CPU is to perform arithmetic and logical operations on data taken from memory, or on information entered through some input device such as a keyboard or scanner. The CPU is controlled by a list of software instructions, called a **computer program**. Software instructions entering the CPU originate in some form of memory storage device, such as a floppy disk or a CD/ROM. These instructions then pass into the computer's random access memory where each instruction is given a unique address.

The CPU can access specific pieces of data in RAM by specifying the address of the data it wants. As a program is executed, data flows from RAM through a unit of wires called the **bus**, which connects the CPU to RAM. The data is then decoded by a processing unit called the instruction decoder, which interprets and implements software instructions. From there, the data is passed to the arithmetic logic unit, ALU, which performs calculations and comparisons. The ALU performs operations such as addition, multiplication, and conditional tests on the data, sending the resulting data back to RAM or storing it for further use.

The term Von Neumann architecture refers to a computer design model that uses memory to hold both instructions and data. The separation of storage from the central processing unit is implicit in this architecture. The term stored program computer is used to mean a computer of this design.

The term Von Neumann architecture came from mathematician John Von Neumann, who wrote about a general purpose stored computing machine in 1946. While Von Neumann's work was pioneering, a number of others also contributed to this idea, including many at the Moore School of Electrical Engineering at the University of Pennsylvania. Various members of the group working on the early computer, ENIAC, were among this group, as well as scientist Alan Touring, who presented a paper on the technology in 1946.

The earliest computing machines had fixed programs. In principle, a desk calculator is a fixed program computer. It can do basic math, but cannot be used for some other purpose, such as a word processor. Some very simple computers still use this design for simplicity and economy. To change the program of such a machine, you have to rewire, restructure, or even redesign the machine. The earliest digital computers were not so much programmed, as they were designed. Reprogramming, when it was possible at all, was a very tedious process, starting with flow charts and paper notes, followed by detailed engineering designs, and then by implementing physical wiring changes.

The idea of the stored program computer changed everything. This was implemented by creating an instruction set, expressed as on/off bits, and designing a decoder circuit that interpreted each instruction, and set logic circuits in such a way as to respond to the instruction.

This allowed a computation to be defined as a series of instructions, which came to be known as a computer program. By treating those instructions in the same manner as data , a stored program machine can be reprogrammed just by loading a different series of instructions from secondary to primary storage. In this arrangement, the computer becomes tremendously more flexible.

The ability to treat instructions as data is what makes assemblers, higher level language compilers, and other automated programming tools possible. A higher level language is actually just a text file. A compiler is a program that reads this type of text file and generates machine language instructions from it. The instructions generated are expressed as the zeros and ones of binary code, which can be loaded to memory to program the computer.

However, as powerful as the Von Neumann architecture is, there are a few drawbacks to it. For example, the separation between the CPU and memory leads to what is known as the Von Neumann bottle neck. Under some circumstances, when the CPU is required to perform minimal processing on large amounts of data , the data transfer rate is very small in comparison with the rate at which the CPU itself can work. This can cause serious limitations in effective processing speed, as the CPU is continuously forced to wait for vital data to be transferred to or from memory. As CPU speed and memory size have increased much more rapidly than the data transfer rate between the two, this bottleneck has become more and more of a problem. Modern object oriented programming is less geared to pushing vast amounts of data back and forth than earlier languages such as Fortran. Yet internally, this is still what computers spend a great deal of time doing.

In addition, self initiated program modifications, such as those implemented by a stored program machine, can be quite harmful, either by accident or design. In some simple stored program computer designs, a malfunctioning program can damage itself, other programs, or even the entire system, leading to a crash. A buffer overflow is one common example of such a malfunction. A buffer overflow occurs when a process attempts to store more data in a given area of memory, than is allocated to the storage of the data. The result is that the data overwrites adjacent memory locations. The overwritten data may corrupt variables or other program instructions, leading to harmful results.

The important point for you to remember about the computer's central processing unit is that CPU architecture reflects the nearly universal adoption of John Von Neumann's stored program concept. This leads to CPU designs that include an arithmetic logic unit, a control unit, and a program counter, that indicates where the next instruction is located in memory.

Another important point to remember is that programs are expressed in the same binary zeros and ones that encode data, and they are stored in the same memory. This makes it possible to reprogram machines simply by loading new instructions to memory.

Podcast 12
COMPUTER PRINTER TECHNOLOGY

Several different technologies support today's computer printers. These technologies fall into two categories: impact and non-impact printers.

Impact printers

One of the oldest types of printers still in common use are **dot matrix printers**. These impact printers use a collection of small pins to strike a ribbon coated with ink, causing the ink to transfer to the paper at the point of impact. Dot matrix printers are divided into two groups, serial dot matrix printers and line dot matrix printers.

The **serial dot matrix printer** is quite common. In this type of printer the print head moves in a horizontal direction, and a circuit sends electrical signals to it which force the appropriate pins to strike against the inked ribbon, making dots on the paper to form the desired characters. The distance between the row of dots in a column determines vertical printing resolution, and the speed at which the print head is moved across the line determines the horizontal printing resolution.

In a **line dot matrix printer**, printing wires span the entire line of print. Whereas a serial dot matrix printer might have a print head of as few as nine wires, a line dot matrix printer has a set of such wires for the entire width of the line. Thus it can print an entire line at once, and nothing moves except the paper, so its print speed can be faster. This type of printer is more complex and expensive and is not very common.

Dot matrix printers with printlines as narrow as 12 to 30 characters can be made inexpensively and are often used in devices such as calculators and credit card purchase machines, to produce receipts on paper tape.

Another type of impact printers was used in mainframe computer environments before laser printers were perfected. A **chain printer** circulates a metal chain of letters in front of the printline at high speed. Hammers at each print position strike when the letter to be printed at that position rotates in place. Chain printers were expensive, noisy, and maintenance-prone, and have almost entirely been replaced with newer technologies.

Impact printers can do something that non-impact printers cannot: they can print multi-part forms. These types of forms have more than one sheet of paper, separated by carbon paper or coated with a chemical that discolors on impact.

Non-impact printers

In non-impact printers no mechanical part of the printing element touches the paper when creating the printed image. Ink jet, thermal, and laser jet printers are all non-impact printers.

Ink jet printers use a series of tiny nozzles to spray drops of ink onto the paper. Continuous jet printer technology was developed in the nineteen seventies by IBM. This technology produces a continuous stream of ink droplets that are sent either directed onto the printing media or into a gutter for ink recirculation. The direction of the droplets is controlled by giving the droplets an electrical charge and using another electric field to deflect them.

"Drop on demand" inkjet printer technology was also developed in the 1970's to counter the high cost of continuous stream printers. Drop on demand printers eject ink droplets only when the droplets are needed to print. This eliminates the hardware required for continuous stream ink jet printing, and the printer can be much less expensive. The droplets might be formed by a piezo-electric nozzle in which a crystal changes shape when an electric charge is sent to it, forcing a droplet of ink to be ejected at the paper surface.

In the 1980's Canon developed an alternative printer technology, the bubble jet printer. This is a drop on demand ink jet printing method in which ink drops are ejected from each printing nozzle by the rapid growth of an ink vapor bubble on the surface of a tiny heated element.[1]

The most popular ink jet printers use a serial printing process. Similar to dot matrix printers, **serial ink jet printers** use print heads with a series of nozzles arranged in a vertical column. The print head is moved horizontally to print a line, the paper is then advanced, and the print head then moves back across the paper, printing the next line in reverse.

Ink jet printers have three advantages. These include quiet operation, the capability to produce color images with photographic quality, and low initial printer purchase price. The great disadvantage of ink jet printers is the cost to maintain them. Comparing cost per page, ink jet printers are many times more expensive than

[1] Hewlett Packard independently developed a similar heat-based printing technology. The thermal printer does not use ink at all, but instead requires the use of special paper with a heat-sensitive chemical coating. When heated by the vertical line of nine dot-shaped printing elements, which are selectively switched on as the print head moves across the line, the chemical turns dark. Dots darkened in this way produce the same type of matrix-formed characters as dot matrix and ink jet printers. This type of printer is widely used to print single-part receipts at grocery store self-checkout machines and credit-card equipped gasoline pumps on paper tape, because they can print much faster than dot matrix printers and require less maintenance.

laser printers to operate. For example, a color ink jet printer may cost $80, and replacement ink cartridges, capable of printing only a few hundred pages, may cost as much as $50.

Non-impact laser printers

Computer **laser printers** are non-impact. They rely on the same technology used in office photocopying machines. This process, named electrophotography, was invented in 1938 by Chester Carlson and developed by Xerox in the period 1940-60. It relies on the unique properties of the element selenium, which does not conduct electricity but can be made into a conductor by the light. Of computer printers, laser printers offer the best print quality and highest resolution.

In electrophotography as used in office copiers, a selenium coating on the surface of a rotating metal drum enclosed in a light-shielded container is uniformly charged with static electricity. The selenium holds the charge, since in this condition it acts like an insulator such as plastic or glass. Then the charged surface is exposed to an optical image. Light causes the selenium to become conductive, to allow the static charge to leak off through the metal drum, in proportion to the amount of light reaching each point. The result is a negative latent image on the surface of the drum, expressed in static electricity. The lightest parts of the image now have the least static charge, and the darkest parts retain the most electric charge.

The image on the drum is "developed" by spreading a fine dark powder named "toner" over the surface. This adheres mostly only to the areas that remain charged, which are the dark areas. The image composed of toner powder is transferred to the surface of a sheet of paper which comes into contact with the rotating drum. The toner powder is then permanently fused to the paper pressure and heat. Finally, the excess toner and static charge are cleaned from the surface of the drum to make it ready for the next charging, exposure, and copying process.

Computer laser printers substitute a laser beam and scanner assembly to form the latent image on the selenium drum, bit by bit. A rotating mirror is used to have the laser beam move across each print line, rather than moving the laser itself. The laser beam, its intensity modulated by electrical signals from the printer's controller, is directed through a lens onto the rotating mirror, which reflects the laser beam across the drum. Computer laser printers are now mass-produced and use the same kind of tiny semiconductor lasers found in CD players. Laser printers typically cost less than $200. A black and white toner cartridge for a laser jet printer may cost about $50 to $100, but, unlike replacement print cartridges in an ink jet printer, can typically produce several thousand printed pages.

LED is an abbreviation for "light emitting diode," a semiconductor device that generates light directly without the use of a heated wire as an incandescent light bulb use. LEDs are used in some computer printers instead of a laser to selectively discharge the static charge on the selenium surface of the rotating metal drum. A row of LEDs extends the full width of the rotating drum, and no mechanical movement is involved in such a printer, simplifying the mechanism.

Laser printers and LED printers are a cost effective solution for applications such as desktop publishing that require high print quality with high speed output. Unlike dot matrix printers, these printers cannot print multiple part forms. But they can quickly print multiple copies of the same image, in instances where multiple copies of output are required. Since they can easily print graphics, they can combine the lines and nomenclature of a form at the same time that they insert particular information. So these printers can easily reproduce a tax form as they print name, address, and other information. These types of printers are widely used for form-letter generation for marketing and other business purposes.

You now know the differences between impact and non-impact computer printers, the essential characteristics of dot matrix, ink jet, and laser jet printers, and the general way in which each of these operates.

Lecture Slides 2

Introduction to Information Technology

Computer Hardware

Course resources

PowerPoint presentations, podcasts, and web links for readings are available at www.ambriana.com > IT Workbook

Print slides at 6 slides per page

Homework, quizzes and final exam are based on slides, lectures, readings and podcasts

Slide 2-2

Main topics

- Four "ages" of computing machinery
- Four generations of electronic computers
- Overview of computer hardware
- Evolution of internal storage (memory)
- Evolution of external storage (tape, disk)
- The stored program concept

Slide 2-3

A little computer history

- Pre-mechanical age: 3000 BC – 1450 AD
- Mechanical age: 1450 – 1840
- Electromechanical Age: 1840 – 1939
- The Electronic Age: 1940 – present

Here's a web link to information about these "ages":
History of Information and Technology

Slide 2-4

Electromechanical age – 1840 - 1939

- Herman Hollerith (1860-1929)
- Data on punched cards
- Electrical, not electronic

Here are some web links:
IBM keypunch machine
IBM mechanical card processing collator

Slide 2-5

Generations of the electronic computer age

- **First generation 1939-54: vacuum tubes**
- Second generation 1954-59: transistors
- Third generation: integrated circuits
- Fourth generation (now): microprocessors
- Fifth generation? Who knows?

Slide 2-6

First generation experiments: 1939 - 1950

- Atanasoff / Berry, Iowa State, (1939)
- Howard Aiken, Harvard Mark I (1944)
- Eckert / Mauchly, ENIAC, 1946, first commercial computer, UNIVAC, 1950
- Parts adapted from **radio** technology
- Thousands of calculations per second

Slide 2-7

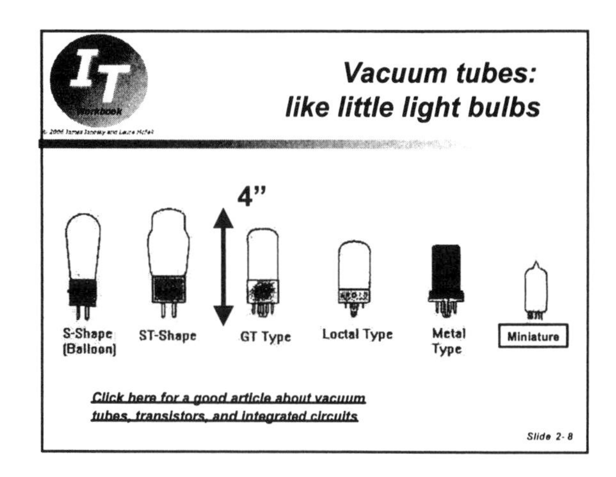

"programming" = rewiring!

The beginnings of modern information technology

- Computers commercialized in 1950 by Univac, 1952 by IBM
- Early computers were **mainframes**
- Expensive and large, not very powerful
- The prediction was that only a very few would be built and sold…

Slide 2-12

IT
Workbook
First commercial computers: 1950 - 1952
Univac I at the U.S. Census Bureau
Click for history
Slide 2-13

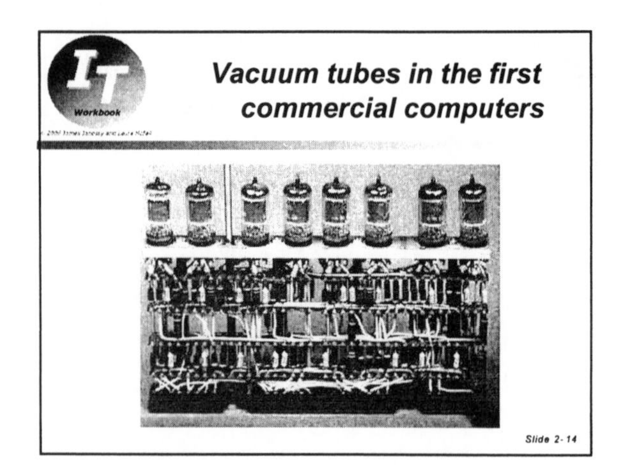
IT
Workbook
Vacuum tubes in the first commercial computers
Slide 2-14

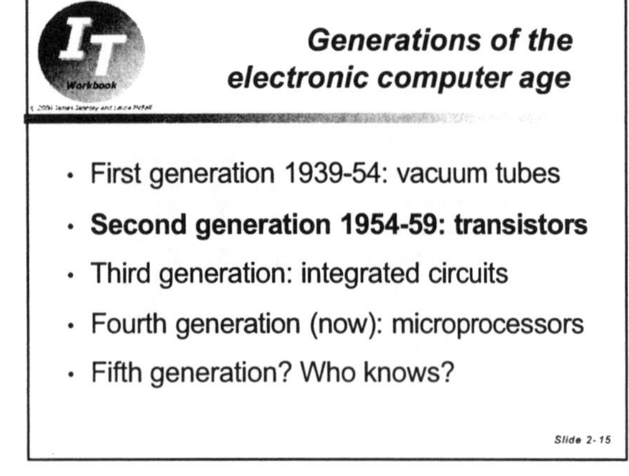
IT
Workbook
Generations of the electronic computer age
First generation 1939-54: vacuum tubes
Second generation 1954-59: transistors
Third generation: integrated circuits
Fourth generation (now): microprocessors
Fifth generation? Who knows?
Slide 2-15

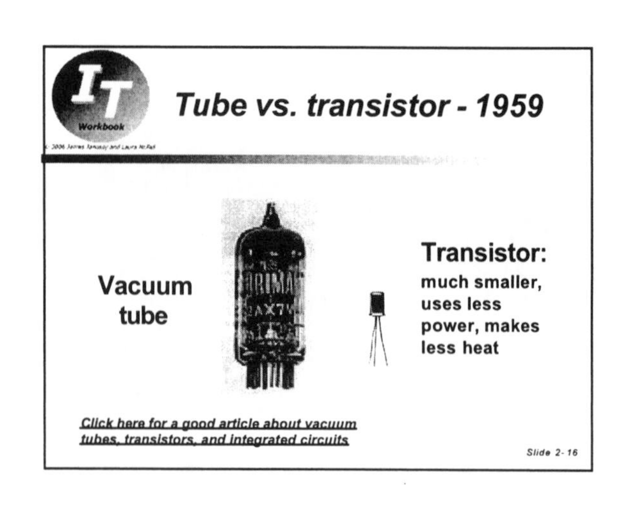
IT
Workbook
Tube vs. transistor - 1959
Vacuum tube
Transistor:
much smaller, uses less power, makes less heat
Click here for a good article about vacuum tubes, transistors, and integrated circuits
Slide 2-16

IT
Workbook
Transistors in computer circuits - 1960
A few separate (discrete) transistors wired on a circuit board with other electronic components
Better than tubes, but low circuit density
Slide 2-17

IT
Workbook
IBM 360 family of computers - 1968
Slide 2-18

Generations of the electronic computer age

- First generation 1939-54: vacuum tubes
- Second generation 1954-59: transistors
- **Third generation: integrated circuits**
- Fourth generation (now): microprocessors
- Fifth generation? Who knows?

Slide 2-19

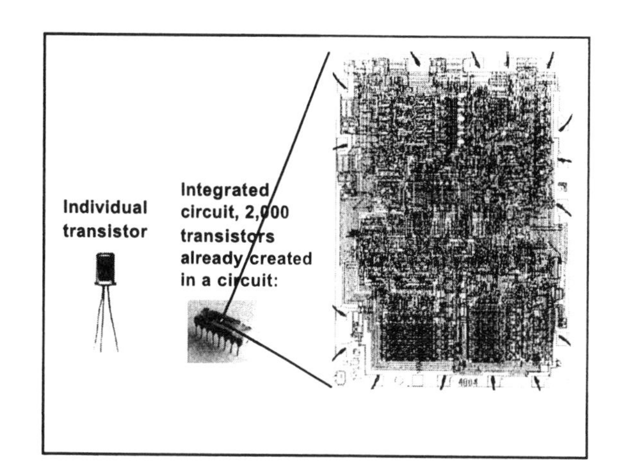

Generations of the electronic computer age

- First generation 1939-54: vacuum tubes
- Second generation 1954-59: transistor
- Third generation: integrated circuits
- **Fourth generation (now): microprocessor**
- Fifth generation? Who knows?

Slide 2-21

Very large scale IC integration

- Now may have up to half a billion transistors all wired and in place forming a CPU
- Heart of a modern PC
- Assembly is about 1.5" square

Slide 2-22

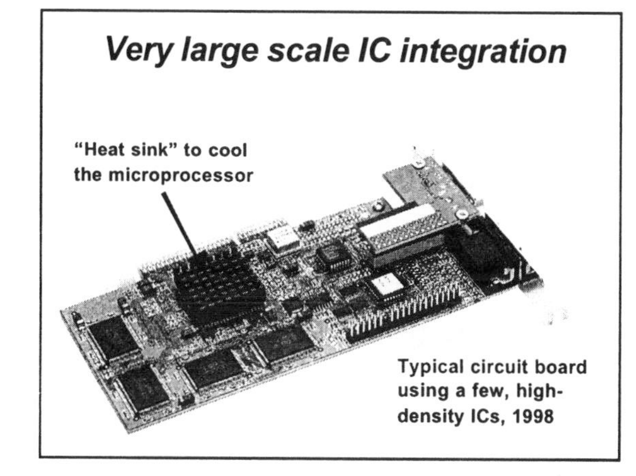

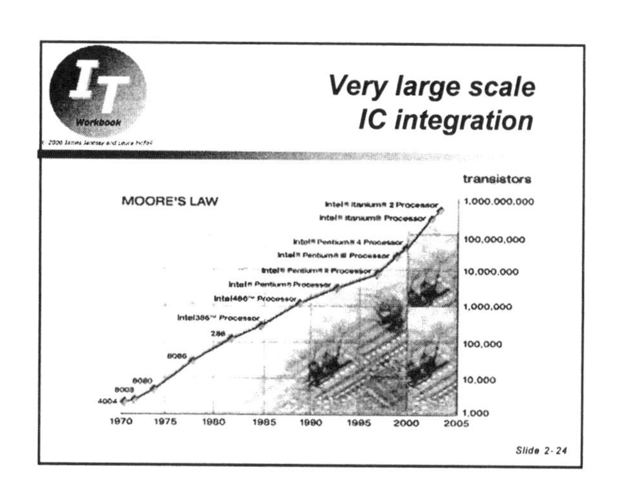

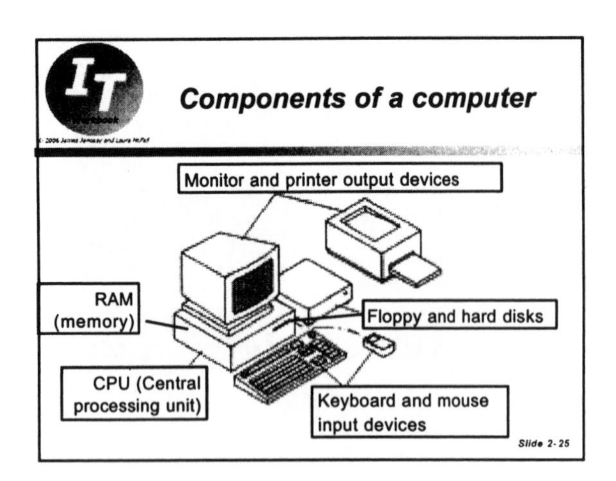

Input and output devices – I/O devices

- Getting information into the computer's memory so it can be operated on
- Getting software into memory (stored program concept)
- Keyboard, mouse, display, printer
- Several technologies have existed...

Slide 2-26

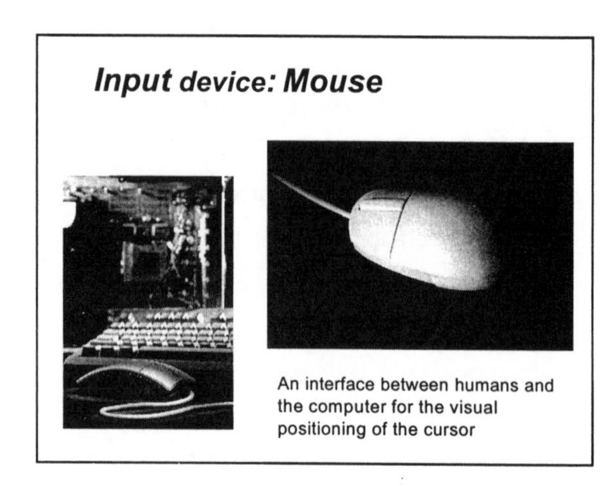

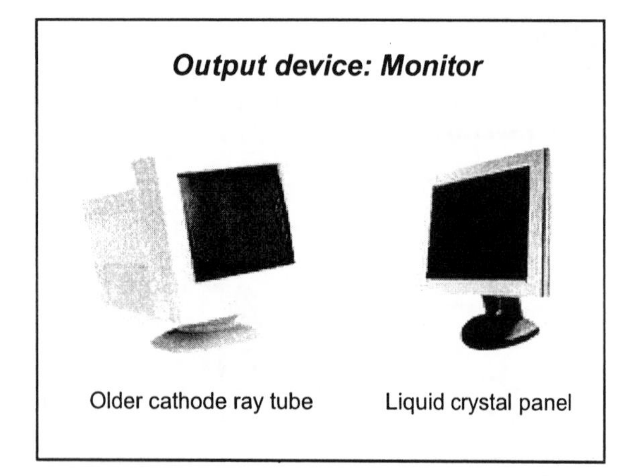

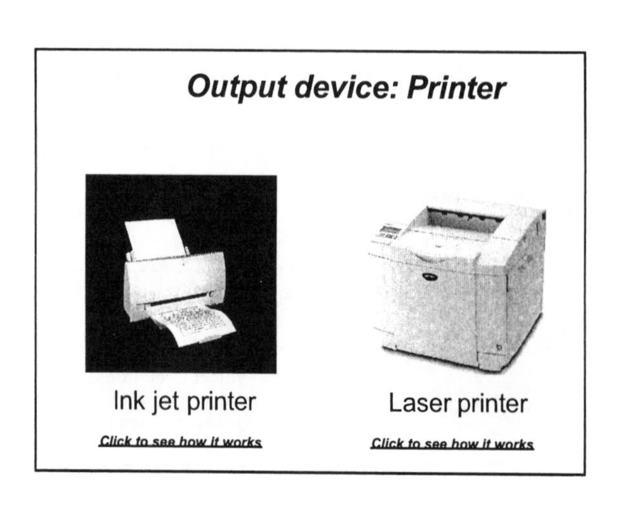

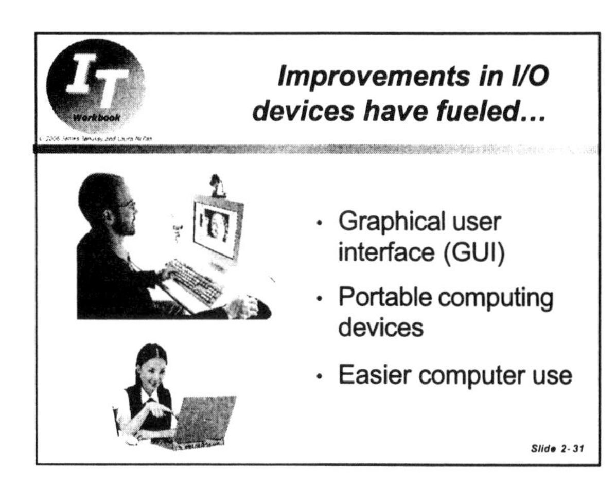

Internal storage: the computer's memory

- Data being operated on is stored here
- With the stored-program concept, the software being executed is stored here
- Organized as "cells" each with its own "address"
- First cell is cell 0, then cell 1, 2, 3, etc.

Slide 2-32

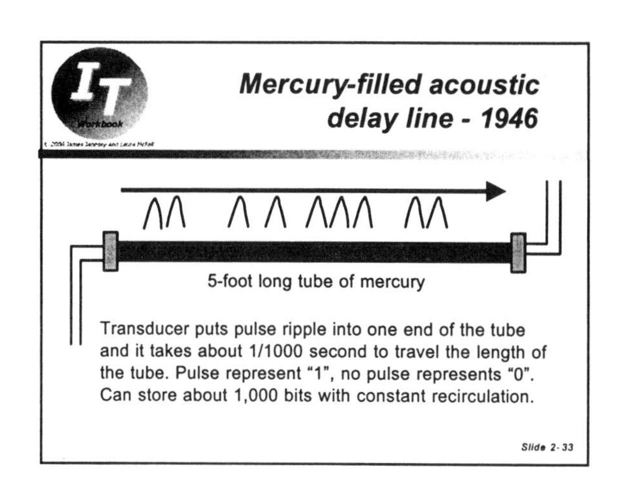

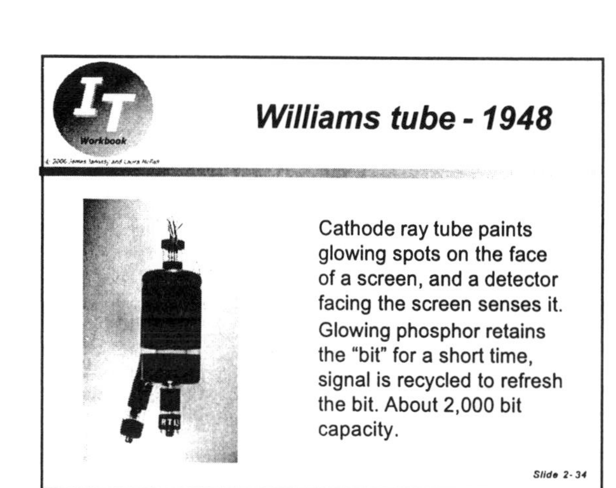

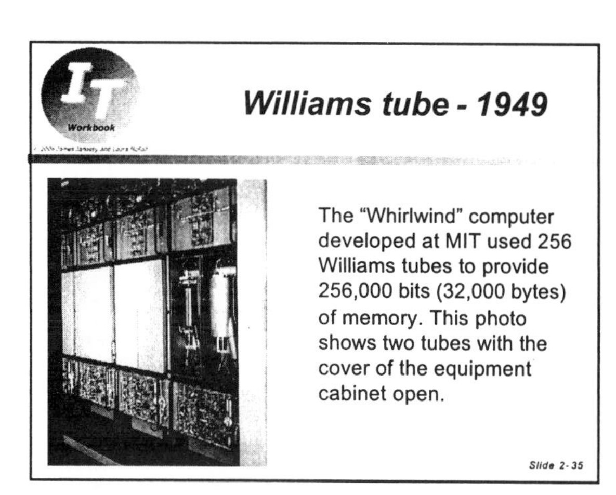

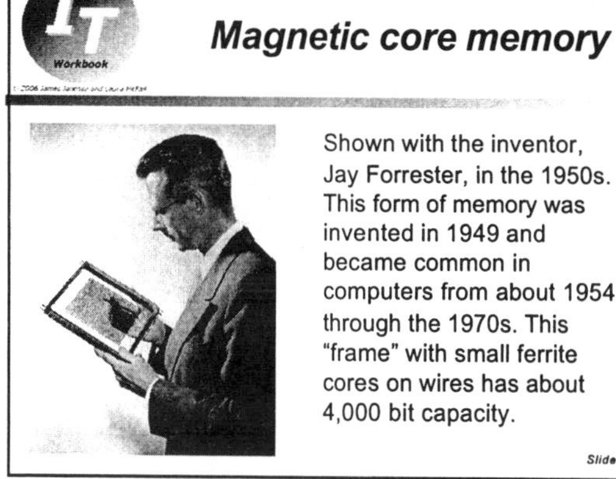

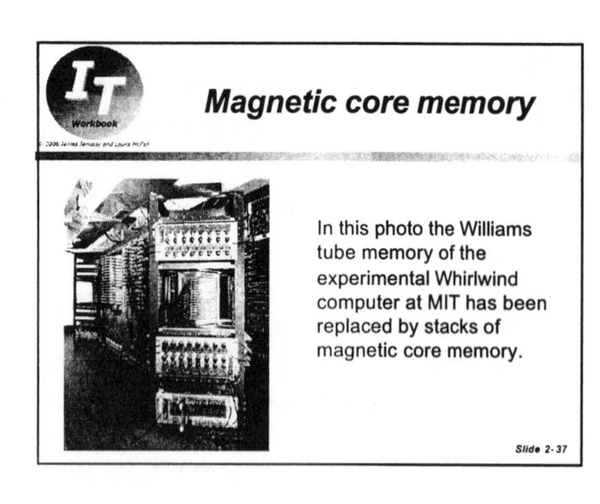

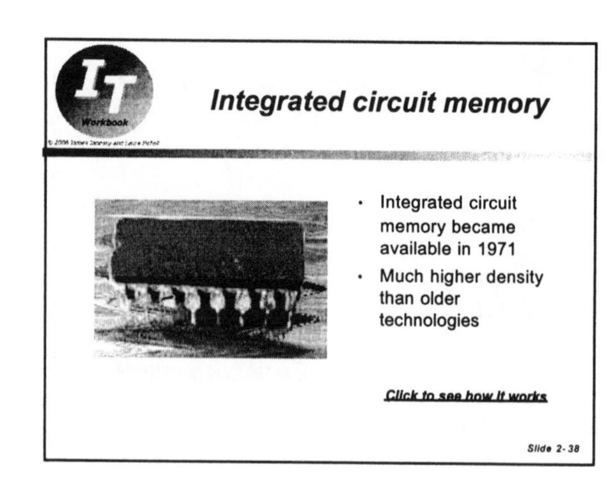

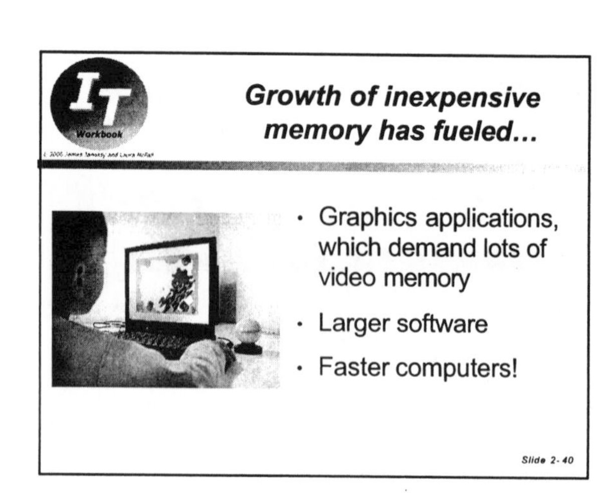

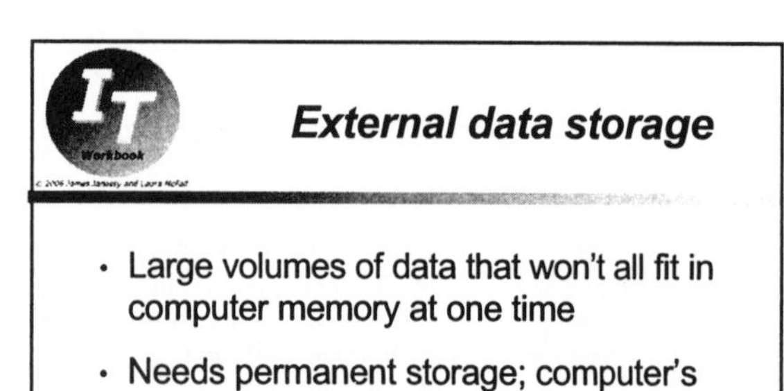

Punched cards

- Carryover from IBM's pre-electronic days
- Low density, 8 bytes/inch
- Slow to read, hundreds of bytes/second

Slide 2-42

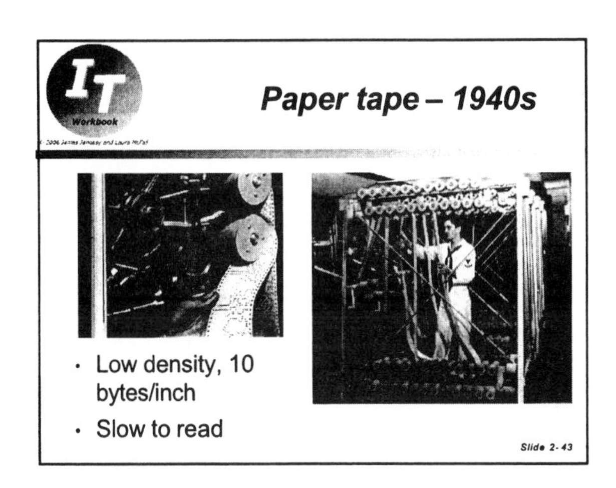
Paper tape – 1940s
Low density, 10 bytes/inch
Slow to read
Slide 2-43

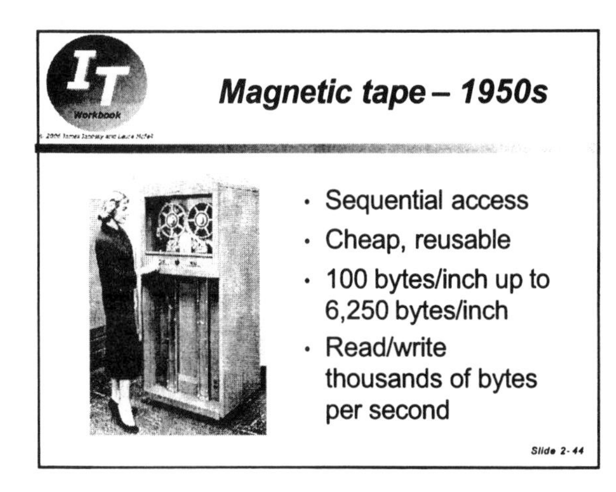
Magnetic tape – 1950s
Sequential access
Cheap, reusable
100 bytes/inch up to 6,250 bytes/inch
Read/write thousands of bytes per second
Slide 2-44

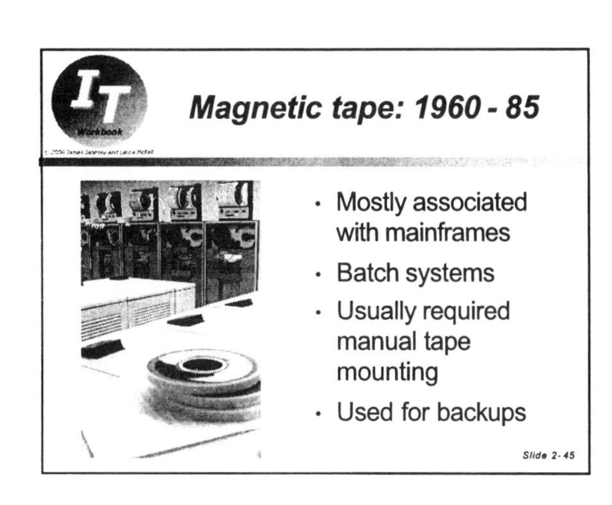
Magnetic tape: 1960 - 85
Mostly associated with mainframes
Batch systems
Usually required manual tape mounting
Used for backups
Slide 2-45

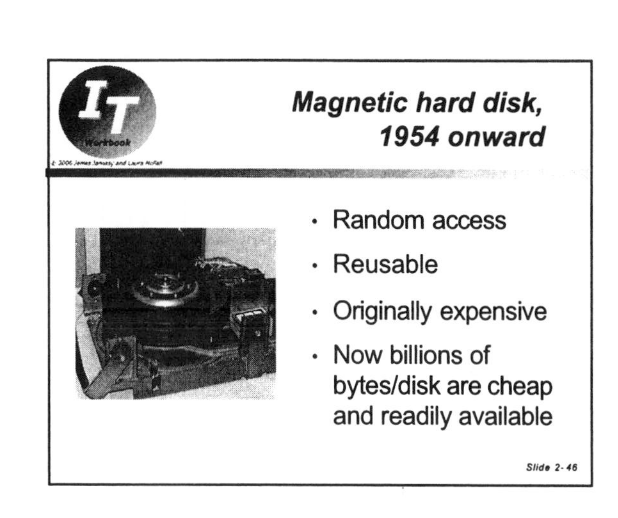
Magnetic hard disk, 1954 onward
Random access
Reusable
Originally expensive
Now billions of bytes/disk are cheap and readily available
Slide 2-46

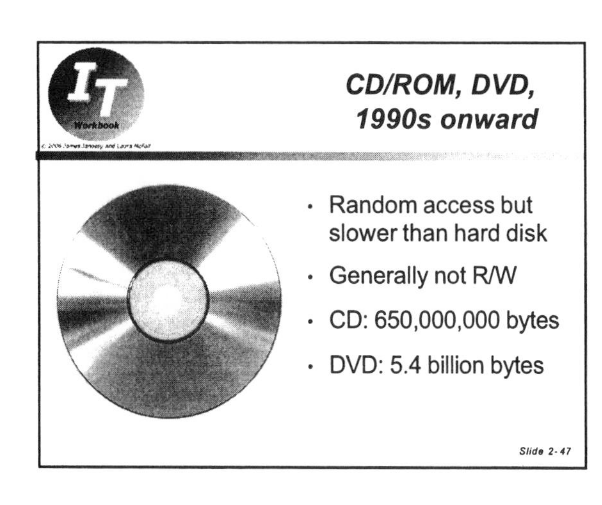
CD/ROM, DVD, 1990s onward
Random access but slower than hard disk
Generally not R/W
CD: 650,000,000 bytes
DVD: 5.4 billion bytes
Slide 2-47

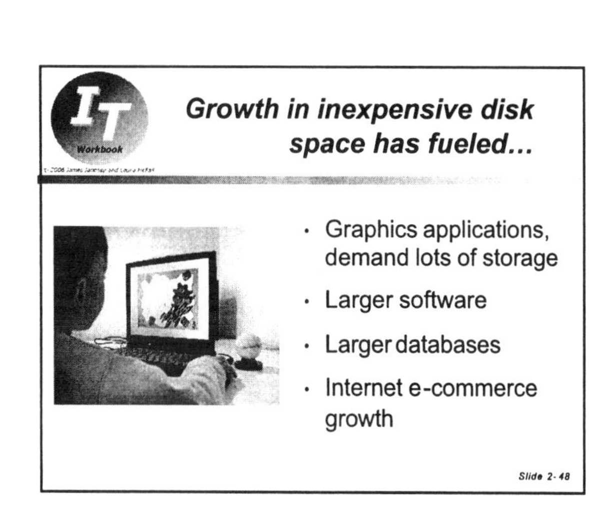
Growth in inexpensive disk space has fueled...
Graphics applications, demand lots of storage
Larger software
Larger databases
Internet e-commerce growth
Slide 2-48

The ASCII Coding System

Text information is encoded electronically in 8 bits per character (byte) according to this standard scheme.

Dec	Hex	Char
0	00	null
1	01	soh
2	02	sot
3	03	eot
4	04	
5	05	
6	06	ack
7	**07**	**bell**
8	08	back
9	09	htab
10	**0A**	**LF**
11	0B	vtab
12	0C	feed
13	**0D**	**CR**
14	0E	
15	0F	
16	10	
17	11	
18	12	
19	13	
20	14	
21	15	nak
22	16	
23	17	
24	18	
25	19	
25	1A	
27	**1B**	**esc**
28	1C	
29	1D	
30	1E	
31	1F	

Dec	Hex	Char
32	20	sp
33	21	!
34	22	"
35	23	#
36	24	$
37	25	%
38	26	&
39	27	'
40	28	(
41	29	)
42	2A	*
43	2B	+
44	2C	,
45	2D	-
46	2E	.
47	2F	/
48	30	0
49	31	1
50	32	2
51	33	3
52	34	4
53	35	5
54	36	6
55	37	7
56	38	8
57	39	9
58	3A	:
59	3B	;
60	3C	<
61	3D	=
62	3E	>
63	3F	?

Dec	Hex	Char
64	40	@
65	41	A
66	42	B
67	43	C
68	44	D
69	45	E
70	46	F
71	47	G
72	48	H
73	49	I
74	4A	J
75	4B	K
76	4C	L
77	4D	M
78	4E	N
79	4F	O
80	50	P
81	51	Q
82	52	R
83	53	S
84	54	T
85	55	U
86	56	V
87	57	W
88	58	X
89	59	Y
90	5A	Z
91	5B	[
92	5C	\
93	5D	]
94	5E	^
95	5F	_

Dec	Hex	Char
96	60	`
97	61	a
98	62	b
99	63	c
100	64	d
101	65	e
102	66	f
103	67	g
104	68	h
105	69	i
106	6A	j
107	6B	k
108	6C	l
109	6D	m
110	6E	n
111	6F	o
112	70	p
113	71	q
114	72	r
115	73	s
116	74	t
117	75	u
118	76	v
119	77	w
120	78	x
121	79	y
122	**7A**	z
123	7B	{
124	7C	|
125	7D	}
126	7E	~
127	7F	

Hexadecimal digits each abbreviate 4 bits. "0" means the bit is off and "1" means the bit is on. Anything capable of unambiguously representing two different "states" or settings can be used to encode a text message in ASCII. Lights, holes punched in paper, magnetized spots and even smoke signals have been used to convey information encoded this way.

Hex	Binary
0	0000
1	0001
2	0010
3	0011
4	0100
5	0101
6	0110
7	**0111**

Hex	Binary
8	1000
9	1001
A	**1010**
B	1011
C	1100
D	1101
E	1110
F	1111

How 8 bits form a "byte" to represent the character "z"

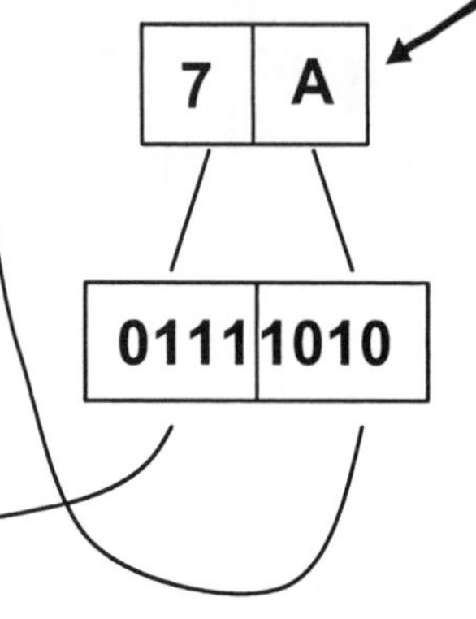

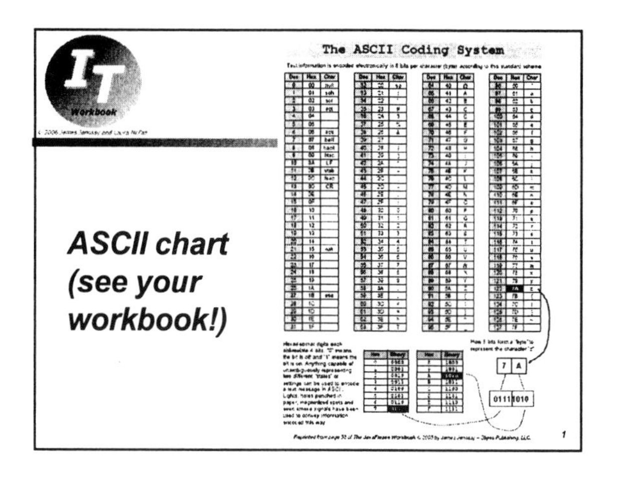
The ASCII Coding System
ASCII chart
(see your workbook!)
1

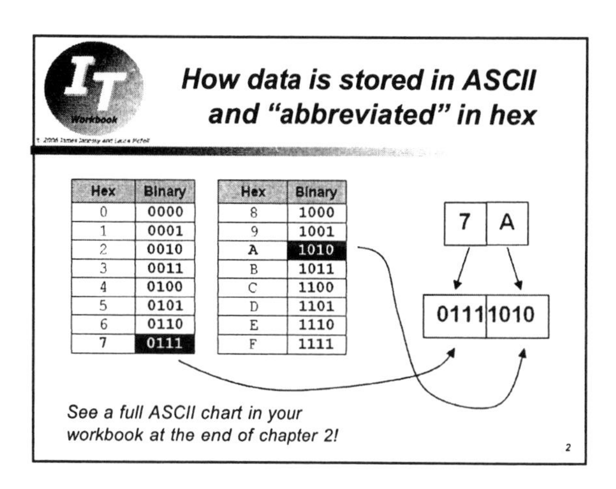
How data is stored in ASCII
and "abbreviated" in hex
Hex Binary
0 0000
1 0001
2 0010
3 0011
4 0100
5 0101
6 0110
7 0111
Hex Binary
8 1000
9 1001
A 1010
B 1011
C 1100
D 1101
E 1110
F 1111
7 A
0111 1010
See a full ASCII chart in your
workbook at the end of chapter 2!
2

A simple ASCII example:
say "Hello, Jim" in ASCII
H e l l o, Jim
3

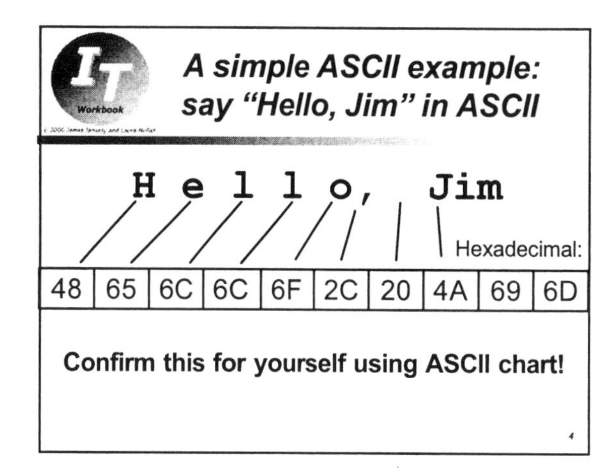
A simple ASCII example:
say "Hello, Jim" in ASCII
H e l l o, Jim
Hexadecimal:
48 65 6C 6C 6F 2C 20 4A 69 6D
Confirm this for yourself using ASCII chart!
4

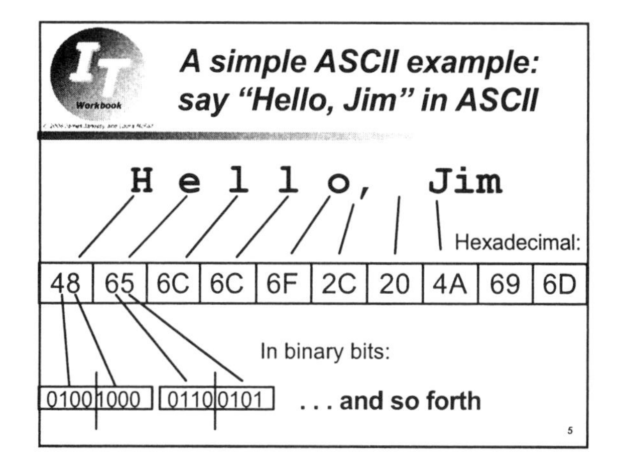
A simple ASCII example:
say "Hello, Jim" in ASCII
H e l l o, Jim
Hexadecimal:
48 65 6C 6C 6F 2C 20 4A 69 6D
In binary bits:
0100 1000
0110 0101
. . . and so forth
5

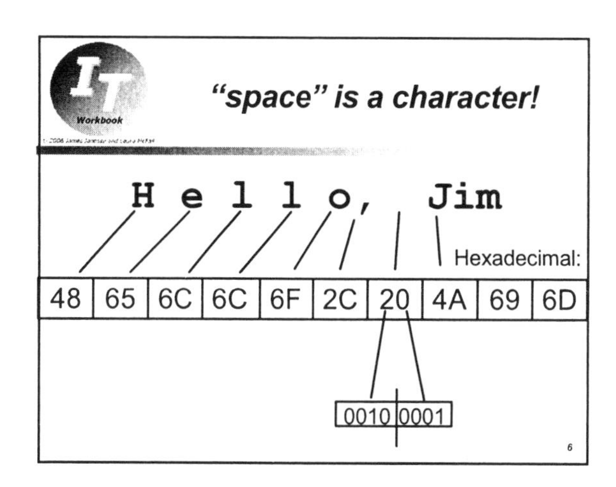
"space" is a character!
H e l l o, Jim
Hexadecimal:
48 65 6C 6C 6F 2C 20 4A 69 6D
0010 0001
6

All bits set to 0: "null"
All bits set to 1: "high values"

- The bit pattern 00000000 has all bits set to zero. ***This not a space!*** It is called the "null" value.
- Hexadecimal is written **X'nn'** where 'nn' is the hex value. Null is **X'00'**
- **X'FF'** is 11111111, high-values

1

Common measures in computer-geek jargon

Prefix	Multiplying number	Common word
pico-	.000000000001	trillionth
nano-	.000000001	billionth
micro-	.000001	millionth
milli-	.001	thousandth
kilo-	1,000	thousand
mega-	1,000,000	million
giga-	1,000,000,000	billion
tera-	1,000,000,000,000	trillion

2

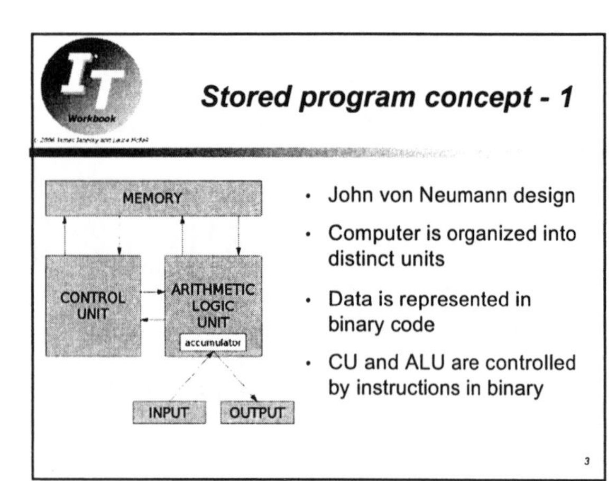

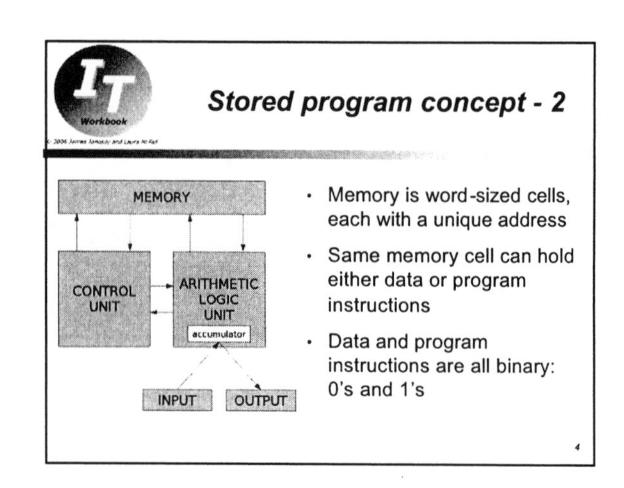

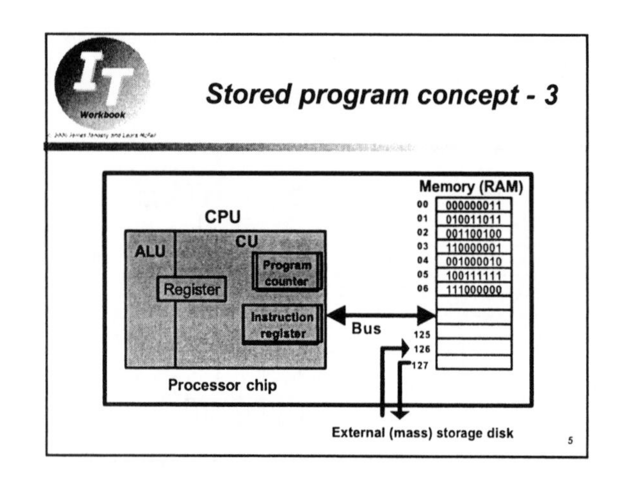

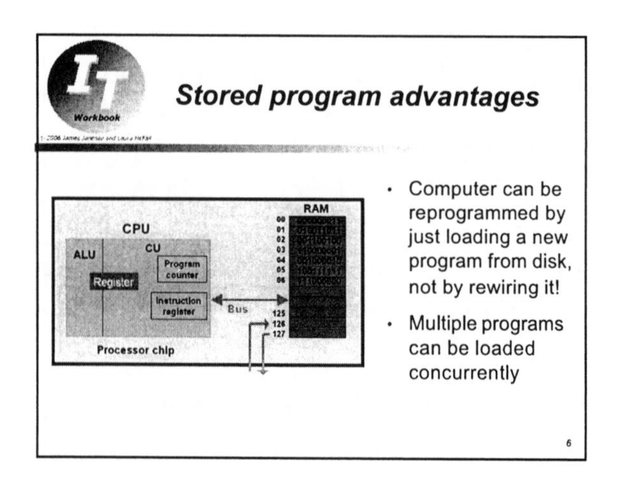

IT Workbook Assignment 2.1

Student name:

Write your answers to the questions below ***within the box****. In each case, please choose your words carefully to answer the specific questions. Avoid simply copying passages from your readings to these answers.*

Using the 8-bit ASCII encoding scheme, a solid box to show a bit that is turned on, and a blank box to show a bit that is turned off, encode "23 Skidoo" in the following series of boxes:

1)

Describe the premechanical, mechanical, electromechanical, and electronic periods of information technology in terms of the **years** cover by each and identify some of the **elements of technology** used for **information storage** in each period.

2) Premechanical era:

3) Mechanical era:

4) Electromechanical era:

5) Electronic era:

IT Workbook Assignment 2.2

Student name:

Write your answers to the questions below within the box. In each case, please choose your words carefully to answer the specific questions. Avoid simply copying passages from your readings to these answers.

Identify and briefly describe the **several major components** of a modern desktop computer:

1)

Describe what the **IBM punched card** (Hollerith card) was and the purpose it served even before computers were invented.

2) what the punched card was:

3) purpose the punched card served before computers were invented:

IT Workbook Assignment 2.3

Student name:

Write your answers to the questions below *within the box*. *In each case, please choose your words carefully to answer the specific questions. Avoid simply copying passages from your readings to these answers.*

The **mercury delay line**, **Williams tube**, **magnetic core memory**, and **RAM chips** are four types of electronic memory that have been used in digital computers. Briefly describe how each stored data and the period of time in which it was employed in computing machinery:

1) Mercury delay line:

2) Williams tube:

3) Magnetic core memory:

4) RAM chips:

Describe some of the **secondary data storage technologies** that preceded magnetic tape. For each, describe the advantage of magnetic tape as a data storage medium over the older technology.

5)

IT Workbook Assignment 2.4

Student name:

Write your answers to the questions below within the box. In each case, please choose your words carefully to answer the specific questions. Avoid simply copying passages from your readings to these answers.

Vacuum tubes replaced **mechanical devices, transistors** later replaced vacuum tubes, and **integrated circuits** later replaced discrete transistors as the components of computer circuitry. In each case of one technology superceding another, explain what the **advantage** of the newer technology was over the old.

1) Vacuum tubes replacing mechanical devices:

2) Transistors replacing vacuum tubes:

3) Integrated circuits replacing discrete transistors:

Both **magnetic tape** and **magnetic disk** data storage house information outside of a computer (secondary storage), and preserve the information even when the computer is turned off. Describe the advantage(s) that magnetic disk data storage offers over magnetic tape data storage:

4)

IT Workbook Assignment 2.5

Student name:

Write your answers to the questions below within the box. In each case, please choose your words carefully to answer the specific questions. Avoid simply copying passages from your readings to these answers.

The **stored program concept** for computing machinery was described by John von Neumann in 1945. Describe and explain this concept is and explain what it contributed to making computers flexible and powerful.

1) Describe and explain the stored program concept:

2) explain what the stored program concept contributed to making computers flexible and powerful:

Describe the six categories of digital computers based on their size and computing power, and the number of concurrent users that each can support:

Category of computer	Size/power	Number of concurrent users
3) Personal digital assistant (PDA)		
4) Notebook/laptop/desktop		
5) Workstation		
6) Minicomputer		
7) Mainframe		
8) Supercomputer		

IT Workbook Assignment 2.6

Student name:

Write your answers to the questions below *<u>within the box</u>*. *In each case, please choose your words carefully to answer the specific questions. Avoid simply copying passages from your readings to these answers.*

Describe how a computer **ink-jet printer** works, and what it is best suited to:

1) how the ink-jet printer works:

2) what the ink-jet printer is best suited to:

Explain how a **CD/ROM** represents the binary-encoded information of computer software and data files:

3)

IT Workbook Assignment 2.7

Student name:

Write your answers to the questions below within the box. In each case, please choose your words carefully to answer the specific questions. Avoid simply copying passages from your readings to these answers.

Describe how a computer **dot matrix printer** works, and its primary use:

1) how a dot matrix printer works:

2) the primary use of a dot matrix printer:

Compare the data storage capacity of a CD/ROM and a DVD, in terms of **bytes**, and in terms of **bits**:

	Bytes	Bits
3) CD/ROM		
4) DVD		

IT Workbook Assignment 2.8

Student name:

Write your answers to the questions below within the box. In each case, please choose your words carefully to answer the specific questions. Avoid simply copying passages from your readings to these answers.

Describe in detail how a **monochrome (black and white) computer laser printer** works, and the specific thing that the laser does in this process:

1) how a laser printer works:

2) the specific thing that the laser does in the laser printer:

Explain what task a **computer laser printer** can do well, that a dot-matrix printer can't do well:

3)

Identify the kind of **printing task** that a computer laser printer cannot perform at all:

4)

IT Workbook Assignment 2.9

Student name:

Write your answers to the questions below within the box. In each case, please choose your words carefully to answer the specific questions. Avoid simply copying passages from your readings to these answers.

"**Moore's Law**" of increasing computer power is not a law, but an informal prediction. Describe the progression it states in the number of transistors on a computer chip from 1970 onward:

1) 1970:

2) 1975:

3) 1980:

4) 1985:

5) 1990:

6) 1995:

7) 2000:

8) 2005:

The terms **kilo**, **mega**, **giga**, and **tera** are all commonly used with computer memory, disk, and CD/ROM storage. Each of these is roughly a power of 10, meaning that a number stated with one of these terms is to be multiplied by some power of 10. Provide the correct interpretation of each of these terms by indicating the "multiplying power of 10" each indicates, in number and word form:

	Multiplying number	*Common word description*
9) kilo-	1,000	thousand
10) mega-		
11) giga-		
12) tera-		

Magnetic disk capacity has increased tremendously since IBM's PC's was released in 1981. At that time, a 10 Mb hard disk was available for the machine. Now, tiny 60 Gb hard disks are used in the Apple Ipod to store digitized sound files. How many times larger in storage capacity is this type of Ipod hard disk than the original 10 Mb IBM hard drive?

13)

IT Workbook Assignment 2.10

Student name:

Write your answers to the questions below within the box. In each case, please choose your words carefully to answer the specific questions. Avoid simply copying passages from your readings to these answers.

In June, 2006, the U.S. Department of Veterans Affairs had to inform the public that an employee had taken data home on a laptop to continue working with it, and that his house had been burglarized and the laptop stolen (see www.eweek.com/articles2/0,1895,1972946,00.asp) . The laptop contained confidential information on over 26,000,000 veterans. This incident highlights a potential danger to everyone given how much information about us is stored in computers, and how much information can be stored on modern hard disks, CD/ROMS, and DVDs. A major danger in loss of personal information is the possibility of **identity theft**. Explain what identity theft is, and some steps that individuals should take to minimize this possibility:

1) what identity theft is:

2) steps you should take to minimize the potential for identity theft:

Sensitive data stored on portable computers or on portable storage media can be safeguarded from unauthorized access by a process of **encryption**. PGP ("pretty good privacy") is a popular method of encrypting data. Explain what data encryption is, and how an individual or organization can obtain and make use of **PGP** to protect stored data:

3) what data encryption is:

4) how to obtain and use PGP to safeguard data:

Computer Software

Software consists of the instructions that tell the computer hardware what to do. The computer hardware circuits stand ready to process instructions, which are simply on/off pulses of electrical energy. These instructions are formed by a programmer. Application, general purpose packages, system, and utility software exist, developed using different programming tools, which are… software!

Systems software manages the computer hardware's basic functions and the human/computer interface common to all programs you might run. The most common systems software on desktop computers is the graphical user interface (GUI) of PCs and MacIntosh computers. This software is called the **operating system** (OS). On PCs it is Windows, and on the Mac and many other computers it is Unix. Many other operating systems exist on other classes of computers, such as MVS on IBM mainframe, Linux on minicomputers and network servers, and scores of specialized operating systems on other types of computers.

Application software is the software that provides end-user functionality, such as word processing, spreadsheets, internet browsing, and supports major business processes such as payroll, financial accounting, purchasing, inventory control, manufacturing, and myriad other functions. Each of these is an "application." You can buy the software for some common applications such as word processing over the counter in a retail store. Other applications, such as custom software solutions for a particular business are usually designed and programmed by personnel within a business, often to gain a competitive advantage.

In this chapter, we'll consider the characteristics of the various types of software, some of the history of operating systems, and the ethical issue of software internet piracy.

Learning objectives

Chapter 3
Computer Software

This chapter, in combination with the web readings, web links, and podcasts available at www.ambriana.com is focused on providing the knowledge to achieve these learning objectives:

- Understand what computer software is, and the four major types of software
- Be able to distinguish the four generations of computer programming languages and their general characteristics, including procedural or non-procedural orientation
- Recognize the differences between application software and operating system software, and the relationship between these
- Understand the process by which computer software is created, including the role of source code, assemblers and compilers, and how software is executed
- Understand what computer utilities are used for, such as a disk defragmenter program
- Be able to explain the difference between a command-line interface and a graphical user interface (GUI)
- Understand the distinction in ownership and use between shareware, freeware, cripple ware, demo software, and public domain software
- Know what software piracy is, be able to recognize the several forms it may take, and the dangers and deficiencies that it poses.

Podcast Scripts – Chapter 3

These readings of the **Information Technology Workbook** are also accessible in audio form as .mp3 files for free download. Visit the workbook web site, www.ambriana.com. You can listen to this material on your own Ipod or .mp3 player or computer, read it, or both listen and read.

Podcast 13
COMPUTER SOFTWARE

Computers dominate every aspect of modern life, and range in size and complexity from large mainframe computers, used by major business organizations, to the small personal computers that have become common household items.

To function, computers must be programmed using the language on which their circuits are based. This **machine language** consists of electronic signals represented by on and off electrical charges. We commonly refer to these patterns by the binary number symbols one and zero.

A computer interprets groups of these binary numbers as ASCII text data, or numeric quantities, or sound or visual data, or as instructions. Sets of related instructions form computer programs, which constitute a basic element in the concept of software.

Computer software is defined as computer programs, procedures, rules, associated documentation, and data pertaining to the operation of a computer system.

Four basic **types** of software exist:

1. application software
2. general purpose software
3. system software
4. utility software.

Let's consider each type of software individually. All four types of software are available for and used with personal computers.

Application software performs specialized functions such as payroll or student records keeping, or other useful work not directly related to the operation of the computer itself.

General purpose software can let you perform certain common office or administrative chores. This type of software includes word processors, spreadsheets, database programs, presentation programs such as Power Point, and other **personal productivity program packages**. These are obtainable in retail computer stores, as off the shelf items.

System software is a collection of programs that manages all the concurrent tasks being performed by a computer, including the execution of application software programs. An **operating system** such as Windows is system software.

Utility software is a program that performs a routine technical task, such as listing or compressing data, copying a file, or sorting data.

A **computer programming language** is also actually software in its own right. We might classify a computer language as system software. A computer language such as C or Java is not generally relevant to a typical computer user, but it is very familiar to a computer programmer . A computer language lets a programmer express the intended logic of an application.

Four generations of computer programming languages exist. These are the means by which computer programs, forming any type of computer software, are actually constructed. Here are the **four generations** of computer programming languages:

1. first generation, **machine** language
2. second generation, **assembler** language
3. third generation, **procedural** languages
4. fourth generation, **non-procedural** languages.

The **first generation** of computer programming languages is the machine language instructions which are the actual on/off patterns of binary instructions a computer executes. This is as low a level as you can get. This language is specific to a given type of computer, and is designed by the engineers who develop the logic gates and circuits of the CPU. Machine language is extremely efficient for a CPU to execute, but very difficult and time consuming for a programmer to write directly. Machine language is also not portable. A given machine language program will work only on the computer chip for which it was intended. Since scores of different computer chips exist, it is undesirable to write general purpose software directly in machine language, if the general purpose software is intended for wide scale distribution.

Second-generation language, called assembler language, corresponds to machine instructions on a one to one basis. That is, one line of assembler code translates into one machine language instruction. Assembler code is easier to write than machine language because acronyms stand for the zeros and ones of machine language. But most programmers still find assembler very cumbersome.

Since computer horsepower now costs much less than a programmer's time, very few employers will pay a programmer to write applications in assembler language because it takes too long, is too error prone, and is too difficult to change when requirements change. Because assembler is essentially a shorthand way of writing machine language code, assembler programs are not portable between different computer chips. For example, a program written in the assembler language for an IBM PC won't work on a Macintosh, and vice versa.

Now let's consider third and fourth computer programming language generations, which are considered "higher level" languages. Compared to machine language and assembler language, higher level languages have the advantages of making it easier and faster to write computer programs. But these generations demand more computer processing power than do machine language or assemblers, for two reasons.

First of all, higher level languages must ultimately produce machine language, a one-time process. The machine language produced must accomplish what the programmer expressed as logical actions in source code statements. Programmers write in **source code**, the syntax of which is designed by the people who create the language. This language is suited to specific types of work. For example, the earliest third generation languages, COBOL and Fortran, were developed in the late 1950's for business and engineering purposes. **COBOL** was designed as the Common Business-Oriented Language, and is somewhat English-like. **Fortran**, on the other hand, stands for "formula translator." Fortran was designed for engineers and scientists; the source code for it is more like equations and formulas. More recently developed languages of this type include PL/1, C, C++, Basic, Pascal, and Java. The software that reads source code as ASCII files and ultimately generates machine language from it is called a **compiler**. Compiling an assembler or machine language program from source code requires computer resources.

But beyond that, the generalized routines composed by higher level language compilers are inherently less efficient than machine language written for a specific situation. A compiler, for example, will ultimately create more lines of machine code to accomplish a specific task than a skilled programmer would to accomplish the same thing in machine language. More lines of code take longer to process.

Despite their execution-time inefficiencies, the use of higher level languages became appealing because it was cost-effective. Computers increased in speed and capacity, rendering the inefficiencies of higher level languages less important. But just as important, the cost of computer programmers rose. It became advantageous to use more productive programming tools, rather than to pay larger and larger salaries to have programmers take longer to work in machine and assembler language.

Third generation languages are procedural in nature. This means that they require the programmer to define specific algorithms, or in other words, specific methodical solutions expressed as step-by-actions, for every process to be performed. A process that is to be repeated, called a loop, must be explicitly defined by the programmer. Thus, a loop to process all rows in a table, must be mapped out in great detail by the programmer. The programmer must initialize variables before starting the loop, and must repeatedly check for the satisfaction of a condition which, when satisfied, will end the loop. Even in third generation languages, computer programming is a very detail-oriented, logically demanding task.

Because national standards were developed for Fortran, COBOL, C, C++, and Java, these languages, and programs written in them, can be used on different manufactures of computers, with identical results. Each different machine has a different compiler, producing the machine language required for that computer, from the standard program statements. This significantly contributes to the longevity of programs written in third generation languages. Even though COBOL has now been superceded by other third-generation languages, hundreds of billions of lines of COBOL code are still in use by computer installations across the world.

Fourth generation languages are known as non- procedural languages. The most well known of these is SQL, which is often called structured query language. In this form of programming language, the programmer specifies what columns of information are wanted, from what tables, and what kinds of selection criteria are to apply. The programmer need not be concerned with explicit actions to input, join (match rows in) tables, select, and display data. Beyond this, SQL provides many powerful functions to handle data conversion and output formatting tasks. The essential elements of SQL are consistent among various relational database products. Learning how to use SQL is a valuable skill.

Other proprietary fourth generation languages exist, owned by one vendor or another. Unlike an open language such as SQL, for which published standards exist, proprietary languages are owned by a single vendor. Using such a language ties an organization closely to that vendor. Widely used proprietary non-procedural fourth generation languages, include SAS and SPSS, both of which are used for statistical analysis of data. These proprietary, non-procedural languages make it possible to invoke complex packaged logic to produce descriptive statistics on sets of data, and to perform complex regression and correlation analyses and display graphic results with just a few statements.

Programs for spreadsheets, word processors, and similar applications are usually written in a third generation language. These are complex software products, designed to present a friendly and intuitive appearance to computer users. These general purpose software products, provide the means for a person with little or no programming background to interact very productively with a computer. The formulas you plug into a spreadsheet can actually be considered a very high level language. They are interpreted by complex source code, written in C, C++ or Java.

Although most programs today are written in third generation languages, interest is growing in languages that even more closely resemble human language. To date, however, even general purpose software that recognizes human speech, or outputs speech, is actually developed in third generation programming languages.

Let's review what you have learned here:

At this point, you should know the difference between different categories of software, and what the different types of programs in each category do:

1. application software
2. general purpose software
3. system software
4. utility software

You should also understand what the different generations of computer programming languages are, and the nature, advantages, and disadvantages of each:

1. first generation, **machine** language
2. second generation, **assembler** language
3. third generation, **procedural** languages
4. fourth generation, **non-procedural** languages.

Finally, you should know and be able to explain the role of a program language compiler and how a computer programmer uses a compiler.

Podcast 14
OPERATING SYSTEMS

Software is divided into two classes, system software and application software. **System software** consists of programs that interact with the computer at a very basic level. This includes operating systems, compilers, and utilities for managing computer resources. **Application software** includes database programs, word processors, and spreadsheets. These are the programs that deliver functionality to end users.

Here, we focus on the system software known as the operating system. The operating system is the most important program running on a computer. Every general-purpose computer has to have an operating system in order to run other programs. Figuratively speaking, applications software sits on top of systems software, because it is unable to run without the operating system and system utilities to support it.

Operating systems perform basic tasks, such as recognizing and accepting input from a keyboard, sending output to the printer or the display screen, keeping track of files on various disks, and controlling peripheral devices such as disk drives, scanners, and printers. By having these services housed in an operating system, none of the application programs need to contain many complex routines that are specific only to the computer hardware in use.

For large computer systems, the operating system has even greater power and responsibility. It acts like a traffic cop, insuring that different programs and users, which are running at the same time, don't interfere with each other's activities. The operating system schedules the running of other programs, and how much of the CPU's resources, and computer's memory, each concurrently executing application program is allowed to use. The operating system is also responsible for security, insuring that unauthorized users do not gain access to the system.

Several types of operating systems exist:

A **multi-user operating system** lets two or more users run programs on the system at the same time. Sometimes this is referred to as a **multi-tasking operating system**, since it provides support for executing more than one application program at the same time. We see this when we are running a word processing program, a spreadsheet program, and accessing the internet, all at the same time. Some of the more complex operating systems allow hundreds, or even thousands, of users to access applications at the same time.

A **multi-processing operating system** supports running a program on more than one central processing unit.

Some older operating systems are batch oriented and do not respond interactively. For example, IBM's older mainframe DOS, which stands for Disk Operating System, was not interactive. It was designed to received instructions from decks of punched cards, on which a user coded special instructions indicating which programs to run, what files to access, and where and how to write outputs. On the other hand, a **real time operating system** responds instantly to input in an interactive manner. For example, **UNIX**, an operating system developed for minicomputers, was designed to communicate interactively with computer terminals. Unix responds to individual commands entered at a terminal, or to process sequences of such commands in a text file known as a batch script.

One older operating system for the PC environment was **Microsoft's DOS** (unrelated to IBM's older mainframe operating system of the same name). Another, more contemporary operating system for PCs, is **Windows XP**. Yet another contemporary operating system for the personal computer environment is a variant of Unix tailored to run on the Apple MacIntosh.

And still another operating system, developed by IBM for personal computers but beset by technical and marketing issues, was **OS2**. OS2 was originally a joint effort of IBM and Microsoft. Its development began in the 1980's but its first releases were hindered by a number of problems. Microsoft eventually abandoned the project and instead pursued the development of its own Windows operating system.

Application programs are designed to run on a specific operating system. The choice of operating system determines the application programs you are able to run. For example, an application program designed to run on Windows XP will not run on a MacIntosh.

As a user, you normally interact with the computer's operating system through a set of commands. For example, the Microsoft DOS operating system supports commands such as copy and rename, for copying files and changing the names of files. In the DOS world, the computer screen acts like a typewriter, and a small prompt appears. The user is expected to know the syntax of a command, and to type it at the prompt and then press the Enter key. These commands are accepted and executed by a part of the operating system called the command line interpreter. Newer operating systems such as Windows provide graphical user interfaces (GUI, pronounced as "gooey") that allow you to initiate actions by pointing and clicking at objects that appear on your screen.

Linux is a free, widely distributed, "open source" operating system that runs on a number of hardware platforms. Open source refers to a program for which the source code is available to everyone, for use and modification, free of charge. Open source code is typically created as a collaborative effort, in which programmers improve the code and share their changes within their community. The core of Linux, called its kernel, was developed by Linus Torvalds to replicate the functionality of Unix. Because it's free and because it runs well on many platforms, including PCs and the Macintosh, Linux has become a very popular alternative to proprietary operating systems such as Microsoft Windows.

The **kernel** of an operating system stays in the computer's memory, so it is important for the kernel to be as small and efficient as possible, while still providing all the essential services required by other parts of the operating system and the application programs it runs. Typically, the kernel is responsible for memory management, process and task management, and disk management.

The term "**proprietary**" means that some entity controls software, and can charge for its use, or impose limits on how you use it. Windows is proprietary because you have to pay Microsoft in order to be able to use it legally. Linux, on the other hand, is not proprietary. Anyone can use it or adapt it to their own purposes. Many open source applications also exist, as legally obtainable alternatives to proprietary software such as Microsoft Office.

Podcast 15
HISTORY OF OPERATING SYSTEMS

Computer operating systems provide a set of functions that are needed by most applications to control the computer's hardware. Early computers had no operating system. Users would arrive at their machines armed with programs and data, frequently on punched paper tape. All of the device control logic, and all of the program logic, would be loaded into the machine, and the machine would set to work. It would process the single program until the program completed execution, or stopped by abnormally ending.

In the early days, an abnormal ending was often dramatic. It caused the machine to completely stop, which we call a computer **crash**. Since computer memory in that period was usually based on magnetic cores, and it was necessary to examine the contents of memory to diagnose the problem, listing out the contents of memory was called a **core dump**.

Later, machines were provided with software libraries, called **run time libraries**, containing device support machine language code. This code was linked to a user's program to assist in operations such as input and output. This library of standard device control routines was the genesis of the modern day operating system. But machines were then still only capable of running a single job at a time.

As computers became more powerful the time needed to actually process a given program decreased. But the downtime required for an operator to manually remove a completed program and set up the machine for the next program remained the same. Since computing machines were expensive, this downtime came to represent precious machine time that was being wasted. A way was needed to speed up the removal of a completed program and the setup and initiation of the next program to be run.

Eventually, run time libraries became programs that started before the first customer job, read in the customer job, controlled its execution, cleaned up after it, recorded its usage, and immediately went on to the next job. Accounting practices, implemented by the runtime libraries, were expanded to keep track of the amount of CPU usage made by each program, the quantity of pages printed, cards read, cards punched, disk storage used, and even operator actions required by jobs, such as changing magnetic tapes. Users could automatically be billed for the running of computer programs based on the resources used.

As time went on, the operating systems that developed from the standard run time libraries were enhanced to be manage an **input queue** (waiting line, or "in box") of program awaiting processing. The operating system was provided with a language that could be used to express a list of programs to be processed, one after another, and to indicate where to obtain the inputs and where to write the outputs of each program. The language was named **job control language** (JCL) and the "job" was the set of programs to be run. A job could be mapped out as a series of program executions, each program execution forming one step of a multiple step job. Job control language became the way for a programmer to state the desired sequence of program executions and the names of the input and output files to be accessed by each program. Job control language for IBM mainframes became quite sophisticated, especially for IBM's premiere mainframe operating system, MVS.

Early operating systems were very diverse, with each vendor producing one or more operating systems specific to their particular hardware. Every operating system, even ones from the same vendor, could have radically different operating procedures. Typically, each time the manufacturer brought out a new machine, it came with a new operating system.

This state of affairs continued until the 1960's, when IBM developed its System/360 series of machines, all of which used the same instruction architecture. The DOS, MVS, VM, and ACP/TPS operating systems were eventually developed for this family of machines. These operating systems are still in modern use, having matured as commercial products for over 40 years. Throughout, IBM maintained compatibility with past versions of these operating systems, so that even programs developed several years ago can be run with the current IBM operating systems with no change. The current version of MVS contains and support a version of Unix as well.

Other companies developed their own proprietary operating systems. Control Data Corporation developed their Scope operating system for batch processing in the 1960's. They developed other operating systems during the 1970's that supported simultaneous batch and timesharing use. **Timesharing** is an interactive form of computer processing in which multiple concurrent users are serviced, each with a small slice of time in rotation. In the late 1970's Control Data worked with the University of Illinois in developing the Plato operating system, which used plasma panel displays and long distance timesharing networks. Plato was remarkably innovative for its time, featuring real time chat, and multi-user graphical games.

UNIVAC, the first commercial computer manufacturer, produced a batch oriented operating system that managed magnetic drums, disks, card readers and line printers. In the 1970's, UNIVAC produced the Real Time Basic system to support large scale timesharing.

Digital Equipment Corporation developed many operating systems for its various minicomputer lines, including their PDP, Tops 10, Tops 20, and VAX time sharing systems. Prior to the widespread use of UNIX, Tops 10 was a particularly popular system in universities and in the early research community where the internet architecture was invented.

With rapidly evolving microchip technology in the 1970's it became possible to package an entire CPU and related circuitry on one chip. These first microprocessor systems made it possible for small home computers to be built. Most of these computers had a built in Basic language interpreter in read only memory. This served as a crude operating system, supporting simple file management operations such as file writing, copying, and deletion, and could load a single program and run it. Most of these single task, single user machines contained minimal amounts of random access memory. They were bought for entertainment and educational purposes, not for business applications. Even the relatively rare word processing applications of the day were mostly self-contained programs that took over the machine completely, as did videogames. Most of these early personal computers did not even have a floppy disk drive, which made using a more capable disk based operating system impossible. For a display device, many attached to a television set.

By the 1980's, low priced cathode ray tube visual displays, memory, hard disks, and faster and more powerful microprocessors became available. It became practical to provide more capable operating systems for personal computers. Microsoft enhanced its MS DOS operating system, which had first been provided as PC DOS on IBM's personal computer in 1981. The Xerox Corporation's Palo Alto Research Center had performed research on a graphical user interface based on icons on the screen and a user manipulated pointing device, known as a mouse. Although Xerox never successfully commercialized this type of human interface, graphical user interfaces were added to operating systems by other venders. Apple's Macintosh operating system was brought to market in 1984, Microsoft Windows was released later in the 1980's, and IBM's OS2 was developed in the 1990s. All of these offered a graphical user interface. IBM's OS2 failed to achieve widespread acceptance as a successful commercial product, but the graphical user interface pioneered by the Mac and Windows have become the expected means for people to deal with computer systems in the modern age.

Operating systems made computers more powerful and capable. In the next podcast of the Information technology Work book, we examine how these powerful resources are sometimes misused, either out of ignorance or willful intent, as we consider software piracy and ethical issues of computer usage.

Podcast 16
SOFTWARE PIRACY

Illegally copied software, called **pirated software**, is readily available to anyone, anywhere. From street vendors in Asia or Europe to unscrupulous computer systems retailers in Chicago or Pakistan, pirated programs are sold for far below their retail prices. Or, in many cases, pirated programs are not sold, but rather copied and shared among friends and associates, with no thought or concern regarding the legality of such an action.

The origin of the software piracy problem lies in the misperception of what you acquire when you "buy" a program. What you are actually buying in the form of property is the physical package and CD/ROM. You don't obtain the software on that media as your possession, you obtain a **license** to use it, in specific ways only. A way that you are most often prohibited from using the software is copying and sharing it with others.

Illegal copies of computer software are circulated in a variety of ways. The most common type of piracy, **soft lifting**, occurs when someone shares a program with someone else not authorized by the licensing agreement to use it. A common form of soft lifting involves purchasing a license to use a single copy of software, then loading the software onto several computers. On college campuses, it is hard to identify software that has not been soft loaded.

People regularly lend programs to their roommates and friends, either not realizing it's wrong, or thinking that the chance of being caught and prosecuted for it is negligible. Soft lifting occurs in businesses and private homes as well.

Unscrupulous hardware dealers commit another type of piracy, **hard disk loading**. Hard disk loading involves loading an unauthorized copy of software onto a computer that is being sold to an end user. This makes the purchase more attractive to the buyer and costs the dealer next to nothing. The dealer usually doesn't provide the buyer with manuals or original CD's of the software. This is how operating systems, such as Windows XP, are frequently pirated.

Renting piracy involves someone renting out a copy of software for temporary use, without the permission of the copyright holder. Although this may seem similar to the way that video CD's are rented, from outlets such as BlockBuster, it is very different. BlockBuster's license agreement with entertainment suppliers includes provision for usage payments based on the number of rentals actually made. The license a software purchaser acquires to use a copy of software does not include this type of provision. The software license you acquire only allows you, the purchaser, to use the software yourself, generally on only one computer at a time. The license you acquire to use the software specifically prohibits renting it to another person. You can generally only transfer the use of the software to another person if you sell it to them, and remove it from your computer system. And even this ability may be prohibited by terms of the licensing agreement.

Unbundling is another form of software piracy. The term OEM is associated with this type of piracy. OEM is an abbreviation for "original equipment manufacturer." An OEM is a company that has a special relationship with software producers. OEM piracy means separately selling software that was originally meant to be included as a package with a specific accompanying product. An OEM buys computers in bulk from a computer manufacturing company and then customizes them, often by adding software licensed from a supplier. The OEM then sells the complete bundle under its own name. **Unbundling piracy** means that software, intended to be sold only as a part of this assembled system, is sold separately to a consumer.

Counterfeiting means producing fake copies of a software package, making it look authentic. This involves providing a product box, CD's, and manuals, all designed to look as much like the original product as possible. Because of their widespread popularity and use, Microsoft products are the ones most commonly counterfeited. A copy of a CD/ROM is made and a photocopy of the manual is produced. Counterfeit software is sold on street corners and at flea markets, and sometimes unknowingly sold even in retail stores. Counterfeit software is typically sold at prices far below the regular retail price.

While the internet greatly increases the opportunity to legitimately sell products and services around the world, it also creates new opportunities to steal software. Unauthorized copying of software formerly required the physical exchange of floppy disks, CD/ROMs, or other hard media. But as the Internet continually gets easier to use, and faster and less expensive, software piracy is made easier too. Hundreds of thousands of web sites now provide downloads of software to anyone. Often, the software provided through these sites has had its program code hacked in order to eliminate copy protection schemes. Sometimes, viruses have been placed into the software to intercept information from a person's computer and transmit it on the internet to parties who may use it for theft.

Illegal distribution of software threatens to undermine the tremendous innovation, number of jobs, and the revenue that the internet currently promises. High technology industry is driving the information revolution, which is the corner stone of the new internet economy. At the heart of these technologies is the concept of intellectual property. Intellectual property includes artistic, graphic, musical, and written works, including computer software. Advances in technology are dependent upon strong intellectual property protection.

While the internet has made it easier for all of us to share information, this doesn't mean we can ignore the laws governing usage of intellectual property. Companies have no incentive to invest in the research and development of technological advances if the return on that investment is diminished by illegal pirating of the resulting products.

Many businesses, as well as private parties, face serious legal risks because of software piracy. Under the law, a company can be held liable for its employees' actions. If an employee installs unauthorized software on company computers or acquires software illegally through the internet while at work, the company can be sued for copyright infringement, which is a form of violation of **intellectual property rights**. This is true even if the company's management is unaware of the employee's actions.

Quite simply, to make or download unauthorized copies of software is to break the law, no matter how few copies are involved. Whether you are casually making a few copies for friends, loaning disks, distributing or downloading pirated software via the internet, or buying a single software program and then installing it on 100 of your company's computers, you are committing copyright infringement. And it doesn't matter if you are doing it to make money or not. All of these actions are illegal and can provoke prosecution.

If you violate **copyright law** and are caught, the copyright owner brings a civil action against you, the owner can seek to stop you from using its software immediately, and can also request monetary damages. The copyright owner may choose between actual damages, which includes the amount it has lost because of your infringement, as well as any profits attributable to the infringement, and statutory damages, which can be as much as $150,000 for each program copied. In addition, the government can criminally prosecute you for copyright infringement. If convicted, you can be fined up to $250,000, or sentenced to jail for up to five years, or both.

It is unwise to engage in software piracy. On an individual basis, it's not clever or smart. If you engage in it at work, you tempt the loss of your job, and a host of expensive legal problems. And with an internet connection on a computer, it is becoming increasingly easy for software vendors to become aware of unauthorized program use, by including logic to make internet transmissions to the vendor to periodically identify the computer using a program. You're far better off to license software legally, and to abide by copyright laws concerning its use.

What can you do as a student to obtain the software you need to be productive? Several alternatives exist. Universities and colleges provide computer labs in which software required for class assignments is already legally licensed and installed. Use the labs if you do not have legal use of the software on your own desktop or laptop. In many instance, schools also have made arrangements for students to be able to legally purchase copies of necessary software at heavily discounted prices. "Educational" versions of software, adequate for completing assigned coursework, but not for commercial use, may also be available. In some cases, vendors such as Microsoft and Adobe provide free "viewing" software that can be used to read or view materials without cost. The free Acrobat Reader, for example, lets you read .pdf files. Microsoft's free PowerPoint viewer lets you view PowerPoint presentations, but not create them. Use the lab to create documents and presentations, and the free reader or viewer to show them or view them on your own computer. And finally, legitimate sites on the web often legally provide entirely free or "shareware" software that replicates the functionality of other products. Become familiar with alternatives products that can accomplish what you need to do. These are all preferable, ethically sound ways to be productive on modern computers without violating copyright laws.

Lecture Slides 3

Introduction to Information Technology

Software and Programming Languages

Course resources

PowerPoint presentations, podcasts, and web links for readings are available at **www.ambriana.com** > IT Workbook

Print slides at 6 slides per page

Homework, quizzes and final exam are based on slides, lectures, readings and podcasts

Slide 3-2

Main topics

- Two main types of software
- Application software – two subcategories
- System software – two subcategories
- Programming language generations
- Operating systems
- Software licensing and piracy

Slide 3-3

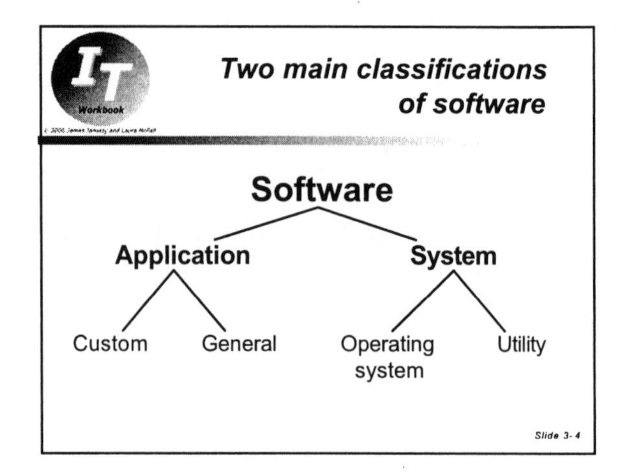

Two main classifications of software

- **Application software:** delivers functionality; directly useful to the end-user; user "sees" it
- **System software:** controls machine devices and interfaces, accesses machine-stored data; user typical does not see it

Slide 3-5

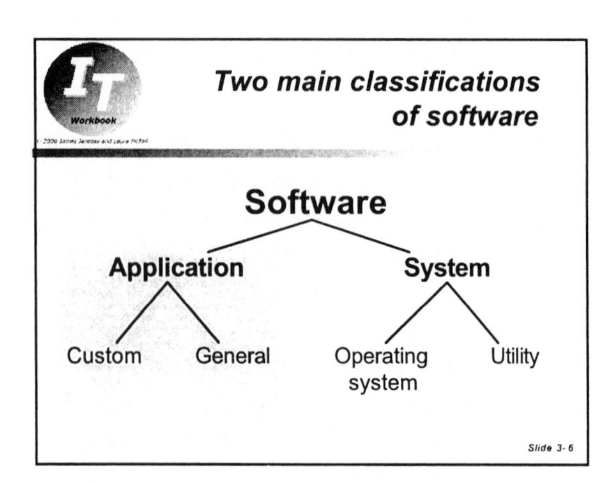

Application software: two subcategories

- **Custom:** such as payroll system, class registration system, air traffic control system; "custom" designed and created
- **General:** word processing, spreadsheet, presentation, desktop database; you can buy it off the shelf, works the same for everybody

Slide 3-7

Can one application program work in different environments?

- Machine "architectures" and operating systems are **different**!
- PC <> Mac <> mainframe <> mini !
- Different CPU instructions, different system processes and commands
- ***Like trying to speak Greek in China!***

Slide 3-8

Application software: How is it developed?

- In the past (prior to around 1990) often developed in COBOL or PL/1; even earlier in assembler language
- Now usually in C, C++, Java, Basic, Visual Basic, or in a proprietary language provided by an Enterprise Resource Planning (ERP) vendor (such as PeopleCode or ABAP)

Slide 3-9

Two main classifications of software

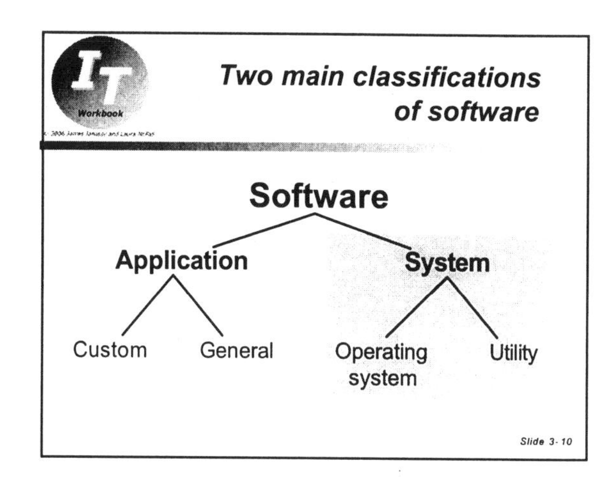

Slide 3-10

System software: two subcategories

- **Operating system:** such as mainframe OS/MVS, ACP/TPF, Unix, MS-DOS, Windows, Linux
- **Utility programs:** programs that do often-necessary chores like sorting, copying, disk formatting, disk defragmentation, data conversion

Slide 3-11

Example of a utility program - how disk "defrag" works

- Typically refers to Microsoft Windows utility called "Disk Defragmenter"
- Designed to solve a problem that occurs because of the way hard disks store data

Slide 3-12

How disk "defrag" works

- Hard disks store data in sectors
- The round, flat disk surface is divided into rings like the rings of a tree
- Divide each ring into pie slices
- Each sector is one pie slice on one ring
- Each sector holds a fixed amount of data, such as 512 bytes

Slide 3-13

How disk "defrag" works

- Read/write arm moves from ring to ring
- To reach a sector, arm moves to the correct ring, waits for the sector to spin by
- **Slow!** (compared to computer)
- Compared to speed of processor and memory, the time it takes for the arm to move and for a sector to spin into place seems like forever!

Slide 3-14

How disk "defrag" works

- Need to minimize arm movement as much as possible!
- Data is stored in sequential sector segments on hard disk
- When disk is new, computer stores data in sequential order, most efficient way to place data on hard disk

Slide 3-15

How disk "defrag" works

- Disk fills up as you install new programs and save documents and files
- When disk is full, erase files to reclaim space
- Files you delete are scattered over the disk
- When you save more files or install more applications, they must be saved in the scattered empty spaces: inefficient use of space, slower to access!

Slide 3-16

How disk "defrag" works

- Disk defragmenter moves files around so every file is stored on sequential sectors
- Good defragmenter also optimizes by moving programs closer to the operating system on the disk
- Goal: minimize arm movement while an application loads

Slide 3-17

How disk "defrag" works

- Done on older disks, defragging increases speed of almost all work
- On a new disk that has never filled up or had any significant number of file deletions, defragmentation will have almost no effect, because everything is sequentially stored already

Slide 3-18

How to defrag in Windows

- Double-click **My Computer**
- Highlight a local hard disk drive by clicking on it once
- Right click the highlighted disk drive
- Click **Properties**
- Click the **Tools** tab, click **Defragment Now**

Click here to learn more about "defragging"

Slide 3-19

System software: How is it developed?

- C, C++, or even assembler; the language must have the ability to deal directly with the hardware
- Language used for utilities has to be compatible with the operating system since it deals with operating system functions

Slide 3-20

What kind of software?

- Word processor, such as Word?
- A program to sort data in a file?
- A photo editor
- DePaul's student system

A link to some reading that helps answer these questions

A quiz to test your knowledge (as part of your homework)

Slide 3-21

Computer programming languages

- **First** generation: machine language
- **Second** generation: assembler
- **Third** generation: Fortran, COBOL, PL/1, C: procedural; C++: object-oriented
- **Fourth** generation: SQL: non-procedural; Java, Visual Basic: event-driven, GUI

Slide 3-22

Computer programming languages – 1st generation

Machine language

```
0001000101000100010100010001010O
0100010100010101100001000001000l
0011100001000111000100010100010O
1100000100100100010011100001000l
0100010000010000100000000100110O
0001000101000100010100010001010O
0100010100010101100001000001000l
0011100001000111000100010100010O
1100000100100100010011100001000l
0100010000010000100000000100110O
0001000101000100010100010001010O
0100010100010101100001000001000l
0011100001000111000100010100010O
1100000100100100010011100001000l
```

- Very hard to write, takes too long to write
- Specific to just one machine
- Inflexible
- Very machine-efficient in memory use and execution speed

Slide 3-23

Computer programming languages – 1st and 2nd generation

Machine language in hex			Assembler
* Microsoft COBOL Code Generator			Page 0001
* Address	Code		T2B.OBJ
*			
00000000	83C404	ADD	SP,04
00000003	C6462500	MOV	BYTE PTR [BP+25],00
00000007	FFE0	JMP	AX
00000009	E90300	JMP	_CHKSTK
0000000F	5B	POP	BX
00000010	3BD8	CMP	BX,AX
00000012	7517	JNE	_CHKSTK
00000014	5A	POP	DX
00000015	59	POP	CX
00000016	3A6E25	CMP	CH,[BP+_CHKSTK]

Slide 3-24

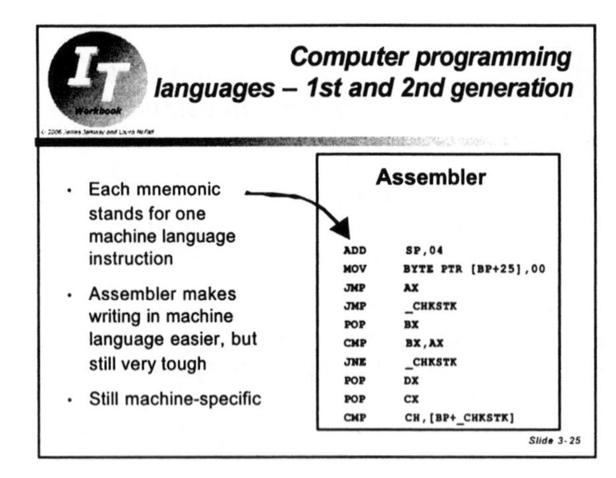
Computer programming languages – 1st and 2nd generation
Workbook
Each mnemonic stands for one machine language instruction
Assembler makes writing in machine language easier, but still very tough
Still machine-specific
Assembler
ADD SP,04
MOV BYTE PTR [BP+25],00
JMP AX
JMP _CHKSTK
POP BX
CMP BX,AX
JNE _CHKSTK
POP DX
POP CX
CMP CH,[BP+_CHKSTK]
Slide 3-25

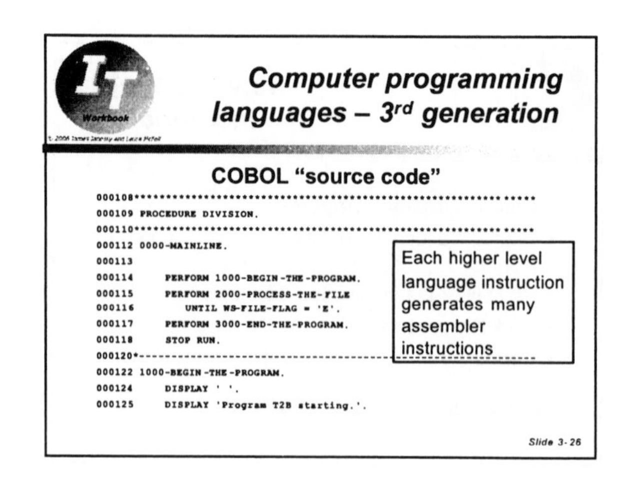
Computer programming languages – 3rd generation
Workbook
COBOL "source code"
000109 PROCEDURE DIVISION.
000112 0000-MAINLINE.
000113
000114 PERFORM 1000-BEGIN-THE-PROGRAM.
000115 PERFORM 2000-PROCESS-THE-FILE
000116 UNTIL WS-FILE-FLAG = 'E'.
000117 PERFORM 3000-END-THE-PROGRAM.
000118 STOP RUN.
000122 1000-BEGIN-THE-PROGRAM.
000124 DISPLAY ' '.
000125 DISPLAY 'Program T2B starting.'.
Each higher level language instruction generates many assembler instructions
Slide 3-26

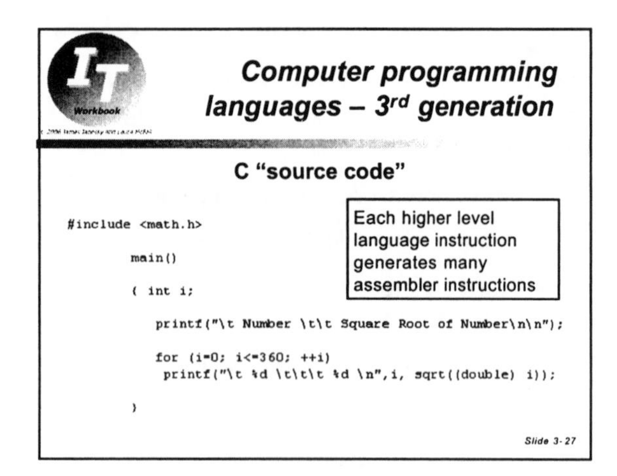
Computer programming languages – 3rd generation
Workbook
C "source code"
#include <math.h>
main()
{ int i;
printf("\t Number \t\t Square Root of Number\n\n");
for (i=0; i<=360; ++i)
printf("\t %d \t\t\t %d \n",i, sqrt((double) i));
}
Each higher level language instruction generates many assembler instructions
Slide 3-27

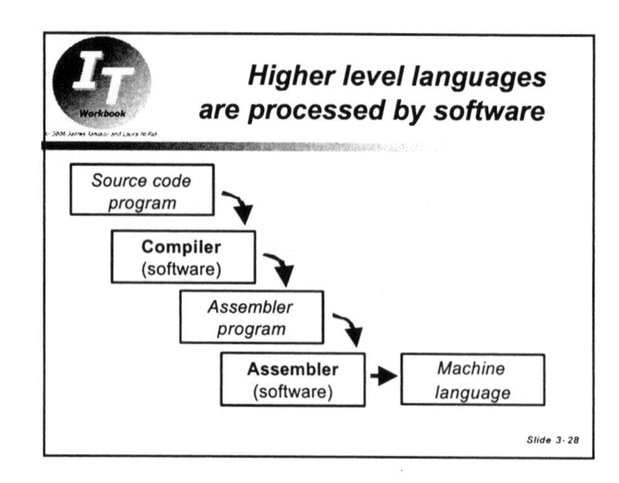
Higher level languages are processed by software
Workbook
Source code program
Compiler (software)
Assembler program
Assembler (software)
Machine language
Slide 3-28

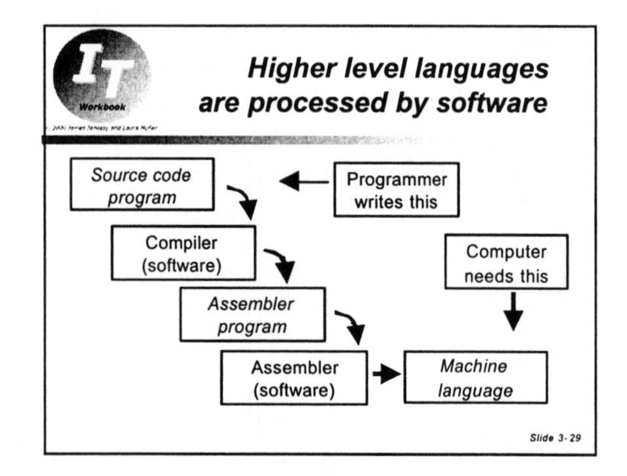
Higher level languages are processed by software
Workbook
Source code program
Programmer writes this
Compiler (software)
Computer needs this
Assembler program
Assembler (software)
Machine language
Slide 3-29

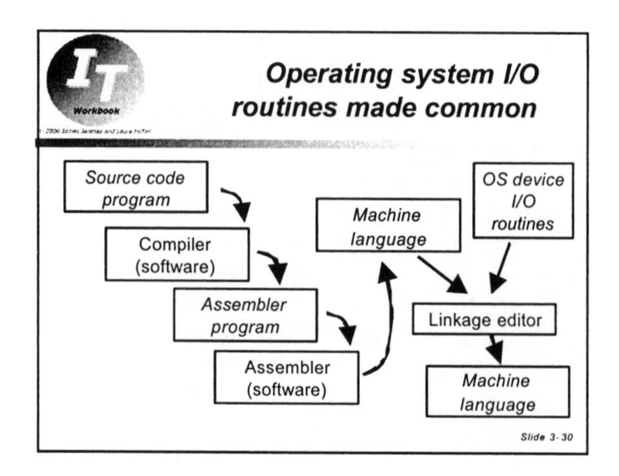
Operating system I/O routines made common
Workbook
Source code program
OS device I/O routines
Machine language
Compiler (software)
Assembler program
Linkage editor
Assembler (software)
Machine language
Slide 3-30

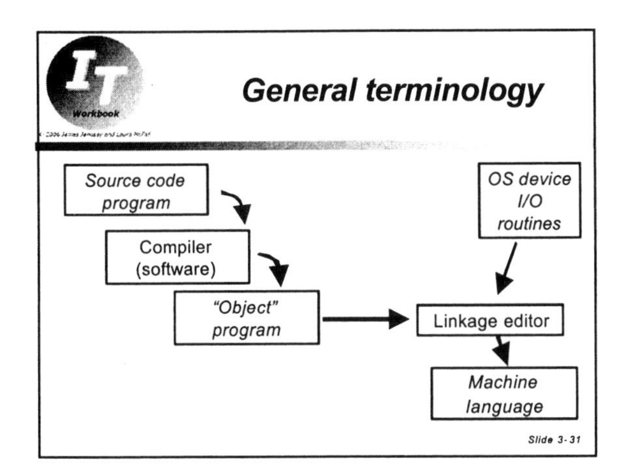
General terminology
Workbook
Source code program
Compiler (software)
"Object" program
OS device I/O routines
Linkage editor
Machine language
Slide 3-31

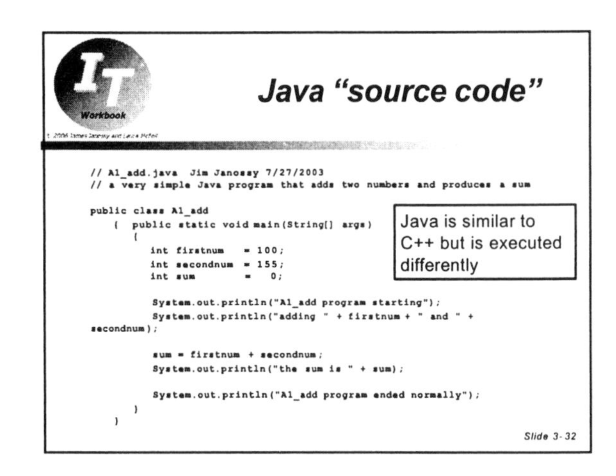
Java "source code"
Workbook
// A1_add.java Jim Janossy 7/27/2003
// a very simple Java program that adds two numbers and produces a sum
public class A1_add
(public static void main(String[] args)
(
int firstnum = 100;
int secondnum = 155;
int sum = 0;
System.out.println("A1_add program starting");
System.out.println("adding " + firstnum + " and " +
secondnum);
sum = firstnum + secondnum;
System.out.println("the sum is " + sum);
System.out.println("A1_add program ended normally");
)
)
Java is similar to C++ but is executed differently
Slide 3-32

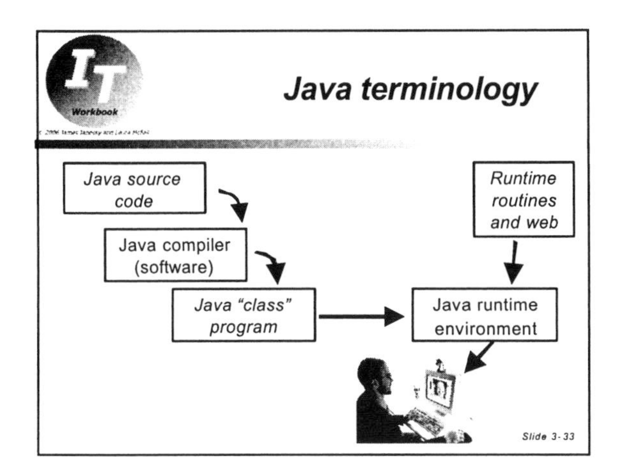
Java terminology
Workbook
Java source code
Java compiler (software)
Java "class" program
Runtime routines and web
Java runtime environment
Slide 3-33

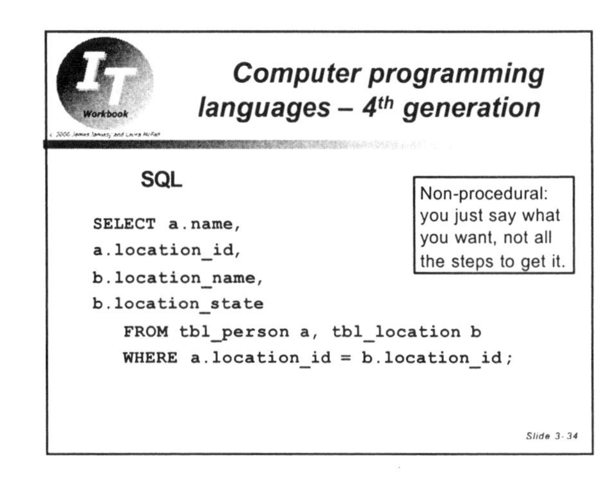
Computer programming languages – 4th generation
Workbook
SQL
SELECT a.name,
a.location_id,
b.location_name,
b.location_state
FROM tbl_person a, tbl_location b
WHERE a.location_id = b.location_id;
Non-procedural: you just say what you want, not all the steps to get it.
Slide 3-34

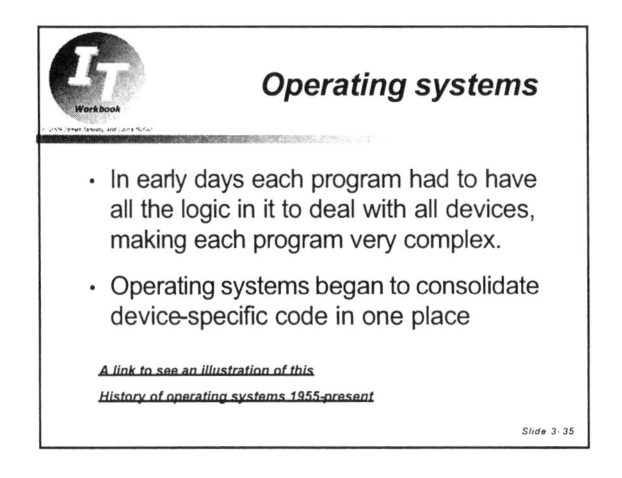
Operating systems
Workbook
In early days each program had to have all the logic in it to deal with all devices, making each program very complex.
Operating systems began to consolidate device-specific code in one place
A link to see an illustration of this
History of operating systems 1955-present
Slide 3-35

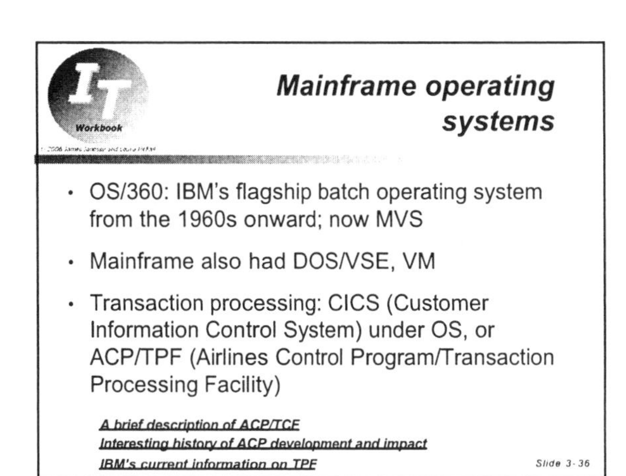
Mainframe operating systems
Workbook
OS/360: IBM's flagship batch operating system from the 1960s onward; now MVS
Mainframe also had DOS/VSE, VM
Transaction processing: CICS (Customer Information Control System) under OS, or ACP/TPF (Airlines Control Program/Transaction Processing Facility)
A brief description of ACP/TCF
Interesting history of ACP development and impact
IBM's current information on TPF
Slide 3-36

Minicomputer operating systems

- Unix – developed at AT&T, then UC Berkeley
- Several proprietary systems, each manufacturer came up with one
- Linux is a Unix clone, very popular!

History of Unix
About Linux

Slide 3-37

Microcomputer operation systems

- CPM – "Control program, microcomputers" early pre-PC era
- MS-DOS, PC-DOS, IBM PCs
- Macintosh GUI OS, now uses Unix
- Windows is Microsoft's GUI OS

About Microsoft Windows
A nice history that covers machines and operating systems

Slide 3-38

Software "types"

- **Retail:** use only if you have paid for it
- **Shareware:** Software is downloadable from the Internet. Try the program for free, but expected to pay if you keep it
- **Freeware:** free for personal use
- **Spyware:** sneaks data about you to the provided of the software

Slide 3-39

What do you get when you "pay for it?"

- **Retail:** use only if you have paid for it
- **Shareware:** Software is downloadable from the Internet. Try the program for free, but expected to pay if you keep it
- **Freeware:** free for personal use
- **Spyware:** sneaks data about you to the provided of the software

Slide 3-40

What do you get when you "pay for it?"

- **Retail:** use only if you have paid for it
- You get a license to use it as specified
- You don't "own" it
- You are not authorized to copy it and give it to others, or to make it available
- Violating the license is copyright infringement!

Slide 3-41

Software piracy and associated problems

- Several ways exist that people acquire software without properly licensing it
- Several downsides to **pirated software**
- Exposure to viruses, lack of documentation, lack of vendor support, serious problems with your employer!

Types of software piracy
Legal issues in copying software without authorization
Internet piracy

Slide 3-42

A0_hello.java

===Program A0_hello compiled on 08/01/03 at 12

```
// A0_hello.java  Jim Janossy  7/28/2003
// This program sends a message to the output

public class A0_hello

   {  public static void main(String[] args)

      {
       System.out.println("Hi there you lucky people!");
      }

   }
```

We'll show all output this way, after the program itself. This illustration is what you find in the LIST.txt file produced when you compile and run a java program using the ***JavaPlease!*** environment. For example, **A0_hello.java** produces **A0_hello_LIST.txt** as output. You can view the output using Notepad, an editor included in all versions of Windows. Don't see it? Click on View, Refresh! *Problems? See Appendix A!*

```
===End of program listing, output follows===================

Hi there you lucky people!

===End of program output====================================
```

Getting started…

This is a first program in Java. I composed it using the Notepad text editor. You can open it in your \JavaPlease folder at A0_hello.java using Notepad. You should do this now to examine it! The program has two lines of comments that start with slashes //. The Java compiler ignores lines like this but we include them to indicate who wrote the program, when they wrote it, and what it is intended to do. All this program does is output the line "Hi there you lucky people!".

Java defines a small set of reserved words and symbols, such as "public", "class", "static", "void", the punctuation symbols { } [ ] ; and the quote " and a few others that have a special meaning. The reserved words and symbols in this program are indicated here in **bold** lettering:

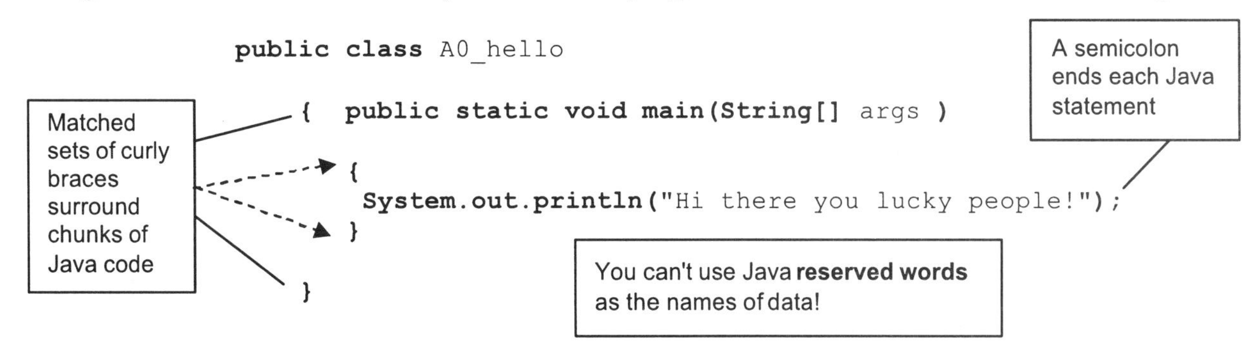

Looked at this way, you can see that only the program name A0_hello, the data name "args", and the phrase we want to print are original to this program. You don't have to understand what all of the reserved words mean in order to start using java to learn about it! But keep in mind that unlike many programming languages in the Windows environment, ***Java is case sensitive.*** This means that in Java words like "string" and "String" are different because of the capitalization. You must capitalize (or not capitalize) correctly in Java or errors will result! Another common source of error is losing track of the matched sets of curly braces that need to surround parts of Java code.

A1_add.java

```
===Program A1_add compiled on 08/01/03 at 12:45==================

// A1_add.java  Jim Janossy 7/27/2003
// a very simple Java program that adds two numbers and produces a sum

public class A1_add

   {  public static void main(String[] args)

      {
         int firstnum    = 100;
         int secondnum   = 155;
         int sum         =   0;

         System.out.println("A1_add program starting");
         System.out.println("adding " + firstnum + " and " + secondnum);

         sum = firstnum + secondnum;
         System.out.println("the sum is " + sum);

         System.out.println("A1_add program ended normally");
      }
   }

===End of program listing, output follows===================

A1_add program starting
adding 100 and 155
the sum is 255
A1_add program ended normally

===End of program output=====================================
```

The **class name** here must exactly match the **file name** under which this program source code is stored on disk

Data declarations

Data item type

To concatenate items in a printing action in Java we use the "+" symbol

The "+" sign means "add" in this context

Performing arithmetic...

In this small Java program we declare two integer number data items, naming them "firstnum" and "secondnum", and we give them initial values of 100 and 155. We also declare a third integer data item named "sum" which we will use to house the value that results when we add the first two numbers together. The reserved word **int** indicates the data type. Besides int, Java supports other forms of whole number data such as **byte**, **short**, and **long**, as well as real numbers (numbers with decimal points and fractional components) in the form of **float** and **double**. Java also supports character data as **char** and **boolean** true/false data types. Collections of characters, usually called "strings" in other languages, are not supported by Java itself but are handled using a class constructed of Java primitives. The defining word for a string is **String**, in which the first letter is capitalized to denote that it is not an inherent part of the Java language.

Data items are called "variables." Variable names can be any contiguous group of letters or numbers except reserved words or the boolean values "true" or "false". By convention we begin variable names with a lowercase letter. Class ("program") names start with a capital letter.

A2_whileloop.java

===Program A2_whileloop compiled on 08/02/03 at 12:32============

```
// A2_whileloop.java  Jim Janossy 7/27/2003
// a Java program including a "while" (until) loop

public class A2_whileloop

   {  public static void main(String[] args)

      {
         int start   =  1;
         int limit   = 10;
         int current =  0;
         int sum     =  0;

         System.out.println("A2_whileloop program starting");
         System.out.println("sum from " + start + " thru " + limit + ":");
         current = start;

         while (current <= limit)
            {
               sum = sum + current;
               System.out.println("After adding " + current +
                  " the sum is now " + sum);
               current = current + 1;
            }

         System.out.println("The total is " + sum);
         System.out.println("A2_whileloop program ended normally");
      }

   }
```

Actions in a "while" loop will not be executed at all if the "while" question starts off as false!

Data declarations

Matched sets of curly braces surround chunks of java code

DO NOT put a semicolon here!

All of the actions between this set of curly braces are repeated as long as the answer to the question **"is current <= limit ?"** is true.

Looping action, also know as **iteration**

A common problem: putting a semicolon on the **while** line. Java punctuation seems strange to newcomers since it involves heavy use of the semicolon. Java statements are terminated with a semicolon. But the end of the while line is not the end of the statement, so you do not put a semicolon there! If you do put a semicolon there you create an "empty" while loop that may never end, since nothing changes the state of the while variable in the while question!

```
===End of program listing, output follows=====

A2_whileloop program starting
sum from 1 thru 10:
After adding 1 the sum is now 1
After adding 2 the sum is now 3
After adding 3 the sum is now 6
After adding 4 the sum is now 10
After adding 5 the sum is now 15
After adding 6 the sum is now 21
After adding 7 the sum is now 28
After adding 8 the sum is now 36
After adding 9 the sum is now 45
After adding 10 the sum is now 55
The total is 55
A2_whileloop program ended normally

===End of program output=====================================
```

The actions within the loop are executed 10 times

B0_array.java

===Program B0_array compiled on 08/04/03 at 17:18================

```
// B0_array.java  Jim Janossy  8/4/2003
// Introduction to simple, one-dimension arrays

class B0_array
{
   public static void main( String[ ] args )

   {
      char letterval [] = new char [10];
      int i;

// beginning

      System.out.println("B0_array program starting");
      System.out.println(" ");

      letterval[1] = 'H';
      letterval[3] = 'e';
      letterval[5] = 'l';
      letterval[6] = 'l';
      letterval[7] = 'o';

// loop

      for (i=0;  i <=9;  i=i+1)
       {
         System.out.println("letterval [" + i + "] is " + letterval[i]);
       }

// end

      System.out.println(" ");
      System.out.println("B0_array program ended normally");
   }
}
```

Define 10 occurrences of a character variable named **letterval**. These will exist as **letterval [0]** through **letterval [9]**.

This assigns values to selected occurrences of letterval by program action. Char variables are initialized with a single character surrounded by single apostrophes (not the quote). An occurrence of a char variable not assigned a value retains the default value of null. Numeric variables not assigned a data value retain the default value of zero.

Pointer value is an index

```
===End of program listing, output follows===================

B0_array program starting

letterval [0] is
letterval [1] is H
letterval [2] is
letterval [3] is e
letterval [4] is
letterval [5] is l
letterval [6] is l
letterval [7] is o
letterval [8] is
letterval [9] is

B0_array program ended normally

===End of program output=====================================
```

Null values contained by default in occurrences 0, 2, 4, 8 and 9 do not print anything

IT Workbook Assignment 3.1

Student name:

Write your answers to the questions below within the box. In each case, please choose your words carefully to answer the specific questions. Avoid simply copying passages from your readings to these answers.

Provide a concise but complete one-sentence definition for **computer software**:

1)

Identify and describe each of the four **major types** of software:

2)

3)

4)

5)

IT Workbook Assignment 3.2

Student name:

Write your answers to the questions below within the box. In each case, please choose your words carefully to answer the specific questions. Avoid simply copying passages from your readings to these answers.

Explain what **shareware** is, what **freeware** is, and how these differ from each other:

1)

Identify and describe each of the **first four generations of computer programming languages** and explain how each is different from the generation that preceded it:

2)

3)

4)

5)

IT Workbook Assignment 3.3

Student name:

Write your answers to the questions below within the box. In each case, please choose your words carefully to answer the specific questions. Avoid simply copying passages from your readings to these answers.

Visit web link #4 at the Chapter 3 button for the Information Technology Workbook at www.ambriana.com and complete the interactive ***Types of Software Quiz*** by clicking the appropriate radio button at the right of each type of software. When you are finished, click the Check Answers button at the bottom of the web page. This will tell you how many of the 20 items you categorized correctly. If you did not score correct on all 20, use your browser's Back button to go back to the quiz and change come of your indications. Continue correcting and rechecking your answers until you achieve a perfect score (all 20 correct).

When you have achieved a score of 20 correct on this quiz, use the File/Print function of your browser to print a copy of your quiz answers (not just the evaluation and score!). Attach that printed copy to this page.

1) *(attach your printed answer sheet)*

Can a program created for the PC Windows environment work on an Apple Mac, and vice versa? Explain why or why not:

2)

IT Workbook Assignment 3.4

Student name:

Write your answers to the questions below within the box. In each case, please choose your words carefully to answer the specific questions. Avoid simply copying passages from your readings to these answers.

Explain what **crippleware** and **demo software** are, and what **purpose** each of these serve:

1)

Identify the categories (general types) of **general purpose software packages** and give an example of each:

2)

3)

4)

5)

IT Workbook Assignment 3.5

Student name:

Write your answers to the questions below within the box. In each case, please choose your words carefully to answer the specific questions. Avoid simply copying passages from your readings to these answers.

In the modern computer environment, can **application software** exist and run without **operating system software**? Describe each type of software, and explain the **relationship** that exists between them:

1)

A **disk defragger** is an example of utility software. Explain in detail what a **disk defragger** does, and why it is a useful piece of software on a personal computer:

2)

IT Workbook Assignment 3.6

Student name:

Write your answers to the questions below within the box. In each case, please choose your words carefully to answer the specific questions. Avoid simply copying passages from your readings to these answers.

Describe what **machine language** is, what **assembler language** is, and the **relationship** between these two computer programming languages:

1) what machine language is:

2) what assembler language is:

3) the relationship between machine language and assembler language:

Identify a **procedural programming language** and a **non-procedural programming language**, and explain the **key difference** between these generations of programming languages:

4) an example of a procedural language:

5) an example of a non-procedural language:

6) the key difference between these generations of programming languages:

IT Workbook Assignment 3.7

Student name:

Write your answers to the questions below <u>*within the box*</u>*. In each case, please choose your words carefully to answer the specific questions. Avoid simply copying passages from your readings to these answers.*

Describe what **source code** is, **how it is prepared**, and what **processes** must be performed to make an executable program from it:

1) what source code is:

2) how it is prepared:

3) what processes are needed to produce an executable program from source code:

Describe how the **execution of a Java program** differs from the execution of a program written in a language such as COBOL or C:

4)

IT Workbook Assignment 3.8

Student name:

Write your answers to the questions below within the box. In each case, please choose your words carefully to answer the specific questions. Avoid simply copying passages from your readings to these answers.

Explain what a **multi-user operating system** can do, what a **multi-tasking operating system** can do, and what a **multi-programming operating system** can do that causes your descriptions of them to be suitable:

1) multi-user operating system:

2) multi-tasking operating system:

3) multi-programming operating system:

Describe what the human/computer interface of a **command-line oriented operating system** provides, and, in contrast, what a **graphical user interface (GUI)** oriented operating system provides in this area:

4)

IT Workbook Assignment 3.9

Student name:

Write your answers to the questions below within the box. In each case, please choose your words carefully to answer the specific questions. Avoid simply copying passages from your readings to these answers.

Explain what **public domain software** is, and what level of **support** a user can expect of it:

1)

Software piracy is common but damaging and potentially dangerous. In addition to the possible legal consequences of it, identify **four deficiencies and dangers** that are associated with using copied or counterfeit software on your computer:

2)

3)

4)

5)

IT Workbook Assignment 3.10

Student name:

Write your answers to the questions below within the box. In each case, please choose your words carefully to answer the specific questions. Avoid simply copying passages from your readings to these answers.

Identify the six types of software piracy that exist, and provide a concise description of what actions are involved in each type:

1)

2)

3)

4)

5)

6)

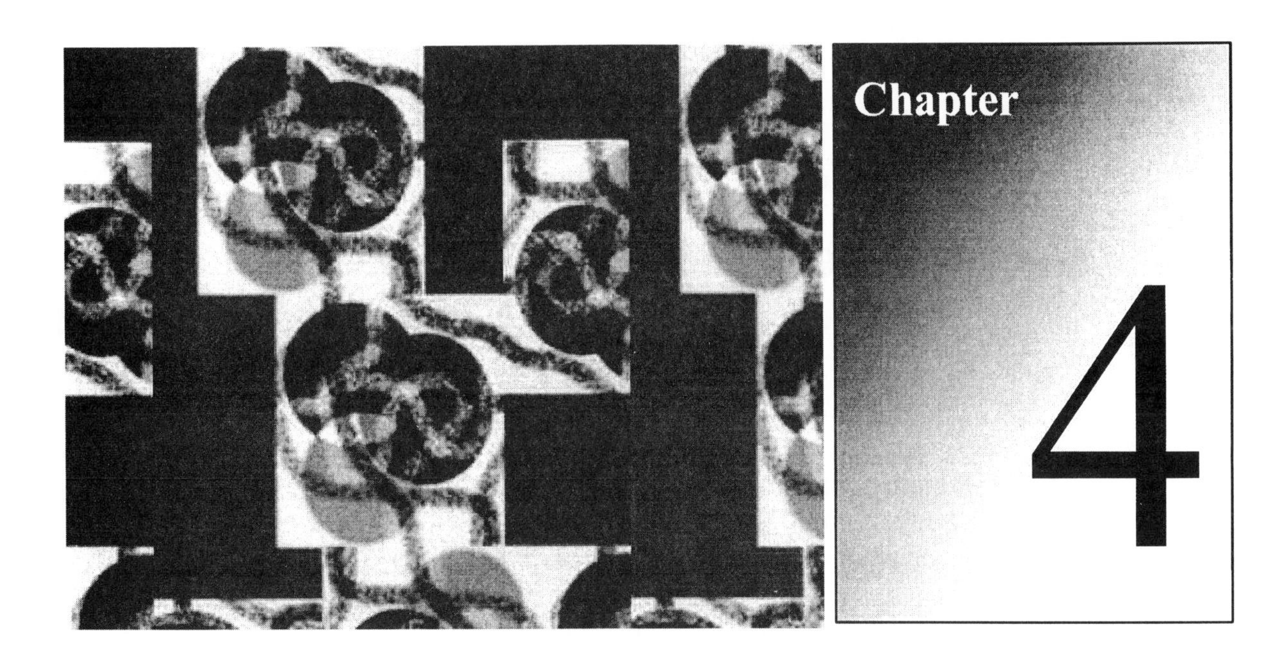

Databases Technology

You may think that any collection of information is a "database." But in the more specific language of information technology **a database is a special arrangement of data that makes it readily accessible**. The best database mechanisms are those that provide rapid, flexible access to data even in ways that had not been entirely anticipated by application database designers.

A database is usually built relying on support software known as a **database management system (DBMS)**. A database management system is complex software that most often serves the purposes of many application systems. That is, much like the operating system, a database management system on machines larger than personal computers is usually a shared resource that supports many other processes running on the computer system.

Three generations of database management system have been invented since computers began to be used commercially. The first two generations are now in only limited use. **Hierarchical databases** such as IBM's IMS system were limited in the types of data structures they could support, but IMS was widely used on mainframe computers, and is still popular in certain large financial applications. **Network databases** were prominent in the 1970s-80s and were more capable than IMS but were also limited in flexibility. Their use waned in the 1990's.

Modern databases are based on relational data concepts. **Relational databases** rely on data arranged in tables accessed by Structured Query Language (SQL). Relational databases free database designers from concerns about how data is physically stored. SQL Server, DB2, and Oracle are modern relational database management systems in common use. In this chapter we'll consider how relational databases work, and some implications of database usage for privacy.

Learning objectives

Chapter 4
Database Technology

This chapter, in combination with the web readings, web links, and podcasts available at www.ambriana.com is focused on providing the knowledge to achieve these learning objectives:

- Be able to distinguish between data and information
- Understand the formats of open and proprietary flat files and the common uses for these data formats
- Know the nature, role, and functionalities of non-relational and relational database management systems
- Understand the progression of technologies comprising the three different generations of database management systems, and be able to describe the capabilities and limitations of each generation
- Be able to describe the goal of data normalization, and why it is important in relational database design
- Understand and be able to distinguish primary keys and foreign keys, and their relationship to the concept of referential integrity
- Recognize the meaning of the symbols on an entity relationship diagram, and how to represent simple relationships between two tables using these symbols
- Understand how data mining is different from simple database queries, the major types of data mining operations, and the principles of fair information practice as it relates to data mining activities.

Podcast Scripts – Chapter 4

These readings of the **Information Technology Workbook** are also accessible in audio form as .mp3 files for free download. Visit the workbook web site, www.ambriana.com. You can listen to this material on your own Ipod or .mp3 player or computer, read it, or both listen and read.

Podcast 17
DATABASE TECHNOLOGY

A **database management system** (DBMS) is a computer program designed to manage a large set of structured data, and to run requested operations on that data. Examples of database management systems include human resources and customer support systems. Originally found only in large organizations with the computer hardware needed to support large data sets, DBMS's have emerged as a standard part of a business operations.

Databases have been in use since the earliest days of electronic computing, but most of the early ones were custom programs written to access custom organizations of data. While modern systems can be applied to widely different databases and needs, these systems were tightly linked to a particular data storage implementation, gaining speed at the expense of flexibility.

As computers grew in capability, this tradeoff became unnecessary. By the mid 1960's a number of general purpose database systems were in commercial use. Interest in standardization began to grow, and in 1971 a standard, known as the CODASYL approach was introduced. Soon a number of commercial products based on CODASYL were available.

IBM developed their own DBMS by 1968, named Information Management System (IMS). It was similar in concept to CODASYL but it used a strict hierarchy for its model of data navigation instead of CODASYL's linked-list network model. Both concepts later became known as **navigational databases** because of the way data was accessed.

IBM employee Edgar Codd was unhappy with CODASYL's navigational approach, notably the lack of a search facility, which was becoming increasingly useful when a database was stored on disk instead of tape. In 1970 he outlined a radically new approach to database construction.

Codd described a system for storing and working with large databases based on mathematical set operators and manipulations. Instead of data being stored in a linked list of freeform rows, Codd's idea was to use a table of fixed-length rows. A linked list system would be very inefficient when storing sparse databases, where some of the data for any one row could be left empty. Codd's relational model solved this by splitting the data into a series of normalized tables, with optional elements being moved out of the main table to other tables, where they would take up room only if needed.

In the relational model, related rows are linked together with a key. For instance, a common use of a database system is to track information about users, names, login information, addresses and phone numbers. In the navigational approach, all of these data would be placed in a single row, and unused items would simply not be placed in the database. But in the relational approach, data would be normalized into a user table, an address table and a phone number table, for example. Rows would be created in these optional tables only if addresses or phone numbers were actually provided.

Linking the information back together is at the heart of this system. In the relational model, some item of information was used as a key, uniquely defining a particular row in a table. When information was being collected about a user, information stored in the related tables would be found by searching for this key in those tables.

Just as the navigational approach required programs to loop in order to collect records, the relational approach requires loops to collect information about any one item. Codd's solution to the necessary looping was a set-oriented language, a suggestion that would later result in development of SQL. Structured English Query

Language, or Sequel, was designed to manipulate and retrieve data. (The name Sequel was later condensed to SQL, because the word "Sequel" was held as a trademark by an aircraft company in the United Kingdom.)

IBM started working on a prototype system, loosely based on Codd's concepts, in the early 1970's. Known as **System R**, it was, unfortunately, conceived as a way of proving Codd's ideas unimplementable. The first quick version was ready in 1974. Work then started on multi-table systems in which all data did not have to be stored in a single large chunk. By 1979, the standardized computer language SQL had been developed and added. Codd's ideas were establishing themselves as both workable and superior to the CODASYL network, pushing IBM to develop a true production version of System R, which was initially named **SQL/DS**. Later, this was renamed **DB2**. Proving E.F. Codd right after all, it revolutionized the world of database management! (Review Codd's 12 rules for relational database management systems on page 108.)

Many people involved with these early databases became convinced of the future commercial success of such systems and formed their own companies to commercialize the work with an SQL interface. These commercialized database versions were all being sold as offshoots of the original products. Microsoft's **SQL Server** is actually a rebuilt version of these original designs. The **Oracle** database management system is another highly successful competing version. **MySQL** is an implementation available for personal use and experimentation but is also capable of supporting actual applications.

One of these original IBM developers, Michael Stonebraker, went on to apply lessons learned to develop new database products, including **PostgreSQL**, now used for global mission critical applications. For example, the .info and .org domain name registries on the World Wide Web use it as their primary data store, as do large companies and financial institutions.

A DBMS can be an extremely complex set of software programs that controls the organization, storage and retrieval of data in a database. One of the basic functionalities a DBMS must provide is a **modeling language** to define the layout of each database hosted in the system, based on the DBMS data model. The three most common organizations are **hierarchical**, **network** and **relational**. A database management system may provide one, two, or all three methods. The most suitable structure depends on the application, transaction rate, and number of inquiries that will be made.

A database query language and report writer is also necessary, to allow users to interactively interrogate the database, analyze its data, and update it. It also controls database security by preventing unauthorized users from viewing or updating the database. For example, an employee database can contain all the data about an individual employee, but one group of users may be authorized to view only payroll data, while others are allowed access to only work history and another only medical data.

A database management system provides a way to interactively enter and update the database, but it may not leave an audit trail of actions or provide the kinds of controls necessary in a multi-user organization. These controls are only available when a set of application programs are customized for each data entry and updating function.

The DBMS can maintain the integrity of the database by not allowing more than one user to update the same row of data at the same time, or by keeping duplicate rows out of a given table in the database. For example, no two customers with the same customer numbers, or key fields, can be entered into the database.

When a DBMS is used, information systems can be changed much more easily as the organization's information requirements change. New categories of data can be added to the database without disrupting the existing system.

Organizations may use one kind of DBMS for daily transaction processing and then move the data into another DBMS better suited for random inquiries and analysis. This is often done when data warehousing is used for analytical reporting. Overall system design decisions are performed by data administrators and systems analysts, while detailed database design is performed by database administrators.

Podcast 18
CREATING A DATABASE

Spreadsheets are the number crunchers of the digital world, and databases are information crunchers. Databases excel at managing and manipulating structured information.

What does structured information mean? Consider the telephone directory. The phone book contains several items of information, name, address, and phone number, about each phone subscriber in a specific area. Each subscriber's information takes the same form. In database language, the phone book is a table that contains a data row for each subscriber. Each subscriber row contains three columns: name, address, and phone number. The rows are sorted alphabetically by the name column, which is called the key.

A database can contain a single table of information, such as the phone book, or many tables of related information. An order entry system for a business, for example, will consist of many tables, including an order table to track orders, a customer table so you can see who placed the order and who to bill, an inventory table, showing the products you have on hand, and so on.

Each of these tables will be linked to one or more of the other tables, so information can be tied together to produce reports or answer questions about the information in the database. Multi-table databases like this are called **relational databases**. They provide exceptional power and flexibility in storing and retrieving information.

To create and maintain a computer database, you need a database program named a database management system (DBMS). Just as databases range from simple, single table lists to complex multi-table systems, database programs also range in complexity.

Some are designed purely to manage single file databases. You can't build a multi-table database with these products. You can create numerous tables for storing different types of information, but there's no way to link information from one table to another. Such programs are sometimes called flat file databases, or list managers. An Excel spreadsheet is such a creature.

Other database programs, called relational database programs RDBMS's, are designed to handle multi-table databases. The most popular desktop relational databases today is Microsoft Access. Lotus Approach and Corel Paradox are similar products. SQL Server, DB2 and Oracle are much heavier duty products for shared server and network use.

A database program provides the tools to design a database structure, create data entry forms to populate the database, validate the data entered and check for inconsistencies, sort and manipulate data in the database, ask the database questions about the data it contains, and produce flexible reports that make it easy to understand the information stored in the database.

Most database programs can be used at many levels. For beginners, templates, sample databases, wizards and experts exist that will do much of the work. If the built in databases don't quite meet the need, it's easy to modify an existing database and to learn to create simple database structures on desktops from scratch. The more powerful database programs enable advanced users to create complete, custom built, application-specific systems.

While a database program can do things across a broad spectrum, such as managing inventory in a supply warehouse to managing personal finances, sometimes it's best not use a database at all, because there's no point in reinventing the wheel. If you want a personal financial manager, you're far better off spending money on a commercially available program than laboring for weeks creating your own version of the same thing. On the other hand, trying to accomplish more complex information management using such a simple approach is counterproductive.

Planning is crucial to creating a database. You must look at the information you want to store and the ways you will want to retrieve it before you start working on the computer. A poorly designed database will hamper anyone attempting to retrieve information in usable form.

Databases range from very simple to very complex. For personal computers, two main types of database programs exit: flat file databases and relational databases. Flat file databases, also called single file or list managers, are as easy to learn as a word processor or spreadsheet program. If you need a database at home, in a class at school or in a small organization, chances are the simple, flat file database will do. A better name for this type of data organization is simply a spreadsheet, although true list managers provide more report creation capabilities than a simple spreadsheet.

With a more complex relational database program, however, you can create a range of databases, from flat file structures to demanding multi-table systems. If you need a database in a small business or large organization, you're likely to need at least some of the features of a more complex relational database.

Whichever type of database program you use, the most crucial step is to design your database structure carefully. The way data is structured will affect every other action. It will determine how easy it is to enter information into the database, how well the database will catch inconsistencies and exclude duplicate data, and how easily information can be retrieved.

One way to see if your database design works is to test it with some sample data. Create a pilot version of your database using a simple tool such as Microsoft Access, feed some test data into it and try manipulating the data to see if you can retrieve data in the ways needed. If your results are not as hoped for, revise your design and come up with a better arrangement of data in tables. Retest your data with more attempts to query the data to see if you can then get what you want.

For example, you may initially enter the full names of people in your database as one column, then discover that it makes more sense to be able to sort the names on last name only. With a little more work, you can break your pilot database structure down further into last name, first name, street address, city, and phone number. With this structure you'll be able to sort your database alphabetically by last name or by city, and you'll be able to extract the data of all those people who live in a particular city.

Creating an efficient table structure often requires breaking columns down into simpler and simpler components. You end up with a table for each "entity" or object type that your database will deal with. For example, instead of placing all the data about each customer and each order placed by the customer in one row, you distribute the information about customers to one table, and orders to another. This eliminates data redundancy, since you no longer repeat the same information about each customer in each order. The technical term for this process is **data normalization**.

In some ways, creating a database that is effective and simple to use is almost anti-intuitive. For example, the initial structure of three columns, name, address, and phone number, seems simpler than the end result of first name, last name, street address, city, and phone number. But while the structure might look more complex, the contents of each column have been reduced to the simplest, most useful components.

Problems can be created by not getting the original design right. But many databases allow minor infractions of database design rules. As long as your table structures provide flexibility and desired output, eliminate redundant or duplicated information, and exclude inconsistencies, the database can work well. And while restructuring the database, if necessary, requires some work, it is later possible to redistribute the information to an even more capable structure.

Podcast 19
DATABASE DESIGN

Database design is the process of producing a detailed model of an organized collection of data. This detailed model must reflect the logical and physical design choices and physical storage requirements needed to express the model in a data definition language such as SQL. Database design can be thought of as the planning of the data structures used to store data. In a relational database these structures are known as **tables**. Once created, this detailed model can be used to create the database itself.

The process of doing database design generally consists of a number of steps carried out by a database designer. The database designer must determine what data will be stored in the database, determine the relationships between the different data elements in the database, and then establish a structure based on these relationships. While this sounds complicated, in essence relational database design consists of identifying the objects ("entities") about which information is being stored, the facts ("attributes") about these objects that we need to record, and the relationships between these objects.

The person performing the database design is most often someone with expertise in the area of database design, not necessarily someone with deep experience in the business area for which the database is being designed. For this reason, it's important that the design be done in collaboration with business analysts who are familiar with the business area, how it functions, and its information needs. Requirements analysis is an important part of the database design process. In **requirements analysis**, business analysts, system engineers and software developers identify the needs of a client.

Requirements analysis takes skill since personnel with business subject area knowledge are frequently unable to clearly identify their system requirements. They may not be accustomed to thinking in terms of the discrete data elements that must be stored in the database. They may understand the information to be stored in the database, but not the database itself.

In performing database design, the designer is develops an understanding of dependencies between data items. For example, can the same customer submit more than one order? (Of course!) All of the orders of a given customer are related to the same customer data. But can different customers order the same items? (Of course!) So the same item may be related to multiple orders, and a given order can specify multiple items. If a customer address changes, it may affect many orders. Understanding the data in these ways is facilitated by use of an **entity-relationship diagram**, which uses consistent notation to document one-to-one, one-to-many, and many-to-many relationships between logical entities.

Once the relationships and dependencies among the various pieces of data have been established, they can be mapped to tables, which store data in rows and columns. Relationships between objects are implemented by using the data values in columns. For example, a row of order data is related to customer by carrying the unique primary key of the customer's data row in every order row. The customer primary key carried in a column in an order table row in this way is called a "foreign key." **Data integrity** exists when all foreign keys exist as primary keys in at least one row in the related table.

Database normalization is applied to minimize the potential for redundant data, and to avoid data inconsistencies during the four data manipulation operations: reads, writes, updates, and deletions. Normalization essentially determines where each fact in the database must be housed in order to avoid redundancies.

Normalized databases possess a design that reflects the true dependencies between entities, allowing quick updates with little risk of introducing inconsistencies. By contrast, a non-normalized database is vulnerable to data anomalies since it stores data redundantly. If data is stored in two locations, but later updated in only one of the locations, the data becomes inconsistent; this is referred to as an **update anomaly**.

Databases intended for online transaction processing are normalized. By contrast, databases intended for analytical processing, such as sales or management reporting, budgeting and forecasting, or financial reporting, are "read only" databases. For such databases, redundant, denormalized data can greatly speed reporting and is not only tolerated, it is emphasized.

Problems can result from poor database design. The data contained in a poorly designed database may be unreliable or inaccurate, and performance may degrade over time. Flexibility of data usage may also suffer, and redundant data may use more system resources than should be necessary. In contrast, a good design meets the following objectives:

- the user's needs are met
- the database is free of data anomalies
- the database is reliable and stable
- the database is easy to use.

Podcast 20
DATA MINING AND ETHICAL ISSUES

Data mining is the practice of extracting previously unknown, potentially useful insights from large databases. It is an umbrella term with varied meanings in different contexts. It is usually associated with a business or organization's need to identify specific trends.

One example of data mining is its use in retail sales. If a store tracks customer purchases and a particular customer buys many silk shirts, a data mining system may make a correlation between the customer's characteristics and silk shirts. Promotional campaigns may then be directed to customers with similar characteristics, and products of a similar type may be promoted to the specific customer. This technique is referred to as market basket analysis.

Data mining technology allows organizations to make effective use of data they have gathered about customers, suppliers, and industry trends, such as names, addresses, buying habits, geographic preferences, price lists, and prices of items on supermarket shelves. But ethical concerns develop if information gathered for one purpose are subsequently used for other purposes. Such uses are offensive to many people, and may violate laws and an organization's stated code of ethics.

Data mining has a wide variety of potential applications. Organizations can use it to detect fraud. Retailers can glean ideas to target promotional campaigns. Medical researchers can identify patterns of symptoms, diseases, and treatment outcomes. The goal in every case is to extract useful insights from a vast body of data.

Computer scientists have defined five major types of data mining operations, associations, sequential patterns, similar time series, classification and regression, and clustering.

Invention of the **association data mining operation** was inspired by attempts to fine-tune interactions with customers in retail trade. By trawling through records from point-of-sale terminals, for example, retailers can discover what types of items sell together, and adjust store layouts and advertising campaigns to profit most effectively from those associations.

While association finds events that occur together from logical collections of events, the **sequential patterns data mining operation** finds common events that occur over a period of time. An example in medical research deals with identifying patterns of diagnoses, symptoms, confirmed diseases, treatments, and outcomes in very large patient records databases. The information has been recorded over a period of many years.

Similar time series data mining operations, as its name suggests, can be used to identify events gathered over a period of time. Applications include identifying companies with similar patterns of growth, and discovering stocks or mutual funds with similar price movements.

Classification and regression operations use existing data to create models of the behavior of variables in databases. A credit card company might use this technique to model characteristics of individuals regarded as good risks for providing credit, based on the total record of many characteristics in its database. Such characteristics might include income, credit history, and type and location of employment.

Clustering data mining operations involve segmenting information in databases into definable, homogeneous groups on the basis of specific characteristics. The basic idea is not new. Clustering has a long history in statistical analysis. What is original in the data mining context is the application of clustering techniques to non-numerical attributes.

How can companies use data mining results? Retail managers who have discovered an unlikely association of items could insure that both don't go on sale at the same time. Japanese banks are using data mining to identify customers most likely to make late credit card payments. Insurance companies mine data on present policyholders to determine the number of claims that a new customer is likely to make and set the cost of their policy accordingly.

But data mining has its down side. For example, Ed Socorro was a sales manager with Hilton Hotels Corporation. Not long after he was hired, a company hired by Hilton to do background checks on new employees reported that Socorro once spent six months in jail. Socorro protested that the background check was wrong, and it was. Still, he was fired. He had actually committed a minor infraction in Illinois, which had supposedly been expunged from his record. This had brought him six months of supervision, a sentence often given for speeding tickets. After that erroneously came up as jail time and Hilton fired him, it took Socorro seven months to find a new job. He eventually filed a lawsuit against Hilton and background checker IMI Data Search, and reached a financial settlement with them.

Socorro learned the hard way about an increasing danger in our ever more networked society: the reliance of corporations and governments on commercially accessible databases that mine the paper trails of our lives. Databases have become remarkably efficient and inexpensive to query. Many employers, schools, and even volunteer organizations now trust them in making decisions about whom to hire and whom to avoid associating themselves with.

But databases are not infallible. They can be misinterpreted or only partially accurate, showing arrests or criminal records that were later wiped clean. And misidentifications can occur, wrongly tarring one individual with the sins of another. The potential for identity theft raises the likelihood that databases will misidentify people and incorrectly report negative things about them behind their backs.

Privacy advocates and civil liberties groups are alarmed. They think some background checks could violate federal employment laws and credit reporting rules that let consumers examine information on file about them. The Internet has made it far easier for anyone to obtain someone else's birthdates and Social Security numbers, and information on lawsuits, divorces, and other personal and potentially embarrassing, but technically public, information.

Of major concern is the misuse of information resulting from the human genome project, which is revealing the connection between specific genetic characteristics and the likelihood of developing DNA tests that could predict debilitating illness decades in advance of onset. Combined with data mining techniques, the resulting information could be misused to discriminate against people having certain characteristics in hiring and health insurance rating, and in social stigmatization.

The possible negative effects of these uses of advanced data mining techniques are the continuing subject of concern to privacy activists and legal authorities. This represents a leading edge of the technology revolution, where it collides with issues of ethics and privacy.

Summary of E. F. Codd's Twelve Rules for Relational Databases

1. All data should be presented to the user in **table form**.
2. **All data should be accessible without ambiguity.** This means each data item can be uniquely identified using the table name, primary key, and column name.
3. **A column should be allowed to remain empty.** "Empty" means containing absolutely nothing, in other words, a null value (hexadecimal X'00'), which is distinct from a space or a number with a value of zero.
4. **The database management system (DBMS) must provide access to its structure through the same tools that are used to access the data.** This is accomplished by storing the structure definition within tables, in the same manner that data is stored, and using SQL statements to define and manipulate the structure.
5. **The DBMS must support a clearly defined language that includes functionality for data definition, data manipulation, data integrity, and database transaction control.** Relational databases use forms of standard SQL, commonly referred to as "structured query language," as their supported comprehensive language.
6. **Data can be presented to the user in different logical combinations called views.** Each view should support the same full range of data manipulation that direct-access to a table has available. *(This is only partially fulfilled in current implementations since update and delete access through logical views conflicts with the common use of logical views to provide read-only access for query and reporting. There is no universal agreement that it is a good idea in any case.)*
7. **Data can be retrieved from a relational database in sets constructed of data from multiple rows and/or multiple tables.** This rule states that insert, update, and delete operations should be supported for any retrievable set rather than for just a single row in a single table.
8. **The user is isolated from the physical method of storing and retrieving information from the database.** Changes can be made to the underlying hardware and storage arrangements without affecting how the user accesses it.
9. **How a user views data should not change when the logical structure (table structure) of the database changes.** This is difficult to implement practically, but database implementation does make some types of changes "transparent" to data users. For example, adding columns to a table does not affect existing uses of columns already present.
10. **Constraints on user input must be provided to maintain database integrity.** Database implementations must at least insure that no part of a primary key is missing (null) and that, if a foreign key is defined in one table, the value in it must exist as a primary key in another table.
11. **A user should be totally unaware of whether or not the database is housed on one computer or distributed across several computers.** This is actually a subcategory of Rule 8. *Implementation of this is inconsistent between commercial DBMS's but is not a major concern in many applications.*
12. **Users should in no way be able to modify the database structure other than through the multiple row database language, SQL.** *In actual commercial implementations, database tables are still accessible to direct manipulation by systems personnel in many implementations, but most end-users are denied access to this type of capability.*

Lecture Slides 4

Introduction to Information Technology

Database Technology

Course resources

PowerPoint presentations, podcasts, and web links for readings are available at
www.ambriana.com > IT Workbook

Print slides at 6 slides per page

Homework, quizzes and final exam are based on slides, lectures, readings and podcasts

Slide 4-2

Main topics

- What is data? How do we make it useful?
- Data storage: flat files and databases
- Database system history and evolution
- Relational database principles, tools, and functionalities
- Relational database and SQL: an example
- Entity Relationship Diagrams
- Data mining: benefits and risks

Slide 4-3

What is data?

- **Data** is nothing more than raw facts
- **Information** is data in a particularly desired, beneficial context
- Desired context of data can vary depending on needs of organization using the data
- **Context provides meaning**

Slide 4-4

Making data useful

- How do we take raw facts and make them into useful data?
- Create a database
- Must understand organization and data needs before creation of database
- Can be very time-consuming
- "I don't know what I want, but this isn't it!"

Slide 4-5

Creating a database

- Database developers first gain clear understanding of organization's needs
- Develop a conceptual model
- Build physical model
- Eliminate redundant data: normalization
- Database maintenance
  - Database up and running efficiently
  - Data backup and restoration
  - Helping users and developers

Slide 4-6

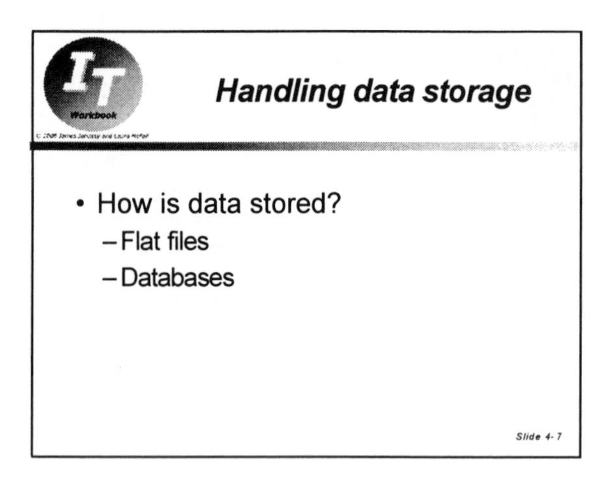
Handling data storage
© 2006 James Janossy and Laura McFall
• How is data stored?
– Flat files
– Databases
Slide 4-7

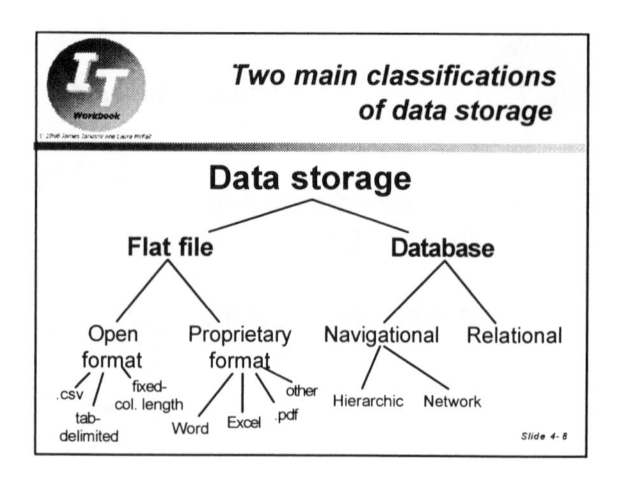
Two main classifications of data storage
Data storage
Flat file
Database
Open format
Proprietary format
Navigational
Relational
.csv
tab-delimited
fixed-col. length
Word
Excel
.pdf
other
Hierarchic
Network
Slide 4-8

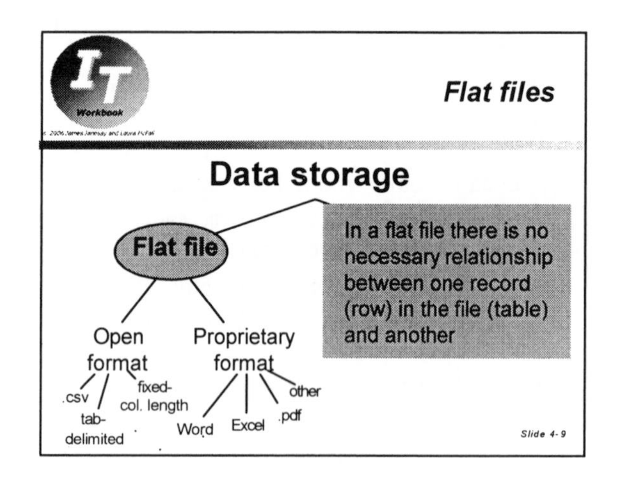
Flat files
Data storage
Flat file
In a flat file there is no necessary relationship between one record (row) in the file (table) and another
Open format
Proprietary format
.csv
tab-delimited
fixed-col. length
Word
Excel
.pdf
other
Slide 4-9

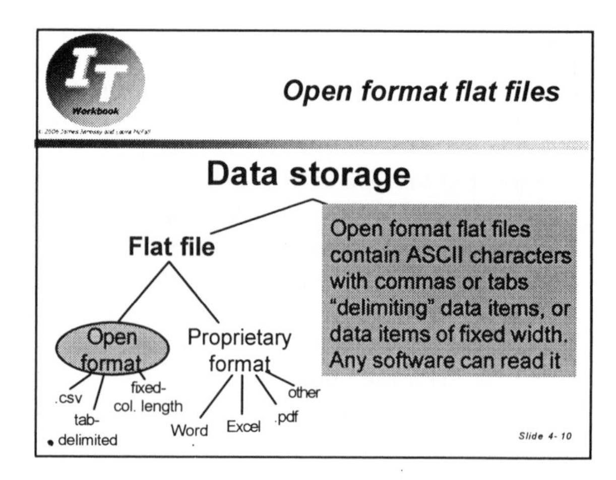
Open format flat files
Data storage
Flat file
Open format flat files contain ASCII characters with commas or tabs "delimiting" data items, or data items of fixed width. Any software can read it
Open format
Proprietary format
.csv
tab-delimited
fixed-col. length
Word
Excel
.pdf
other
Slide 4-10

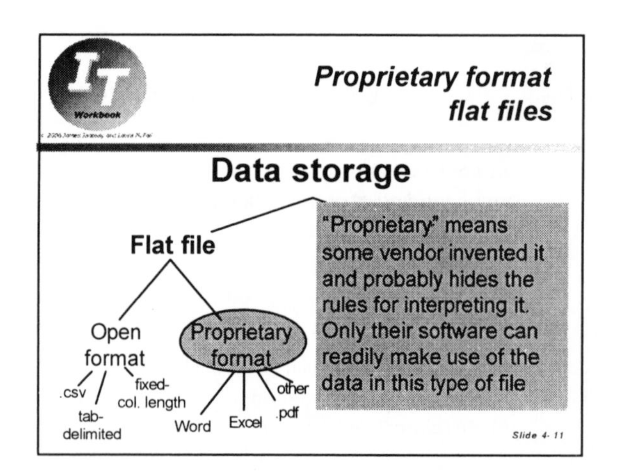
Proprietary format flat files
Data storage
Flat file
"Proprietary" means some vendor invented it and probably hides the rules for interpreting it. Only their software can readily make use of the data in this type of file
Open format
Proprietary format
.csv
tab-delimited
fixed-col. length
Word
Excel
.pdf
other
Slide 4-11

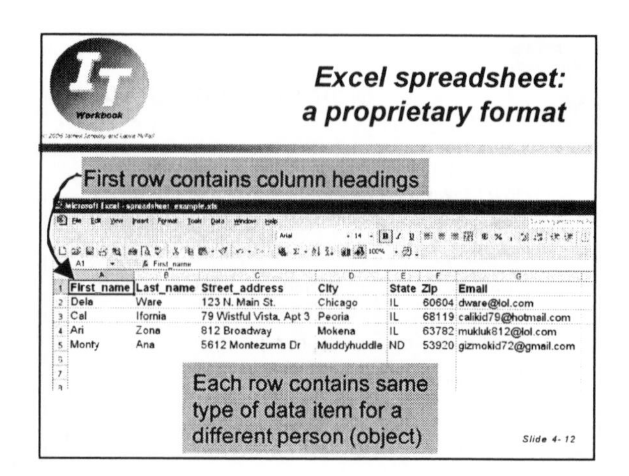
Excel spreadsheet: a proprietary format
First row contains column headings
First_name Last_name Street_address City State Zip Email
Dela Ware 123 N. Main St. Chicago IL 60604 dware@lol.com
Cal Ifornia 79 Wistful Vista, Apt 3 Peoria IL 68119 calikid79@hotmail.com
Ari Zona 812 Broadway Mokena IL 63782 mukluk812@lol.com
Monty Ana 5612 Montezuma Dr Muddyhuddle ND 53920 gizmokid72@gmail.com
Each row contains same type of data item for a different person (object)
Slide 4-12

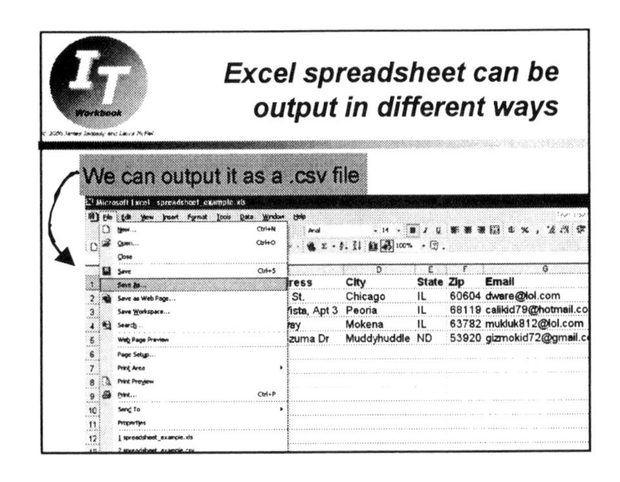
Excel spreadsheet can be output in different ways
We can output it as a .csv file

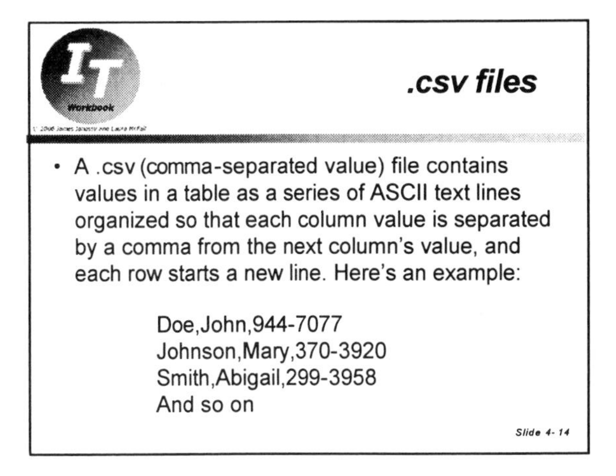
.csv files
• A .csv (comma-separated value) file contains values in a table as a series of ASCII text lines organized so that each column value is separated by a comma from the next column's value, and each row starts a new line. Here's an example:
Doe,John,944-7077
Johnson,Mary,370-3920
Smith,Abigail,299-3958
And so on
Slide 4- 14

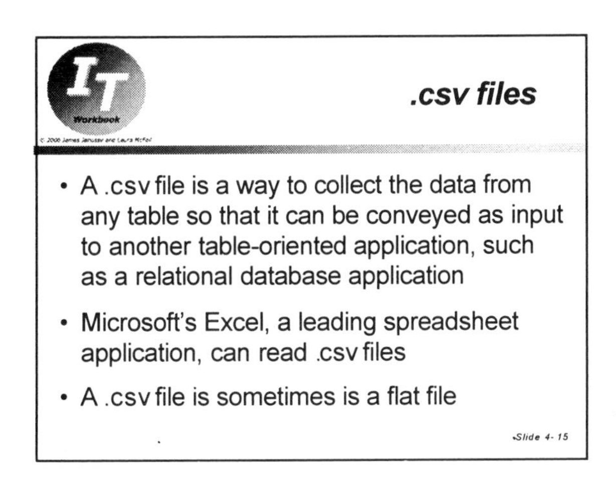
.csv files
• A .csv file is a way to collect the data from any table so that it can be conveyed as input to another table-oriented application, such as a relational database application
• Microsoft's Excel, a leading spreadsheet application, can read .csv files
• A .csv file is sometimes is a flat file
Slide 4- 15

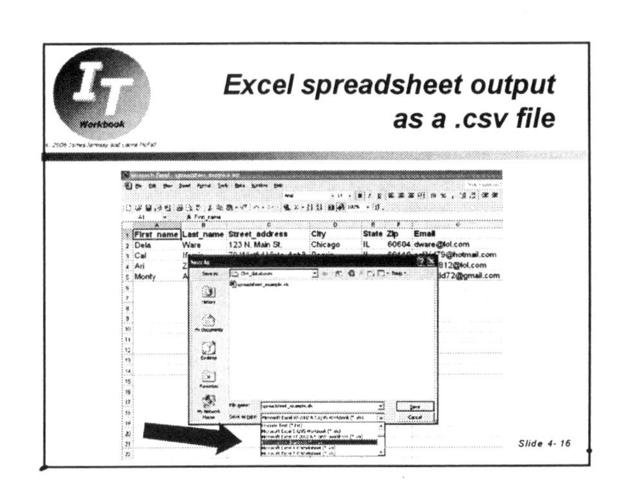
Excel spreadsheet output as a .csv file
Slide 4- 16

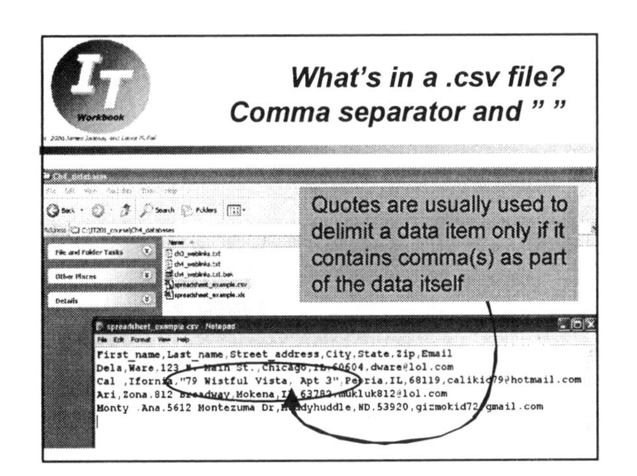
What's in a .csv file? Comma separator and " "
Quotes are usually used to delimit a data item only if it contains comma(s) as part of the data itself

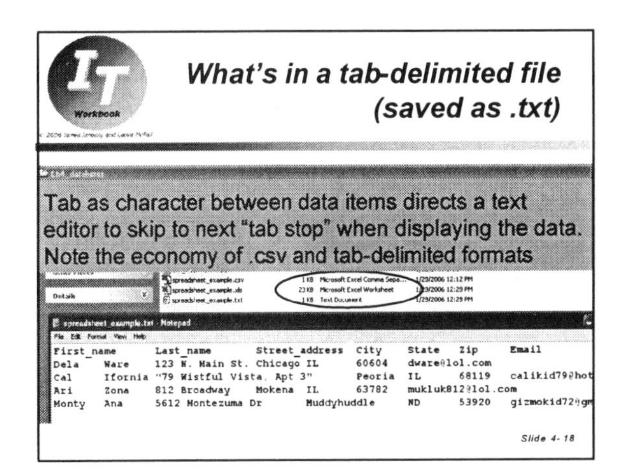
What's in a tab-delimited file (saved as .txt)
Tab as character between data items directs a text editor to skip to next "tab stop" when displaying the data. Note the economy of .csv and tab-delimited formats
Slide 4- 18

Tab-delimited files

- Tab-delimited files use tabs between fields or data items
- Saved as text files (.txt extensions)
- Tabs direct text editor to “skip” to next tab stop when displaying data contained in the file

Slide 4- 19

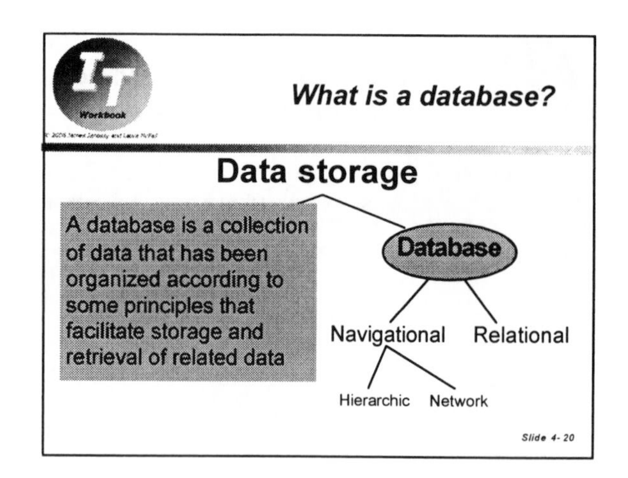

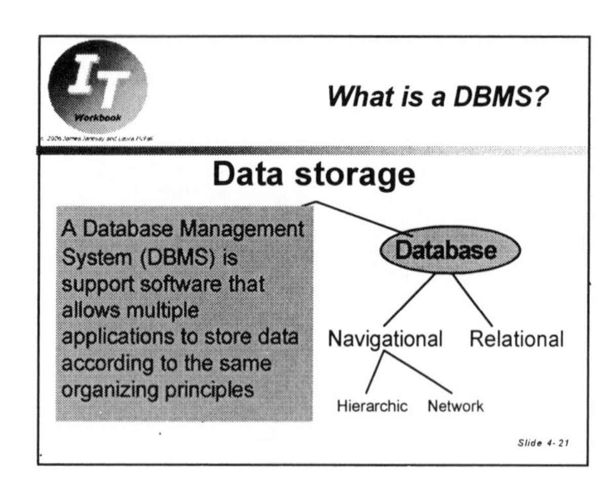

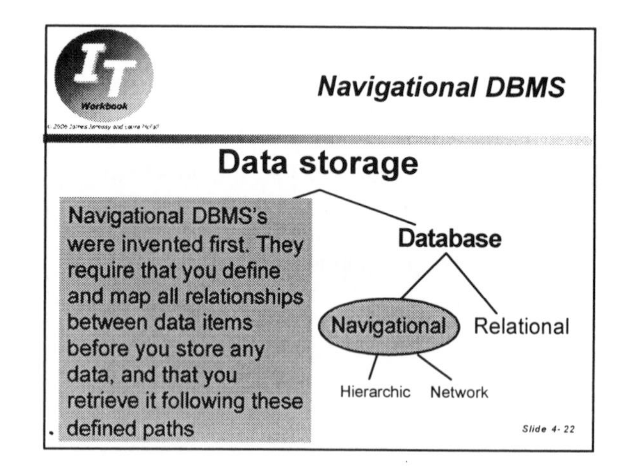

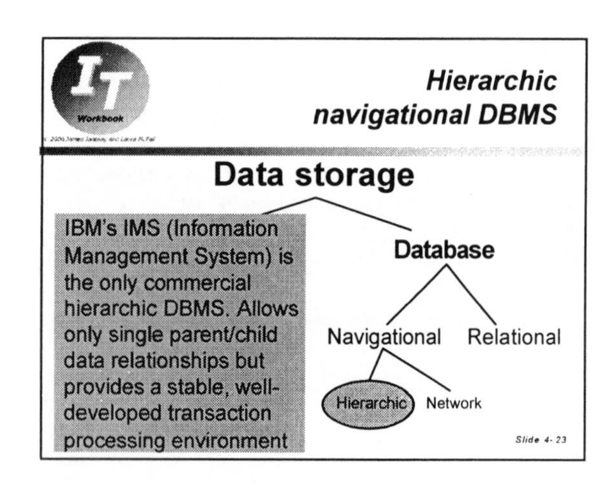

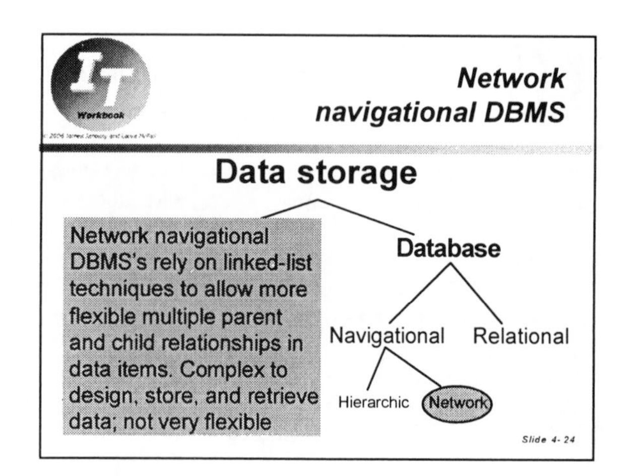

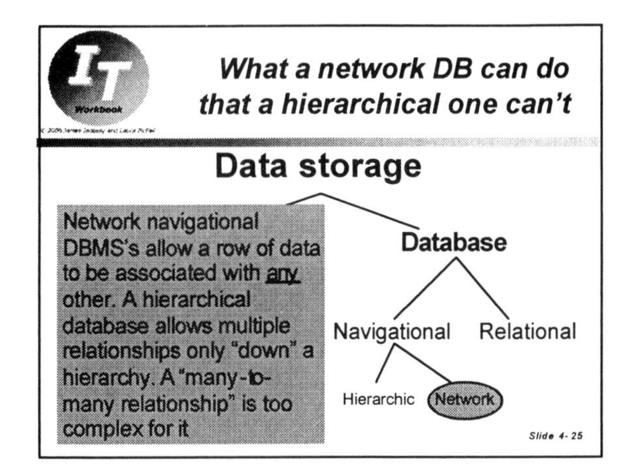
IT
Workbook
What a network DB can do
that a hierarchical one can't
Data storage
Network navigational DBMS's allow a row of data to be associated with any other. A hierarchical database allows multiple relationships only "down" a hierarchy. A "many-to-many relationship" is too complex for it
Database
Navigational
Relational
Hierarchic
Network
Slide 4-25

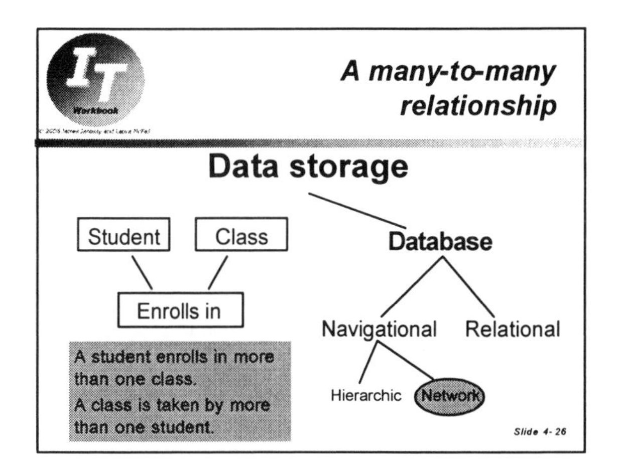
IT
Workbook
A many-to-many
relationship
Data storage
Student
Class
Enrolls in
A student enrolls in more than one class.
A class is taken by more than one student.
Database
Navigational
Relational
Hierarchic
Network
Slide 4-26

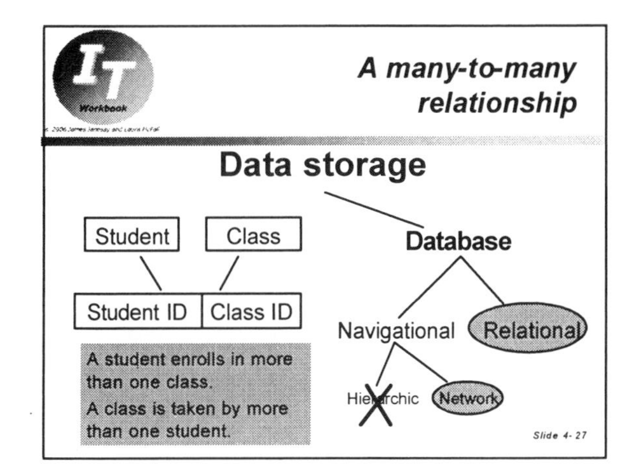
IT
Workbook
A many-to-many
relationship
Data storage
Student
Class
Student ID
Class ID
A student enrolls in more than one class.
A class is taken by more than one student.
Database
Navigational
Relational
Hierarchic
Network
Slide 4-27

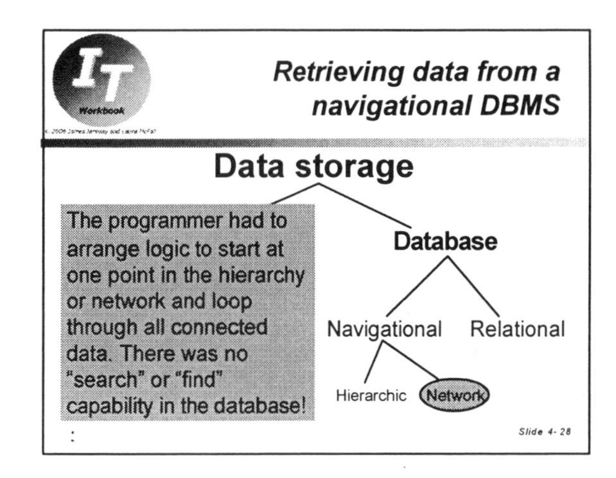
IT
Workbook
Retrieving data from a
navigational DBMS
Data storage
The programmer had to arrange logic to start at one point in the hierarchy or network and loop through all connected data. There was no "search" or "find" capability in the database!
Database
Navigational
Relational
Hierarchic
Network
Slide 4-28

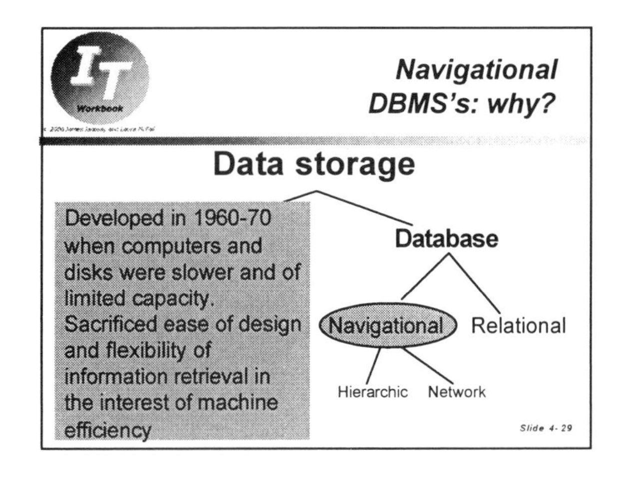
IT
Workbook
Navigational
DBMS's: why?
Data storage
Developed in 1960-70 when computers and disks were slower and of limited capacity. Sacrificed ease of design and flexibility of information retrieval in the interest of machine efficiency
Database
Navigational
Relational
Hierarchic
Network
Slide 4-29

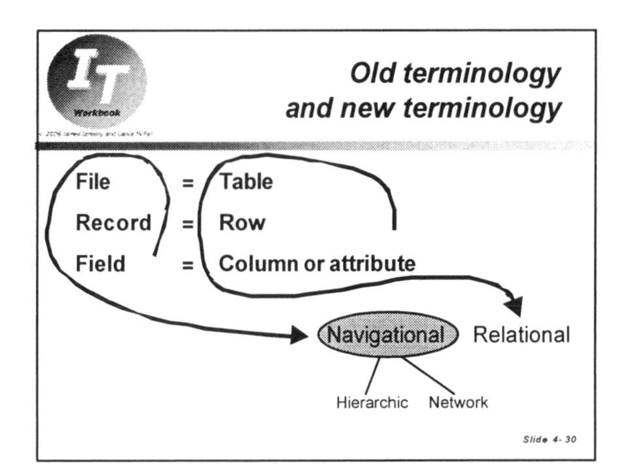
IT
Workbook
Old terminology
and new terminology
File = Table
Record = Row
Field = Column or attribute
Navigational
Relational
Hierarchic
Network
Slide 4-30

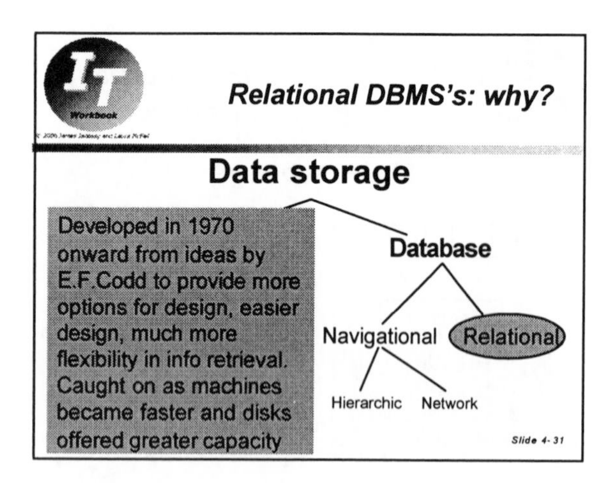
Relational DBMS's: why?
Workbook
Data storage
Developed in 1970 onward from ideas by E.F.Codd to provide more options for design, easier design, much more flexibility in info retrieval. Caught on as machines became faster and disks offered greater capacity
Database
Navigational
Relational
Hierarchic
Network
Slide 4-31

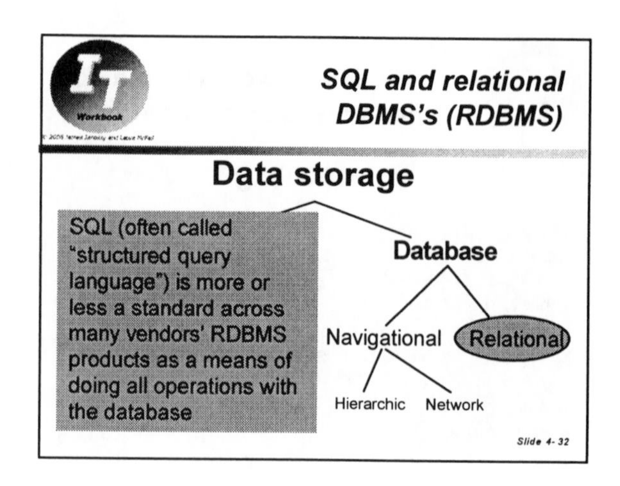
SQL and relational DBMS's (RDBMS)
Workbook
Data storage
SQL (often called "structured query language") is more or less a standard across many vendors' RDBMS products as a means of doing all operations with the database
Database
Navigational
Relational
Hierarchic
Network
Slide 4-32

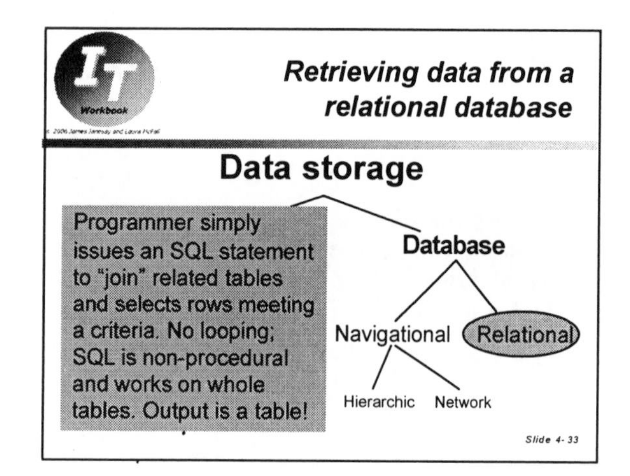
Retrieving data from a relational database
Workbook
Data storage
Programmer simply issues an SQL statement to "join" related tables and selects rows meeting a criteria. No looping; SQL is non-procedural and works on whole tables. Output is a table!
Database
Navigational
Relational
Hierarchic
Network
Slide 4-33

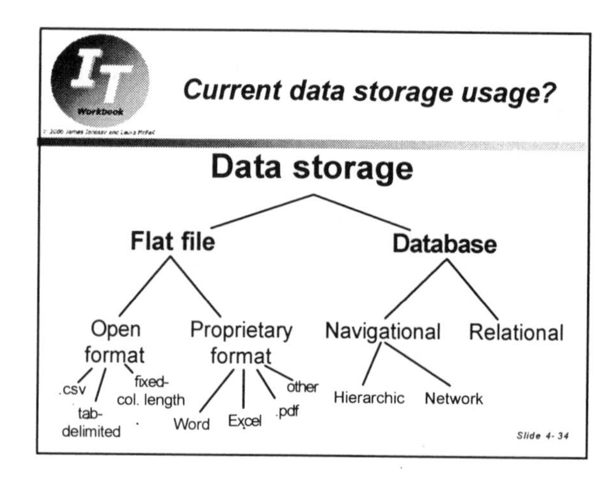
Current data storage usage?
Workbook
Data storage
Flat file
Database
Open format
Proprietary format
Navigational
Relational
.csv
fixed-col. length
tab-delimited
Word
Excel
other
.pdf
Hierarchic
Network
Slide 4-34

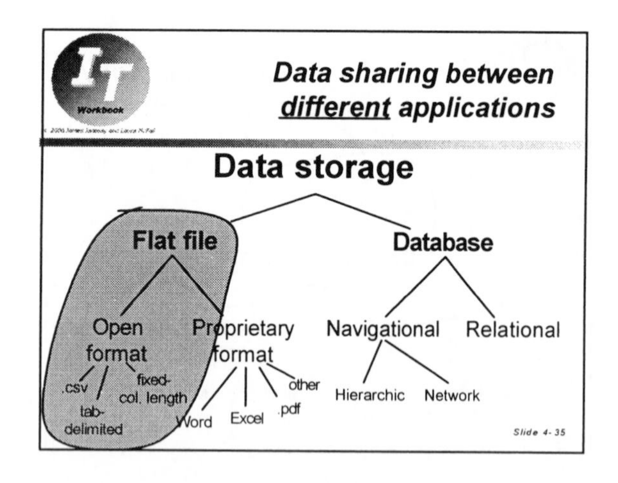
Data sharing between different applications
Workbook
Data storage
Flat file
Database
Open format
Proprietary format
Navigational
Relational
.csv
fixed-col. length
tab-delimited
Word
Excel
other
.pdf
Hierarchic
Network
Slide 4-35

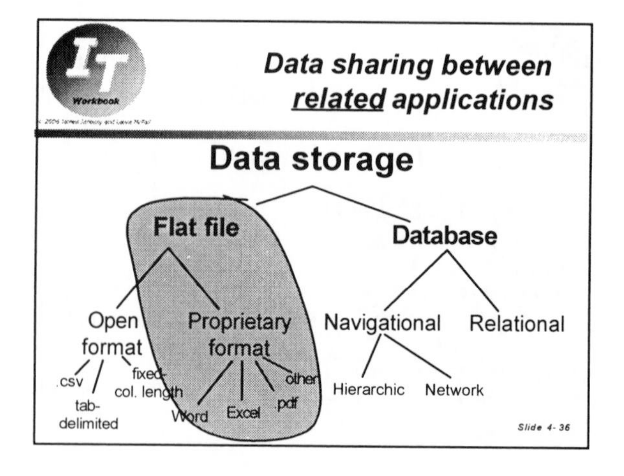
Data sharing between related applications
Workbook
Data storage
Flat file
Database
Open format
Proprietary format
Navigational
Relational
.csv
fixed-col. length
tab-delimited
Word
Excel
other
.pdf
Hierarchic
Network
Slide 4-36

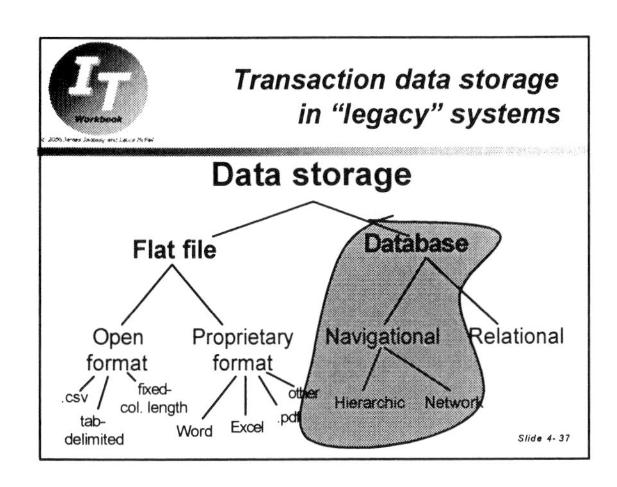

Legacy systems

- A legacy system is an existing computer system or application program that continues to be used because the organization does not want to replace it
- Many people use this term to refer to antiquated systems

Slide 4- 38

Legacy systems

- Considered potentially problematic
- Often run on obsolete (and slow) hardware
- Hard to maintain, improve, and expand; general lack of understanding of the system
- System designers may have left
- Inadequate documentation
- Integration with newer systems may be difficult; new software may use different technology

Slide 4- 39

Legacy systems

Organizations can have compelling reasons for keeping a legacy system:

- Costs of redesigning the system are prohibitive
- System requires close to 100% availability, so it cannot be taken out of service
- The way the system works is not well understood or documented; replacement means "reinventing" it
- The system works satisfactorily

Slide 4- 40

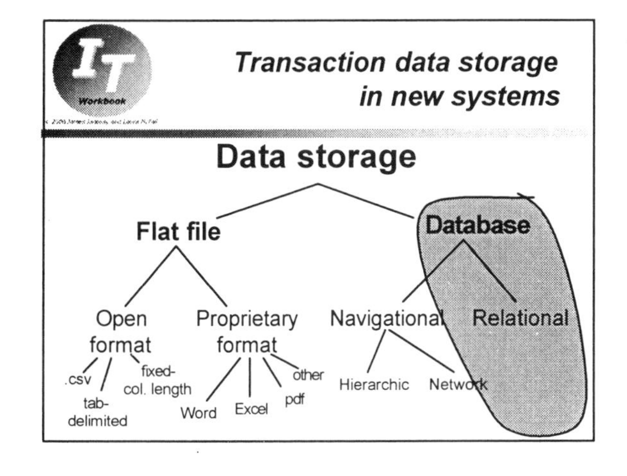

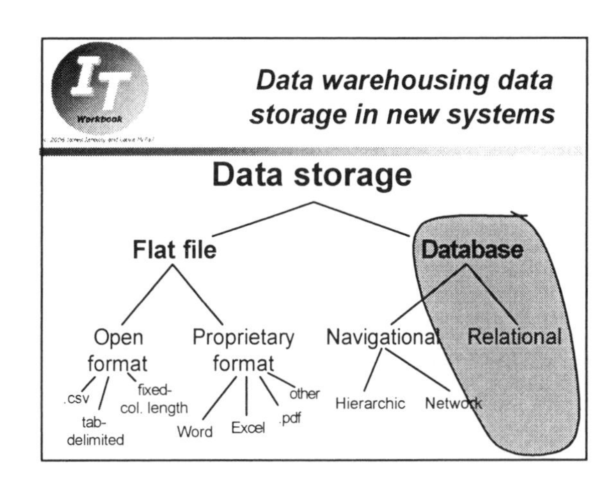

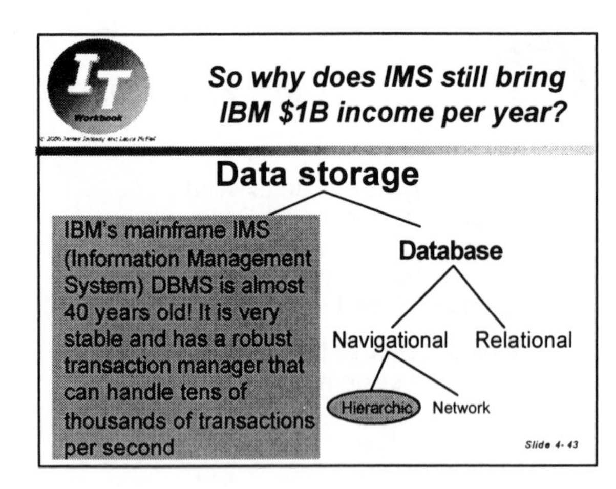

Relational database concepts

E.F. Codd (1923-2003)

- Pioneered by E.F. Codd at IBM's Almaden (California) research lab in 1970s
- Original "Alpha" language to manipulate data was implemented as SEQUEL, renamed "SQL"
- DB2 and Oracle were first implementations

Slide 4- 44

Relational database principles

- Data is stored in sets called tables
- Each table row is info about one object
- Rows can be selected from tables by a criteria based on their content
- Tables can be "joined" and result it a new set (table) of data

E.F. Codd's 12 rules for relational databases

Slide 4- 45

Relational database: Four basic functionalities

- A schema modeling language (a schema file specifies a format for displaying or editing a table or a view in a relational database)
- Data structures optimized to handle large volumes of data in disk storage
- An interactive query, reporting and update language
- A transaction mechanism that insures data integrity with concurrent users

Slide 4- 46

Uses of modern RDBMS

- **Transaction processing:** provide A.C.I.D. services to online applications via SQL INSERT, UPDATE statements
- **Information retrieval:** responsive extraction of information in response to SQL SELECT statements
- **Web publishing** of information

Slide 4- 47

What is A.C.I.D.?

- **Atomicity** - All transactions are either performed completely, or are not done at all; a partial transaction that is aborted must be rolled back
- **Consistency** - The effects of a transaction must preserve required system properties. For instance, if funds are transferred between accounts, a deposit and withdrawal must both be committed to the database, so that the accounting system does not fall out of balance (bank balance transfer)

Slide 4- 48

What is A.C.I.D.?

- **Isolation** - Intermediate stages of a transaction cannot be visible to other transactions
- So, in the case of a transfer of funds between accounts, both sides of the double-entry bookkeeping system must change together for each transaction
- Other transactions will not "see" this transaction until both sides of the bookkeeping system have been changed

Slide 4-49

What is A.C.I.D.?

- **Durability** - Once a transaction is committed, the change must persist, except in the face of a truly catastrophic failure
- If the server room is destroyed you can't expect the software to keep transactions persistent!
- But the system should be designed to be resistant to reasonably traumatic sorts of system failures such as a network link breaking down or perhaps even a disk drive malfunctioning

Slide 4-50

Tools are provided by modern PC-based RDBMS's to...

- Design the structure of a database
- Create data entry forms
- Validate data on entry for inconsistencies
- Sort and manipulate data
- Query the database to extract data
- Produce formatted reports and outputs

Slide 4-51

Three types of data relationships

- A relationship is an association between two tables that enables a relational database to pull together related data that's stored in multiple tables
- There are many kinds of relationships.The three most common are:
  - One-to-many
  - One-to-one
  - Many-to-many

Slide 4-52

Three types of data relationships

- **One-to-many** is the most common relationship, in which the primary key value matches none, one, or many records in a related table

Slide 4-53

Three types of data relationships

- **One-to-one** is probably the least common of the three, where a primary key value matches only one (or no) record. These relationships are almost always forced by business rules and seldom flow naturally from the actual data

Slide 4-54

Three types of data relationships

- **Many-to-many** happens when both tables contain records that are related to more than one record
- A relational database doesn't directly support a many-to-many relationship, so a third table - an "associating table" - must be created
- The "associating table" contains a primary key and a foreign key to each of the data tables. After breaking down the many-to-many relationship, you have two one-to-many relationships between the associate table and the two data tables

Slide 4-55

An example application: Public Library Video Rentals

- 1,000 video titles on the shelf
- 250 regular patrons
- Who has what checked out, when due back in?
- Be able to list all patrons, all titles, titles most frequently checked out, checkout history of each title, and can browse by director, lead actor, viewing time

Slide 4-56

An example application: Public Library Video Rentals

This is a **many-to-many** relationship.

- Each library patron can check out **multiple videos** (at one time, or in sequence)
- Each video can be checked out by **many patrons** (one after another)

Slide 4-57

Isolating video item "attributes"

The video data row:

Title_ID	Title	Director	Lead_actor	Minutes

One row of data for each video, no data redundancy

Slide 4-58

Isolating public library patron attributes

The patron data row:

Patron_ID	Name	Address	Phone_nbr

One row of data for each patron, no data redundancy

Slide 4-59

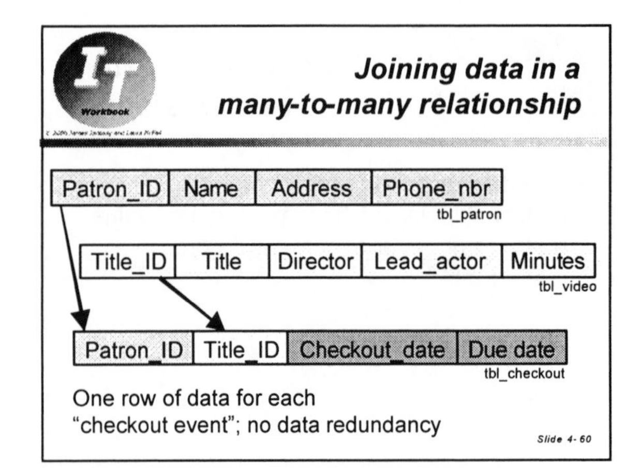

An example application: Public Library Video Rentals

- List all patrons and phone numbers alphabetically by name:

```
SELECT name, phone_nbr
  FROM tbl_patron
  ORDER BY name;
```

Slide 4-61

An example application: Public Library Video Rentals

- List all patrons and phone numbers alphabetically by name:

```
SELECT name, phone_nbr
  FROM tbl_patron
  ORDER BY name;
```

Oops! It would have been better to break the name into constituent parts such as lastname and firstname

Slide 4-62

The database design process

- Break composite columns like name into constituent parts
- Create unique key column for each item
- Eliminate repeating groups (what if we wanted to record all actors in each video?)
- Make sure each attribute depends only on the whole row key

Slide 4-63

These are actually the rules of "data normalization"

- Break composite columns like name into constituent parts
- Create unique key column for each item
- Eliminate repeating groups (what if we wanted to know all actors in each video?)
- Make sure each attribute depends only on the whole row key

More about the process of data normalization

Slide 4-64

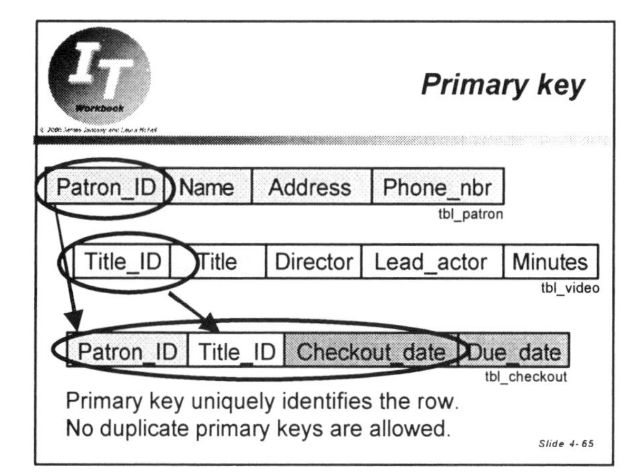

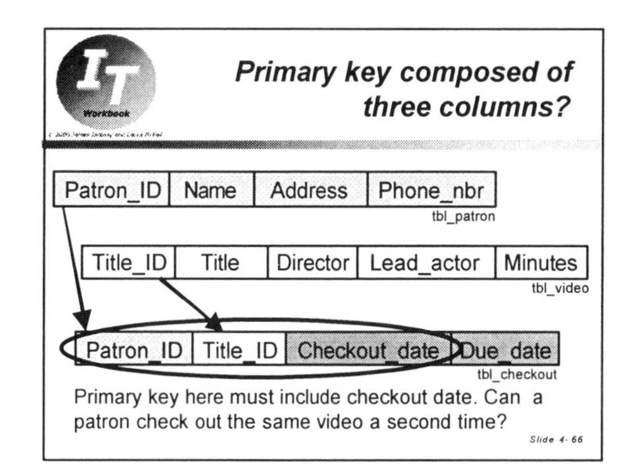

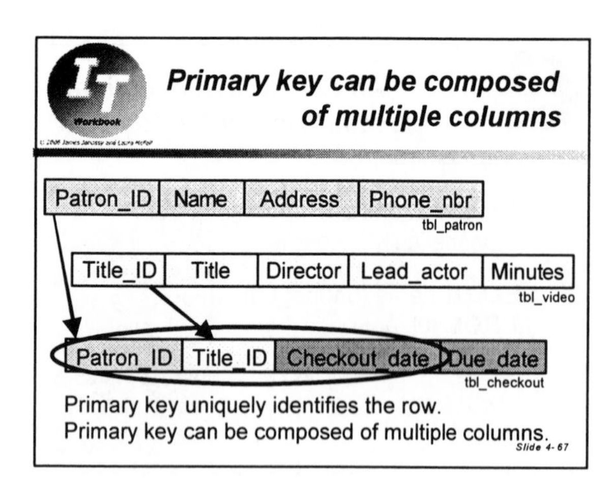

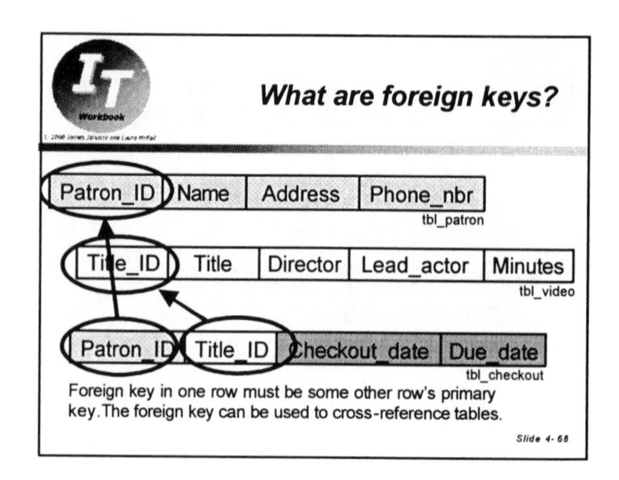

*Select all columns using **

- List all patrons and phone numbers alphabetically by name:

```
SELECT *
  FROM tbl_patron
  ORDER BY name;
```

Slide 4-69

*Select all columns using **

- List just the videos running less than 90 minutes and list them from shortest length to longest length:

```
SELECT *
  FROM tbl_video
  WHERE tbl_video.minutes < 90
  ORDER BY minutes;
```

Slide 4-70

Alias for table name

- List just the videos running less than 90 minutes and list them from shortest length to longest length:

```
SELECT *
  FROM tbl_video A
  WHERE A.minutes < 90
  ORDER BY minutes;
```

This is an "alias" and makes it easier to code the column names

Slide 4-71

Joining multiple tables

- List the titles of videos due back on February 3, 2006 and who has them now:

```
SELECT A.name, B.title_id, B.title, C.due_date
FROM tbl_patron A, tbl_video B, tbl_checkout C
WHERE A.patron_id = C.patron_id
    and B.title_ID = C.title_id
    and C.due_date = "02/03/2006";
```

Slide 4-72

Referential integrity

- Referential integrity is a feature provided by relational database management systems (RDBMS's) that prevents applications from entering inconsistent data
- Most RDBMS's have various referential integrity rules that you can apply when you create a relationship between two tables

Slide 4-73

Referential integrity

- A column in table B contains a foreign key that points to a column in Table A. Referential integrity would prevent you from adding a row to Table B with a column value that is not in A.
- Referential integrity rules might specify that if you delete a row from Table A, any rows in Table B that are linked to the deleted row will also be deleted.
- This is called a **cascading delete**

Slide 4-74

Referential integrity

- Referential integrity rules can specify that whenever you modify the value of a column in Table A that serves as a key, all rows in Table B that carry the value of the column as a foreign key will also be changed
- This is called **cascading update**

Slide 4-75

Referential integrity

- **Patron_ID is a foreign key in tbl_checkout**

  Can you delete a patron's row in tbl_patron if they have ever checked out a video?

Slide 4-76

Referential integrity

- **Title_ID is a foreign key in tbl_checkout**

  Can you check out a video to a patron if that Title_ID doesn't exist in tbl_video?

Slide 4-77

Entity-relationship diagrams - (ERDs)

- An entity-relationship diagram (ERD) is a specialized graphic that illustrates the relationships between entities in a database
- ERDs often use symbols to represent three different types of information
  - Boxes are commonly used to represent entities
  - Diamonds are normally used to represent relationships
  - Ovals are used to represent attributes (columns).

Slide 4-78

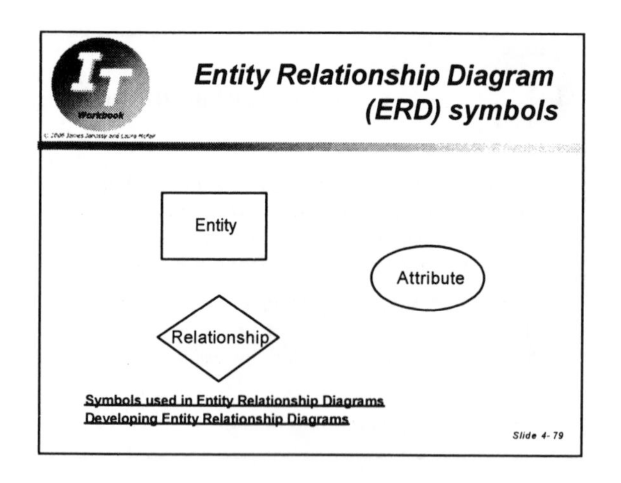
Entity Relationship Diagram (ERD) symbols
Entity
Attribute
Relationship
Symbols used in Entity Relationship Diagrams
Developing Entity Relationship Diagrams
Slide 4- 79

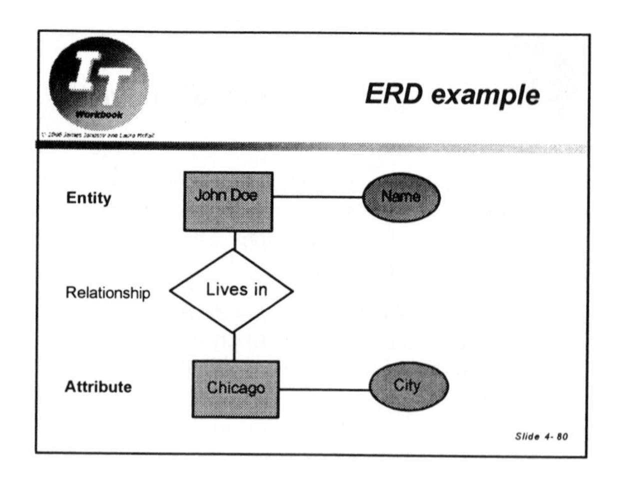
ERD example
Entity
John Doe
Name
Relationship
Lives in
Attribute
Chicago
City
Slide 4- 80

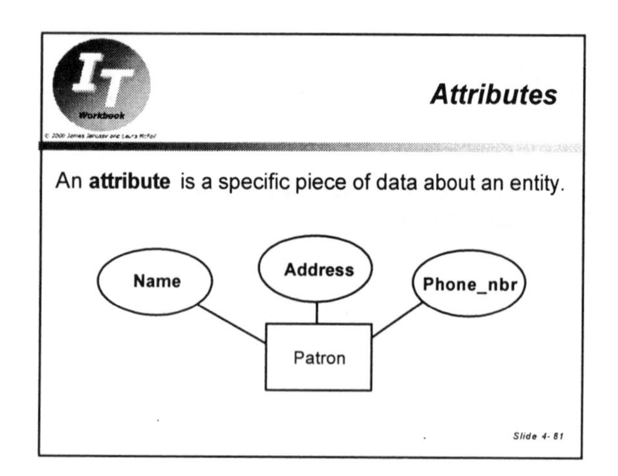
Attributes
An attribute is a specific piece of data about an entity.
Name
Address
Phone_nbr
Patron
Slide 4- 81

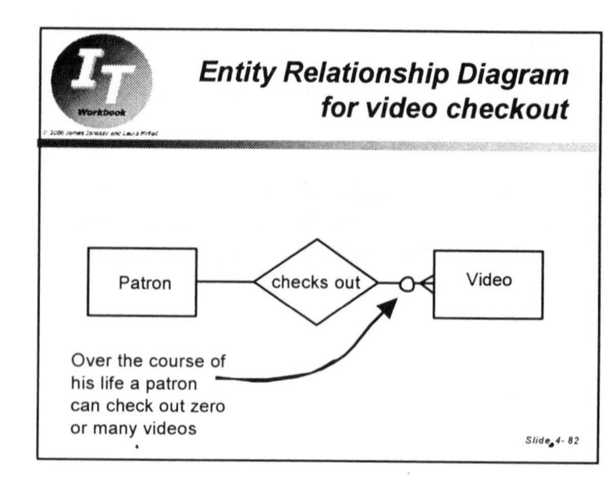
Entity Relationship Diagram for video checkout
Patron
checks out
Video
Over the course of his life a patron can check out zero or many videos
Slide 4- 82

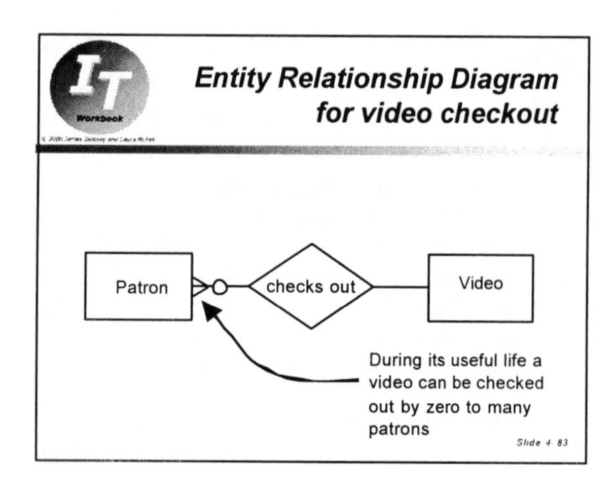
Entity Relationship Diagram for video checkout
Patron
checks out
Video
During its useful life a video can be checked out by zero to many patrons
Slide 4- 83

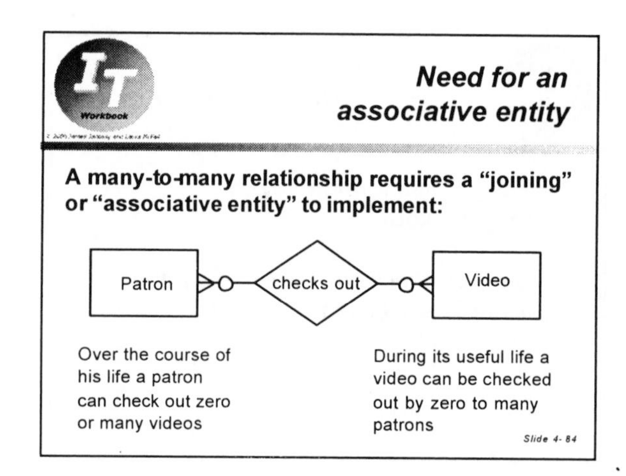
Need for an associative entity
A many-to-many relationship requires a "joining" or "associative entity" to implement:
Patron
checks out
Video
Over the course of his life a patron can check out zero or many videos
During its useful life a video can be checked out by zero to many patrons
Slide 4- 84

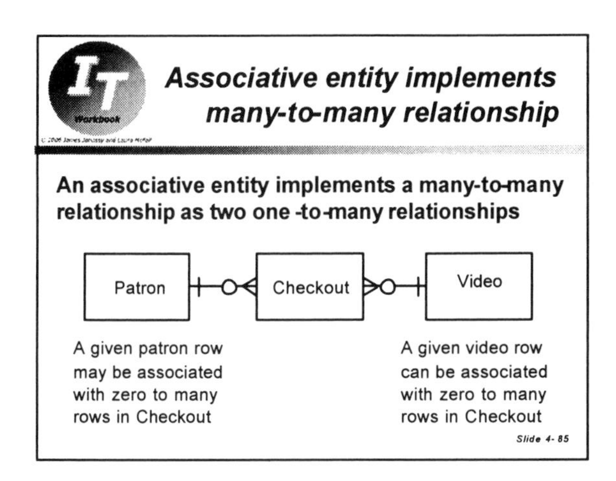

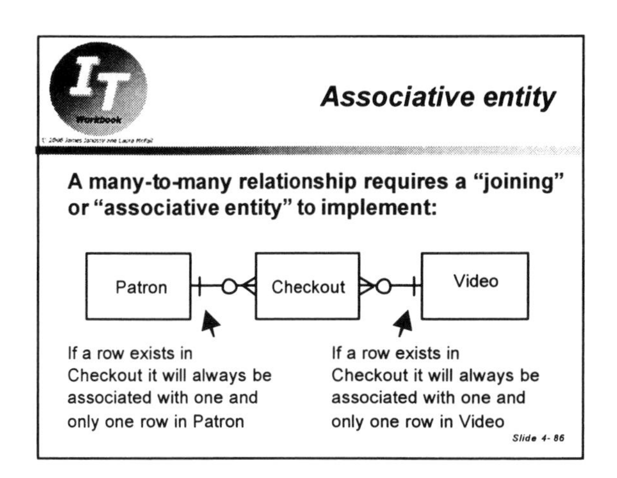

Object-oriented and hypermedia databases

- Object-oriented databases (OODM) are based on object-oriented programming concepts
- Useful for handling complex data
  - pictures
  - audio files
  - video files

Slide 4-87

Object-oriented and hypermedia databases

- Hypermedia databases store data in nodes that can be linked to one another using hyperlinks
- Each node can store any type of data
- World Wide Web is an example of a hypermedia database

Slide 4-88

Data mining benefits and risk

- **Data mining** is automated search for patterns in huge accumulating databases
- Process of applying analytical and statistical methods to data to find patterns
- What can data mining be used for?

Slide 4-89

Data mining: five types of operations

- **Association**
- Sequential patterns
- Similar time series data
- Classification and regression
- Clustering

Similar patterns of things, different people or events

Slide 4-90

Data mining: five types of operations

- Association
- **Sequential patterns**
- Similar time series data
- Classification and regression
- Clustering

Similar outcomes over time, different people or events

Slide 4-91

Data mining: five types of operations

- Association
- Sequential patterns
- **Similar time series data**
- Classification and regression
- Clustering

Similar outcomes over time such as patterns of growth or price movements

Slide 4-92

Data mining: five types of operations

- Association
- Sequential patterns
- Similar time series data
- **Classification and regression**
- Clustering

Models that measure related behavior of variables

Slide 4-93

Data mining: five types of operations

- Association
- Sequential patterns
- Similar time series data
- Classification and regression
- **Clustering**

Segmenting non-numerical data into homogeneous groups

Slide 4-94

Data mining benefits and risk

- **Benefits:** better understanding of complex relationships
- **Risks:** invasion of privacy, differential treatment of people
  - "Preferred shopper" cards
  - Medical data

Forms of data mining and examples of benefits

Slide 4-95

Data "repurposing?"

- Ethical principle: data can be used only for the purpose for which it was collected
- "Re-purposing" data means to use it for things it was not collected for: tempting, but not appreciated, may not be legal
- *Have people opt-in, or opt-out?*

Implications of data mining for fair practices and privacy

Slide 4-96

IT Workbook Assignment 4.1

Student name:

Write your answers to the questions below within the box. In each case, please choose your words carefully to answer the specific questions. Avoid simply copying passages from your readings to these answers.

Provide a **concise description** of a **database management system** (DBMS):

1)

Identify and briefly describe the **four basic functionalities** that a **DBMS** must provide:

2)

3)

4)

5)

IT Workbook Assignment 4.2

Student name:

Write your answers to the questions below within the box. In each case, please choose your words carefully to answer the specific questions. Avoid simply copying passages from your readings to these answers.

Explain why Information Management System **(IMS)**, IBM's original hierarchical (navigational) database management system for mainframe computers, is still so widely used that it generates an estimated $1 billion in revenue for IBM annually:

1)

Identify and discuss the information structure that a **network model navigational database** allows that a **hierarchical model navigational database** does not:

2)

IT Workbook Assignment 4.3

Student name:

Write your answers to the questions below within the box. In each case, please choose your words carefully to answer the specific questions. Avoid simply copying passages from your readings to these answers.

Explain what is meant by **open format flat files**, and identify and describe the three common such formats, and what they are often used for:

1) open format flat files are:

2) one open format flat file is:

3) another open format flat file is:

4) and yet a third open format flat file is:

5) what open format flat files are often used for:

Explain what is meant by **proprietary format flat files**, and identify three such formats of files:

6) proprietary format flat files are:

7) a proprietary format flat file:

8) another proprietary format flat file:

9) and yet another proprietary format flat file:

IT Workbook Assignment 4.4

Student name:

Write your answers to the questions below <u>within the box</u>. In each case, please choose your words carefully to answer the specific questions. Avoid simply copying passages from your readings to these answers.

A typical PC-based database program (DBMS) gives you **several tools**. Identify and briefly describe what six of these tools do:

1)

2)

3)

4)

5)

6)

Explain what **data publishing** is in connection with the World Wide Web and describe an example of it:

7) what data publishing is:

8) an example of data publishing:

IT Workbook Assignment 4.5

Student name:

Write your answers to the questions below within the box. In each case, please choose your words carefully to answer the specific questions. Avoid simply copying passages from your readings to these answers.

The process of **designing a database** can be summarized in four steps. (These steps actually comprise the process of **normalization**.) List and briefly describe each of these four steps:

1)

2)

3)

4)

In terms of a relational DBMS, describe what a **primary key** is, and what a **foreign key** is:

5) a primary key is:

6) a foreign key is:

IT Workbook Assignment 4.6

Student name:

Write your answers to the questions below within the box. In each case, please choose your words carefully to answer the specific questions. Avoid simply copying passages from your readings to these answers.

Explain what a **legacy system** is and describe several **problems** now experienced with them:

1) a legacy system is:

2) problem #1:

3) problem #2:

4) problem #3:

5) problem #4:

Identify and describe four basic functionalities of **relational database management systems**:

6)

7)

8)

9)

IT Workbook Assignment 4.7

Student name:

Write your answers to the questions below within the box. In each case, please choose your words carefully to answer the specific questions. Avoid simply copying passages from your readings to these answers.

Discuss the information access actions a programmer needs to take in acquiring information from a **navigational database** and b) the information access actions a programmer needs to take in acquiring information from a **relational database**:

1) information access actions with a navigational database:

2) information access actions with a relational database:

Explain what **referential integrity** means in terms of relational database management systems, and how it protects the accuracy and completeness of data:

3) referential integrity means:

4) how referential integrity protects the accuracy and completeness of data:

IT Workbook Assignment 4.8

Student name:

Write your answers to the questions below <u>within the box</u>. In each case, please choose your words carefully to answer the specific questions. Avoid simply copying passages from your readings to these answers.

Explain what an **Entity Relationship Diagram** (ERD) depicts, what the **rectangle** symbol on it represents, and what the **diamond** symbol on it represents.

1) what an Entity Relationship Diagram (ERD) depicts:

2) what the rectangle symbol ☐ represents on an ERD:

3) what the diamond symbol ◇ represents on an ERD:

Draw an entity relationship diagram representing that a student can be enrolled in one or more classes, and that the same class can be taken by none or more than one student:

4)

IT Workbook Assignment 4.9

Student name:

Write your answers to the questions below within the box. In each case, please choose your words carefully to answer the specific questions. Avoid simply copying passages from your readings to these answers.

Describe what **data mining** is, and why it is more involved than simply querying a database:

1) what data mining is:

2) why data mining is more involved than simply querying a database:

IBM has identified **five major types of data mining operations**. Provide a brief description of each of these types of operations:

3) associations:

4) sequential patterns:

5) time series:

6) classification and regression:

7) clustering:

IT Workbook Assignment 4.10

Student name:

Write your answers to the questions below within the box. In each case, please choose your words carefully to answer the specific questions. Avoid simply copying passages from your readings to these answers.

Data mining, in the words of Ann Cavoukian, can "in the absence of adequate safeguards, jeopardize informational privacy." (*Data Mining: Staking a Claim on Your Privacy*, January, 1998, www.ipc.on.ca). She lists eight principles of fair information practices, but raises issues with five of them. List the **fair information practices principles on which Cavoukian raises issues**, and briefly discuss each of the issues she raises:

1)

2)

3)

4)

5)

Computer Networks

A **telecommunications system** transmits data from one place to another. It is made up of **transmission media** (cables or radio waves), it connects **devices** such as computers, it **routes signals** with data communications hardware using a common **signaling protocol** (electronic language), and it uses specialized network software to **manage its operation**.

In this chapter we distinguish the characteristics of analog signals and digital signals. The ordinary telephone system handles voice signals in an analog fashion. **Analog signals** mimic the original information varying in strength like the air pressure waves of the voice. Analog signals suffer from serious disadvantages for high speed transmission, which we discuss in the chapter.

Digital signals overcome many of the limitations of analog transmission. Digital signals represent information in discrete values that can be transmitted as a series of coded pulses. **Modems** (modulator/demodulators) adapt the remaining analog components of the voice telephone system to carry digital information. In the modern world transmission media such as **digital subscriber lines** (DSL), modems for cable television systems, **fiber optic cable** and extremely **high frequency radio** (wireless) methods have been developed to provide high speed pathways for digital signals.

Digital communication technologies now make it possible to connect computers in **local area networks** (LAN) within one or more buildings, **wide area networks** (WAN) over distances of many miles, and other forms of interconnections. In the chapter, we examine the components, characteristics, and techniques of various types of computer networks.

Learning objectives

Chapter 5
Computer Networks

This chapter, in combination with the web readings, web links, and podcasts available at www.ambriana.com is focused on providing the knowledge to achieve these learning objectives:

- Be able to identify what a communication system is, and its components
- Understand the difference between analog signals and digital signals, and how analog signals are represented in digital form and vice versa, and the role of a modem
- Be able to describe the analog components of the telephone system, and the digital components, and how multiple digital signals are combined into one transmission stream through multiplexing
- Recognize what broadband communication is, and be able to distinguish the different technologies that can be used to accomplish it
- Understand what a computer network is, and the four types of networks as defined by their size
- Understand and be able to describe the different types of transmission media, computer network topology, and protocols used in computer networks
- Be able to describe the capabilities and differences between Bluetooth and Wi-Fi wireless connection technologies, and the appropriate uses of each
- Understand and be able to explain what routers, gateways, and firewalls are and do in computer networks
- Be able to describe the role of satellite technology, WiMAX, and mesh networks in achieving Internet connectivity for remote areas not otherwise served.

Podcast Scripts – Chapter 5

These readings of the **Information Technology Workbook** are also accessible in audio form as .mp3 files for free download. Visit the workbook web site, www.ambriana.com. You can listen to this material on your own Ipod or .mp3 player or computer, read it, or both listen and read.

Podcast 21
TELECOMMUNICATIONS

A telecommunications system transmits data from one place to another. It is made up of **transmission media** such as cables, or radio waves. It connects devices such as computers, routes signals with data communications hardware using a common signaling protocol, and it uses specialized network software to manage its operation. Communications systems are either analog or digital.

The ordinary telephone system handles voice signals in an analog fashion. **Analog signals** mimic the original information, varying in strength like the air pressure waves of the voice. Analog signals suffer from serious disadvantages for high speed transmission, because any noise or distortion added by amplifying equipment is subsequently indistinguishable from the information in the signal.

Digital signals overcome the limitations of analog transmission. **Digital signals** represent information in the discrete zero and one values of digital ciruitry, which can be transmitted as a series of pulses. Digital signals are not amplified, they are recreated in the process of long distance transmission, so they do not accumulate spurious noise as do analog signals. Even the major connecting signal pathways of the "plain old telephone system" (POTS) were converted to digital technology in the last part of the twentieth century. **Modems**, which are **modulator-demodulator** circuits, adapt the few remaining analog parts of the telephone system to carry digital information over the "last mile," the wire connection between the telephone system's local office and subscriber telephones in most homes.

In the modern world, transmission media such as digital subscriber lines, DSL, modems for coaxial cable television systems, fiber optic cable, and extremely high frequency radio wireless methods have been developed to provide high speed pathways for digital signals. These technologies now make it possible to connect computers in high-speed local area networks within one or more buildings or in wide area networks over distances of many miles.

In this chapter of the Information Technology Workbook we'll introduce you to many of these technologies and the components that make modern computer networking possible. In preparing to learn about computer networking, make sure you review the web links at the Chapter 5 button of the Information Technology Workbook at www.ambriana.com.

Podcast 22
COMPUTER NETWORKS - LAN

Four different categories of computer networks exist. These are:

- local area networks, LANs
- metropolitan area networks, MANs
- wide area networks, WANs
- personal area networks, known as PAN's.

We consider LANs in this podcast, and the other categories in subsequent podcasts.

LAN

A **local area network** is a computer network or system for communicating between computers that covers a small local area such as a home, office, or small group of buildings. The first LANs were created in the late 1970's and were used to links several large central computers at one site. Now, most LANs connect work stations and personal computers.

Each individual computer in a LAN has its own central processing unit with which it runs programs. But it also has the ability to access data and other computer devices anywhere on the LAN. This means that many users can share expensive devices such as laser printers and scanners, as well as data. Users can also use the LAN to communicate with each other by sending e-mail or engaging in chat sessions. LANs are most likely to be based on ethernet, which is the most widespread LAN technology in use today, or WiFi technology.

Ethernet is a communication technique (protocol) for the conveyance of digital signals on wires or cables. WiFi technology is wireless. WiFi allows a person with a wireless-enabled computer or personal digital assistant (PDA) to connect to the Internet when close to an access point. An access point is a low-power radio transmitter and receiver that communicates with a circuit card in appropriately equipped devices.

The defining characteristics of LANs are that they run at much higher data rates than wide area networks, while covering a smaller geographic area. At most, LANs cover a few city blocks, and they usually do not involve the use of leased telecommunication lines. Larger LANs employ redundant links and routers capable of using various techniques to recover from failed links in the network. Modern LANs will often have connections to other LANs via routers, and use leased telephone circuits, to create a wide area network (WAN). LANs also usually have a connection to the Internet. Links to other LANs can be tunneled across the Internet using virtual private network (VPN) technologies.

In the days before personal computers a business or university might have just one central computer, with users accessing it via computer terminals over simple low-speed cabling. Networks such as IBM System Network Architecture (SNA) were aimed at linking terminals, or other mainframes, at remote sites using leased telephone circuits. These proprietary arrangements were actually wide area networks, since they could span long distances.

With the popularization of DOS-based personal computers in the 1980s, users sites grew to have dozens or even hundreds of computers. The initial attraction of networking these was generally to share disk space and laser printers, both of which were very expensive at the time. Much enthusiasm developed for the concept, but many incompatible types of network hardware, cabling, and communications protocol implementations were created by various vendors in the market. It was typical, for example, for each computer vendor to have its own type of network card, cabling, protocol, and network operating system.

Novell Netware finally resolved the situation and helped dispel this chaos. Netware supported forty different competing network card and cable types, and provided a more sophisticated operating system than most of its competitors. NetWare dominated the personal computer LAN business from early after its introduction in 1983 until the mid 1990s, when Microsoft introduced its Windows/NT Advanced Server, and Windows for work groups. Subsequently, hardware became standardized and communication protocols centering on Internet technology assumed the prominence they enjoy today.

Podcast 23
MAN, WAN, AND PAN NETWORKS

Four different categories of computer networks exist. These are:

- local area networks, LANs
- metropolitan area networks, MANs
- wide area networks, WANs
- personal area networks, known as PANs.

We'll consider MANs, WANs, and PANs in this podcast.

MAN

Metropolitan Area Networks are a relatively new class of network. **Metropolitan Area Netwo rks** are large computer networks that span a college campus or a city. MANs typically use a wireless infrastructure, or optical fiber connections, to link their sites.

An **optical fiber** is a transparent thread made of plastic or glass, used for very efficiently transmitting light generated by a modulated laser. Optical fiber circuits can transmit information at much higher data rates than copper wires or coaxial cable. Optical fiber circuits are commonly used in the high capacity backbones of telecommunications systems and, increasingly, even in local networked sites.

MANs serve a role similar to that of an internet service provider (ISP) but for corporate users with large LANs. Three important features set MANs apart from LANs and WANs.

1. **The MAN network size falls somewhere between LANs and WANs.** Many MANs cover an area the size of a city, although they can be as small as a group of buildings, or as large as the northern half of Scotland.
2. **A MAN is not typically owned by a single organization.** Rather, the MAN may be owned by a consortium of users or by an Internet Service Provider, which sells the service to subscribing users.
3. **A MAN frequently acts as a high speed network, allowing the sharing of regional resources.** For example, a university may have a MAN joining many of their LANs, which are situated around a site with an area measured in square kilometers. From the MAN, the university could have links to the Internet.

Older MAN technologies are in the process of being displaced in most areas by Ethernet-based MANs, such as Metro Ethernet. MAN links between LANs have also been built without cables using either microwave, radio, or infra red communication links.

WAN

Now, let's consider wide area networks, called WANs. A wide area network is a computer network that covers a wide geographical area and can involve a vast array of computers. WANs are used to connect local area networks together, so that users and computers in one location can communicate with users and computers far distant. WANS often connect computers through public networks such as the telephone system or via satellite.

The most widely known example of a WAN is the Internet, which is a huge public network. But some WANs are built for one particular organization and are private. Others, built by internet service providers such as Yahoo and Google, provide connections from an organization's LAN to the Internet.

WANs are most often built using high capacity communication lines leased from a provider of telecommunications services such as AT&T or Sprint. At each end of the leased line, a router connects to the LAN on one side, and to a hub within the WAN on the other. Network protocols provide data transport and addressing functions.

PAN

Now let's consider Personal Area Networks, called PANs. A **personal area network** is a computer network used for communicating among small computerized devices within a very short distance of one another. Telephones and personal digital assistants devices (PDA) are commonly linked using PAN technology.

The most common PAN technology is Blue Tooth, a wireless protocol operating at a much lower transmission rate than WiFi, over shorter distances. Personal area networks may also be wired with the Universal Serial Bus (USB) on modern computers.

PANs can be used for communication among the personal devices themselves, which is known as **intrapersonal communication**, or for connecting to a higher level network and the Internet, which is known as an **uplink**.

Podcast 24
NETWORK CONFIGURATIONS

A network configuration, also referred to as **network topology**, is the specific physical arrangement of the elements in a network. These elements can be computers, printers, and scanners, and other electronic devices. A given element has one or more links to other elements in the network, and the links can appear in a variety of different shapes, or configurations.

The simplest connection is a one-way link between two devices. This type of connection, such as between an individual computer and a printer, is not regarded as a network connection. Several types of actual network configurations exist, connecting more than two devices. The most common of these are bus, star, ring, mesh, and wireless mesh topologies. We'll consider each of these topologies in a separate podcast.

Podcast 25
BUS NETWORK

A network configuration, also referred to as network topology, is the specific physical arrangement of the elements in a network. These elements can be computers, printers, and scanners, and other electronic devices. A given element has one or more links to other elements in the network, and the links can appear in a variety of different shapes, or configurations. In this podcast we consider bus network topology.

Bus network topology is a network configuration in which all network elements are connected together by a single communication line, called a bus. Several common instances of the bus architecture exist including one in the motherboard of most computers.

Bus networks are the simplest way to connect multiple computers, but they often have problems when two computers want to transmit at the same time, on the same bus. This result in a "collision" of signals. Systems that use bus network architecture must implement a scheme for collision handling or collision avoidance. One way to deal with the situation is to have each workstation circuit check the bus first before trying to transmit (Carrier Sense, Multiple Access, or CSMA) or to include a bus master circuit that controls access to the shared bus resource.

A bus network is passive as far as workstations are concerned. This means that the computers on the network are not responsible for moving the signal along. Each of the computers on the bus simply listens for a signal addressed to it, and ignores signals addressed to other computers.

Bus networks are easy to implement and extend, and are well suited for temporary networks because they are quick and easy to set up. Bus topologies are typically the cheapest network topology to implement. In addition, they are robust, since the failure of one computer on the network doesn't affect the operability of other computers on the network.

However, bus networks are difficult to administer and troubleshoot. They are limited in cable lengths and the number of computers supported. A cable break can disable major portions of the network, and maintenance costs may be higher than in other topologies. Bus network performance degrades as additional computers are added or as traffic on the network gets heavy. In addition, bus networks offer low security because all computers on the bus can see all data transmissions, and one virus circulating in the network can affect all computers on the network.

Podcast 26
STAR NETWORK

A network configuration, also referred to as network topology, is the specific physical arrangement of the elements in a network. These elements can be computers, printers, and scanners, and other electronic devices. A given element has one or more links to other elements in the network, and the links can appear in a variety of different shapes, or configurations. In this podcast we consider star network topology.

Star networks are one of the most common computer network topologies. In its simplest form, a **star network** consists of one central computer that acts as a router to transmit messages to the others. All computers on the star network communicate with others by transmitting to, and receiving from, just the central computer. No computer workstation connects directly to any other computer workstation. Thus the failure of a transmission line linking any computer workstation to the central computer will result only in the isolation of only that computer workstation.

Compared to other network types, star networks are easy to implement and extend. Star networks are well suited for temporary networks because they allow quick setup. Star networks are reliable and do not suffer from problems with data collisions.

But the disadvantages of star topology includes limited cable length and number of supported computers, and possible higher maintenance costs. Failure of the central computer can disable the entire network, and one virus in the network can affect all computers.

Podcast 27
RING NETWORK

A network configuration, also referred to as network topology, is the specific physical arrangement of the elements in a network. These elements can be computers, printers, and scanners, and other electronic devices. A given element has one or more links to other elements in the network, and the links can appear in a variety of different shapes, or configurations. In this podcast we consider ring network topology.

In a **ring network** every computer has exactly two branches connected to it. Each computer in the ring acts as a repeater, receiving every transmission, accepting transmission addressed to it, and forwarding other transmission along the ruing. Because of the active nature of each computer workstation on the ring, ring networks can span greater distances than other physical topologies.

But ring networks tend to be inefficient when compared to star networks because data must travel through more points before reaching its destination. For example, if a given ring network has eight computers on it, to get from computer 1 to computer 7, data must travel from computer 1 through computers 2, 3, 4, 5, and 6 to finally reach its destination at computer 7. But because of this, if one computer fails, a portion of the full network also fails, since a part of the ring is disconnected by the failure.

Ring networks are often the most expensive topology. Due to the need for extra cabling to complete the ring, ring networks canoften be difficult to install, and are more dependent on the physical placement of each device in the network.

The most popular example of the ring topology is a **token ring network**. In this type of ring network, special signals named tokens traverse the ring at all times. A computer that wishes to send a message to another computer in the network waits to receive an unused token and then attaches its message to that token and forwards it on the ring.

Podcast 28
MESH NETWORK

A network configuration, also referred to as network topology, is the specific physical arrangement of the elements in a network. These elements can be computers, printers, and scanners, and other electronic devices. A given element has one or more links to other elements in the network, and the links can appear in a variety of different shapes, or configurations. In this podcast we consider mesh network topolo gy.

In **mesh network topology**, several computers exist, with two or more paths between each. This concept is applicable to wireless and wired networks. Let's consider a wired mesh network first.

In a full mesh wired network, a direct wired link exists between each computer and each other computer. These direct connections provide a high amount of reliability. Mesh networks can still operate even when a computer breaks down or a connection line fails. But wired networks designed with this topology are usually very expensive. Typically, only special purpose situations, such as military applications, can warrant the high wired mesh network cost to gain this ultimate degree of reliability.

Now, let's consider a wireless mesh network. Wireless mesh networking is mesh networking implemented over a wireless LAN. This type of mesh network is relatively inexpensive, and very reliable and resilient, since each computer workstation, here called a "node," needs only to transmit as far as the surrounding computer nodes. Each computer acts as a repeater to transmit data from nearby computers to peers, resulting in a network that can span large distances if computers are naturally spread out across the geography.

Wireless mesh networks are especially appealing for coverage over rough or difficult terrain. The multiple communication paths allow for continuous connections and automatic reconfiguration around blocked paths as a transmission hops from computer to available adjacent computers. Wireless mesh networks may involve either fixed or mobile devices.

Wireless mesh networks are diverse, and provide reasonably priced communication in difficult environments, such as tunnels and oil rigs, and for battlefield surveillance and other high speed mobile video applications. Since a wireless mesh network has the potential to be much less expensive than other types of networks, many wireless community network groups are already creating wireless mesh networks.

Podcast 29
BLUETOOTH WIRELESS

The various parts of a computer network are a community of electronic devices. They communicate with each other using a variety of wires, cables, signals and light beams, and an even greater variety of connectors and plugs. When any two devices need to talk to each other, they have to agree on several things. How much data will be sent? How will they speak to each other? All participants in an electronic discussion need to know what the bits mean and whether the message they received is the same one sent. This coordination of understanding requires the development of an agreed upon set of commands and responses, known as a communication protocol.

Two **wireless protocols** have been developed and popularized for use in local area networks. They are named Bluetooth, after a Danish king, and WiFi. We consider Bluetooth in this podcast, and WiFi in the subsequent podcast.

Bluetooth is a wireless networking technology designed to connect devices over short distances. Bluetooth is inexpensive, automatic, and low powered, yet it can connect up to eight devices simultaneously.

Even with low power requirements, Bluetooth doesn't need line of sight between communicating devices, and the walls in your house won't stop a Bluetooth signal. This makes it useful for controlling several devices in different rooms.

Bluetooth transmits data at the rate of 742,000 bits per second via low-power radio waves. This is much slower than WiFi but adequate for the purpose of connecting wireless computer keyboards and mice, for connecting personal digital assistants to computers and each other, and for portable telephone technology. Bluetooth communicates on a frequency band in the 2.4 gigahertz range (2.4 billion cycles per second), which has been set aside by international agreement for the use of industrial, scientific and medical devices. This is the same ultra high frequency band used by early cordless telephones, microwave ovens, garage door openers, and some radar systems.

Bluetooth technology facilitates small area networking by eliminating the need for user intervention to establish a connection, and by keeping power requirements for communication very low, extending battery life and limiting range.

As a networking device standard, Bluetooth defines standards at two levels, the physical level and the communication protocol level. It provides agreement at the physical level by establishing the radio frequencies to be used and how devices are to automatically switch frequencies hundreds of times a second to avoid interference ("spread spectrum" communication). It provides agreement at the communication protocol level by establishing agreement between devices on when bits are sent, how many bits will be sent, and how the devices in communication can be sure that the message received is the same as the message that was sent.

At the **physical level** Bluetooth defines how its use of radio frequencies will minimize the chance for interference with other devices, since a number of devices that you may already use take advantage of the same frequency band. Insuring that Bluetooth and these other devices don't interfere with one another was a crucial part of the Bluetooth design process. One of the ways Bluetooth devices avoid interfering with other systems is by sending out very weak signals. The low power limits the range of a Bluetooth device to about thirty feet, reducing the chances of interference between your computer system, your portable telephone, and your neighbor's garage door opener.

A more significant way that Bluetooth minimizes the chance for interference relies on a technique called spread spectrum frequency hopping. Spread spectrum frequency hopping makes it very rare for more than one device to be transmitting on the same frequency at the same time by constantly changing the frequency used for transmissions. Bluetooth's transmitters change frequencies 1,600 times per second! Since every Bluetooth transmitter uses spread-spectrum transmitting automatically, it's highly unlikely that two transmitters close enough to be in communication with each other will be on the same frequency at the same time. In the unlikely event that a transmission does encounter interference, the error detection and retransmission mechanism also built into Bluetooth compensates for it.

Now, let's consider some of the aspects of the **communications protocol level** as defined by Bluetooth. When Bluetooth-capable devices come within range of one another, an electronic conversation automatically takes place to determine whether they have data to share or whether one needs to control the other. You don't have to press a button or give a command to initiate this sequence of data exchanges. Once the initializing data exchanges take place, the Bluetooth devices form a network on their own.

Bluetooth systems automatically create a personal area network (PAN) that may fill a room, or may encompass no more distance than that between the cellular phone on your belt and the headset on your head. Once a PAN is established, the members in the network hop radio frequencies in unison so they stay in touch with one another and avoid interfering with other PANs that may be operating nearby.

Bluetooth has been very successful for its intended purpose. Applications ranging from PDA connectivity, to cordless keyboard and mice for desktops and laptops, to digital camera connectivity to computers are just a few of the modern applications of Bluetooth technology.

Podcast 30
WiFi WIRELESS

The various parts of a computer network are a community of electronic devices. They communicate with each other using a variety of wires, cables, signals and light beams, and an even greater variety of connectors and plugs. When any two devices need to talk to each other, they have to agree on several things. How much data will be sent? How will they speak to each other? All participants in an electronic discussion need to know what the bits mean and whether the message they received is the same one sent. This coordination of understanding requires the development of an agreed upon set of commands and responses, known as a communication protocol.

Two **wireless protocols** have been developed and popularized for use in local area networks. They are named Bluetooth, after a Danish king, and WiFi. In a previous podcast we considered Bluetooth. In this podcast consider WiFi.

WiFi is a wireless networking technology designed to allow computers to communicate in a local area network over distances of hundreds of feet. Despite it's common misinterpretation WiFi doesn't mean "wireless fidelity." WiFi is a trademarked designation licensed by the WiFi governing body to products that pass compliance tests demonstrating that meet compatibility standards.

WiFi products are widely available in the market, and, significantly, WiFi has a global set of standards, so that the same products can work in different parts of the world. This makes it unique among communications offerings. Cellular phone systems, for example, often use different standards in different countries, and cell phones used in the United States cannot be used in Europe, whereas the same WiFi devices can be used in both places.

WiFi uses stronger radio frequency signals than does Bluetooth, and it communicates at much higher speeds. Whereas Bluetooth communicates at less than one megabit per second, WiFi systems transmit and receive at either 11 megabits per second for older systems, or 54 megabits per second for more modern ones. This is more than 70 times faster than Bluetooth.

Originally intended to be used for mobile devices and LANs, WiFi is now often used for Internet access at fixed locations. WiFi lets a person with a wireless enabled computer, or personal digital assistant, connect to the Internet when in proximity of a WiFi access point. The geographical region covered by one or several access points is called a hot spot.

WiFi allows LANs to be deployed without cabling to each computer, significantly reducing the cost of network deployment and expansion. Using WiFi, it's possible to install LANs in places where new cables cannot be run, such as historic buildings. But some technical concerns do exist with WiFi usage.

For one thing, as a result of the higher signal strength, power consumption for WiFi is high as compared to Bluetooth, making battery life and heat generation more of a consideration. Security of data transmission is an even more serious concern with WiFi systems. The most common WiFi wireless encryption standard has been shown to be breakable, even when correctly configured. While most newer wireless products support the improved WiFi protected access protocol, many first-generation access points cannot be upgraded in the field and will have to be replaced to support it.

And although WiFi networks are designed to have a limited range, WiFi signals do easily pass through ordinary building walls. WiFi signals are strong enough to communicate between houses and apartments. This can pose a problem in high density areas such as large apartment buildings. When several residents in the area establish WiFi access points, it is common for a WiFi equipped computer in a house or apartment to detect a strong enough signal from a neighbor to erroneously establish a connection with some one else's access point.

Although WiFi technology can be used for telephones, it seems unlikely that it will compete directly against cellular technology for mobile phone usage. WiFi phones have a limited range by comparison to cellular technology, so setting up enough access points to support mobile phone usage, would be too expensive to be practical.

In addition, cellular technology bounces a connection from cell phone tower to cell phone tower while maintaining one solid connection to a traveling user. Current WiFi offerings can't do this, since they connect to only one access point at a time. Once you're out of range of one hot spot, the connection will drop. You would need to be reconnected to the next access point each time you passed out of the range of one and into the range of another.

Technical and security concerns notwithstanding, WiFi has considerable potential for local area network implementation. It has already grown in popularity to the point that most laptop computers sold today include WiFi wireless communications circuitry built in. Hot spots in restaurants, schools, and libraries are common, and many provide free high speed connection to the internet.

The standards for WiFi continue to be refined, both in communication speed and security. Both Bluetooth and WiFi will most likely be a part of the local area networking picture for many years to come, as more and more devices achieve inexpensive and reliable unattended connectivity through their use.

Podcast 31
ROUTERS

A number of components in addition to computers are critical to the effective operation of a computer network. Among these components are routers, gateways, proxy servers, firewalls, and the communications protocol used by the computer network to send messages across the network. Also among these components is specialized software to control the devices making up the network, in the form of network operating systems. In this podcast we consider routers.

A **router** is a specialized computer that forwards **packets** of data, that is, pieces of messages, over packet switching networks. A router is connected to at least two networks, commonly two LANs or WANs, or a LAN and its Internet service provider's network. Routers are located at gateways, which are the places where two or more networks connect. Routers use message headers and forwarding tables maintained by the routers in a network to determine the best path for forwarding data packets through the network.

A **message header** is a special part of the overhead associated with a message. The message header carries information about the message and its destination. Think about a message header like an envelope in which an ordinary paper letter is sent by snail (postal) mail. The envelope is not the message, but it contains information, such as the recipient's address, needed to get the message to the proper destination. In network transmissions, a message header is part of the data packet and contains information about the transmission. The header, which can be accessed only by the operating system or by specialized programs, may also contain the date the message or file being sent was created and the size of the transmission.

In e-mail, for example, the first part of a message contains controlling data,, such as the subject, origin and destination e-mail addresses, and the path an e-mail takes. The header contains information about the e-mail client. As the e-mail travels to its destination, information about the path it took will be appended to the header by the routers that handle the message in transit.

Routers use communications protocols to communicate with each other. **Communication protocols** are agreed upon formats and procedures for transmitting data and acknowledging its receipt.

Podcast 32
GATEWAYS AND PROXY SERVERS

A number of components in addition to computers are critical to the effective operation of a computer network. Among these components are routers, gateways, proxy servers, firewalls, and the communications protocol used by the computer network to send messages across the network. Also among these components is specialized software to control the devices making up the network, in the form of network operating systems. In this podcast we consider gateways and proxy servers.

Gateway

A gateway is a node, or device on a network, that serves as an entrance to another network. In large organizations, the gateway is the computer that routes traffic to the outside network, and may, for example, serve (deliver) web pages. In small organizations or private residences the gateway is the ISP that connects the user to the Internet. In large organizations, the gateway often acts as a proxy server and a firewall.

The gateway is also associated with a router, which uses headers and forwarding tables, to determine where data packets are to be sent, and a data switch, which provides the actual path for the packet, in and out of the gateway.

Proxy Server

What is a proxy server? A **proxy server** is a server sitting between a client application, such as a web browser, and a real server. It intercepts each request to the real server, to see if it can fulfill the request itself. If not, it forwards the request to the real server. A Proxy server can also serve as a firewall by effectively hiding the true network address of its users, enhancing the security of a local area network.

Proxy servers have two main purposes. They can improve network performance, and they can enhance control over what is visible to the users of a network, and what external users on the Internet can "see" of an internal network.

Proxy servers can dramatically improve performance for groups of users because they save the results of all requests for a certain amount of time. Consider one user accessing the World Wide Web through a proxy server. The proxy server saves the requested page. If another user now requests the same web page, the proxy server can serve it, instead of getting a new copy of the web page across the internet. Since the proxy server is often on the same network as the user, this is a much faster operation.

Proxy servers can also be used to filter requests. For example, a company might use a proxy server to prevent its employees from accessing a specific set of web sites. Real proxy servers support hundreds or even thousands of users. Major online services, such as AOL, for example, employ an array of proxy servers. A proxy server can also block incoming requests generated from the Internet, keeping the network free of attempts to infiltrate it.

Podcast 33
FIREWALLS

A number of components in addition to computers are critical to the effective operation of a computer network. Among these components are routers, gateways, proxy servers, firewalls, and the communications protocol used by the computer network to send messages across the network. Also among these components is specialized software to control the devices making up the network, in the form of network operating systems. In this podcast we examine what fire walls are.

Firewalls are a first line of defense in a computer network in protecting private information. A **firewall** is a system designed to prevent unauthorized access to or from a private network. Firewalls can be implemented in hardware or software, or a combination of both. Firewalls are frequently used to prevent unauthorized Internet users from accessing private networks such as intranets.

Like the Internet itself, intranets are used to share information. Intranets are networks based on internet technology, but belonging to an organization, and accessible only by the organization's members or employees. Secure intranets are now the fastest-growing segment of the Internet, because they are much less expensive to build and manage than private networks that were based on proprietary technology and protocols.

All messages entering or leaving an intranet must pass through the firewall, which examines each message. The firewall can prevent messages that don't meet specified security criteria from being delivered. The firewall can do this simply by dropping the message and not passing it on.

Several types of firewall techniques exist. One of these is a packet filter, which examines each data packet entering or leaving the network and accepts or rejects it based on user-defined rules. Packet filtering is fairly effective and transparent to users, but difficult to configure.

Another type of firewall is an application gateway, which applies security mechanisms to specific applications, such as FTP and Telnet servers. This is very effective, but it can cause a degradation in performance.

Still another firewall technique is the circuit level gateway, which applies security mechanisms when an Internet connection is established. Once the connection has been made, packets can flow between hosts without further checking.

In practice, many firewalls employ two or more of these techniques in combination. The administration of the controls for a firewall is generally the responsibility of the security or network group of an organization. Windows/XP provides a simple firewall that you can activate to gain protection against unauthorized access to your computer, especially if you have a continuous Internet connection.

Podcast 34
COMMUNICATION PROTOCOLS AND NETWORK OPERATING SYSTEMS

A number of components in addition to computers are critical to the effective operation of a computer network. Among these components are routers, gateways, proxy servers, firewalls, and the communications protocol used by the computer network to send messages across the network. Also among these components is specialized software to control the devices making up the network, in the form of network operating systems. In this podcast we consider communication protocols and network operating systems.

Communication protocols

A **communication protocol** is an agreed upon format for transmitting data between two nodes or devices. Protocols determine the type of error checking to be used, the data compression method, if any, how the sending device will indicate that it has finished sending a message, and how the receiving device will acknowledge that it has received a message.

A variety of standard protocols have been developed. Each protocol offers advantages and disadvantages. Some protocols are simpler than others, some are more reliable, and some are faster. Protocols can be implemented in either hardware or software. Your computer must support the correct protocols if you want to communicate with other computers.

TCPIP is an abbreviation for **Transmission Control Protocol, Internet Protocol**, the suite of communication protocols used to connect host computers on the Internet. TCPIP uses several protocols, the two main ones being TCP, which is Transmission Control Protocol, and IP, Internet Protocol. TCPIP is built into the UNIX operating system and is used by the Internet, making it the standard for transmitting data over networks. Even network operating systems that have their own protocols, such as Netware, also support TCPIP.

Transmission Control Protocol is a "transport" level protocol and deals only with packets. TCP enables two hosts to establish a connection and exchange streams of data. TCP guarantees delivery of data and also guarantees that packets will be delivered in the same order in which they were sent.

Internet Protocol is a "network" level protocol and is the common element in the public Internet. It is used for communicating data across a packet-switched network and provides widely-recognized addressing services. Whereas TCP reliably gets data packets to their site destination, Internet Protocol supports their distribution across disparate types of local area networks using the unique IP addressing mechanism.

Network operating systems

Network operating systems, abbreviated NOS, are operating systems that include special functions for connecting computers and devices to a LAN. Some operating systems, such as UNIX, have networking functions built in. The term network operating system, however, is generally reserved for software that enhances a basic operating system by adding networking features. Novell Netware and Windows/NT, are examples of an NOS.

You have now completed all of the podcasts for Chapter 5 of the Information Technology Work book. Make sure you review the material at the web links provided at the work book web site at the Chapter 5 button at the Information Technology Workbook at www.ambriana.com.

Lecture Slides 5

Introduction to Information Technology

Communications Networks

(C) 2006 Jim Janossy and Laura McFall, Information Technology Workbook

Course resources

PowerPoint presentations, podcasts, and web links for readings are available at
www.ambriana.com > IT Workbook

Print slides at 6 slides per page

Homework, quizzes and final exam are based on slides, lectures, readings and podcasts

Slide 5-2

Main topics

- Communications systems
- The telephone network
- Overview of computer networks
- Categories of networks
- Network configurations
- Wireless technologies
- Network components

Slide 5-3

What is communication?

- Exchange of information
- Humans communicate with sound and visual cues including text and images
- Sound and sight have limited range
- We "impress" intelligence on energy that can be quickly transported great distances and decoded by the receiver

Slide 5-4

What is a communications system?

A collection of

- individual communications networks
- transmission systems
- relay stations
- tributary stations
- data terminal equipment (DTE)

capable of **interconnection** and **interoperation** to form an integrated whole

Slide 5-5

What is a communications system?

These are the five common characteristics of every communications system:

- serve a common purpose
- are technically compatible
- use common procedures
- respond to controls
- operate in unison

Slide 5-6

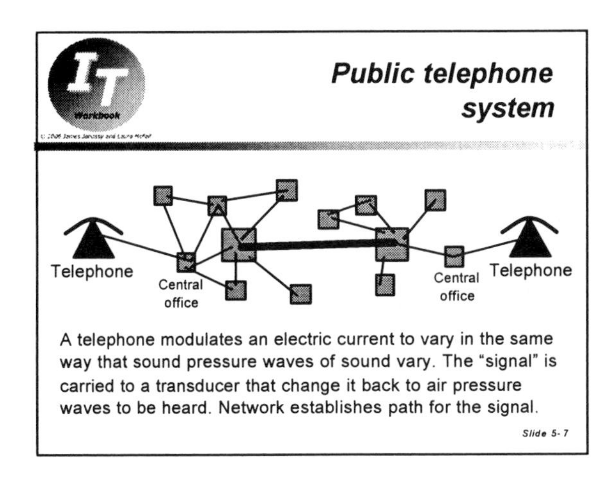
Public telephone system
Telephone
Central office
Central office
Telephone
A telephone modulates an electric current to vary in the same way that sound pressure waves of sound vary. The "signal" is carried to a transducer that change it back to air pressure waves to be heard. Network establishes path for the signal.
Slide 5-7

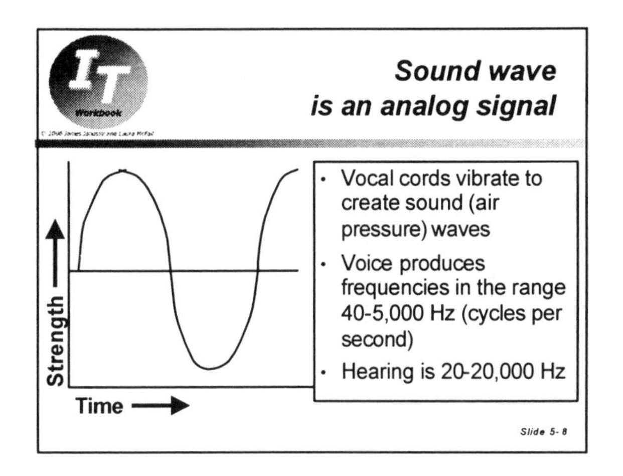
Sound wave is an analog signal
Vocal cords vibrate to create sound (air pressure) waves
Voice produces frequencies in the range 40-5,000 Hz (cycles per second)
Hearing is 20-20,000 Hz
Strength
Time
Slide 5-8

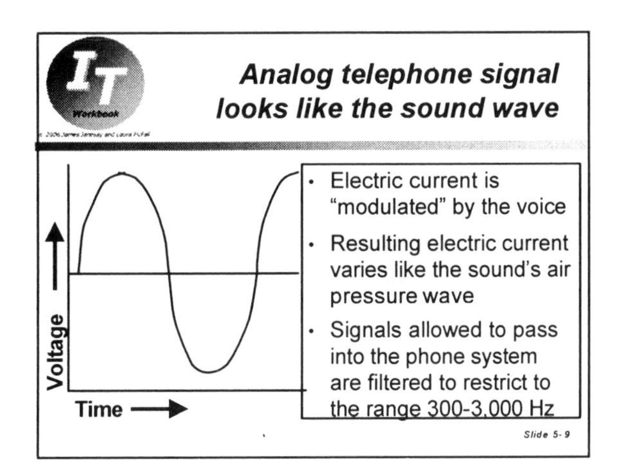
Analog telephone signal looks like the sound wave
Electric current is "modulated" by the voice
Resulting electric current varies like the sound's air pressure wave
Signals allowed to pass into the phone system are filtered to restrict to the range 300-3,000 Hz
Voltage
Time
Slide 5-9

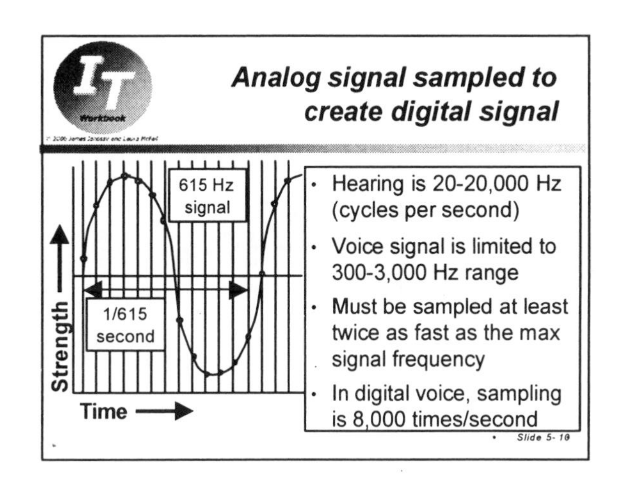
Analog signal sampled to create digital signal
615 Hz signal
1/615 second
Hearing is 20-20,000 Hz (cycles per second)
Voice signal is limited to 300-3,000 Hz range
Must be sampled at least twice as fast as the max signal frequency
In digital voice, sampling is 8,000 times/second
Strength
Time
Slide 5-10

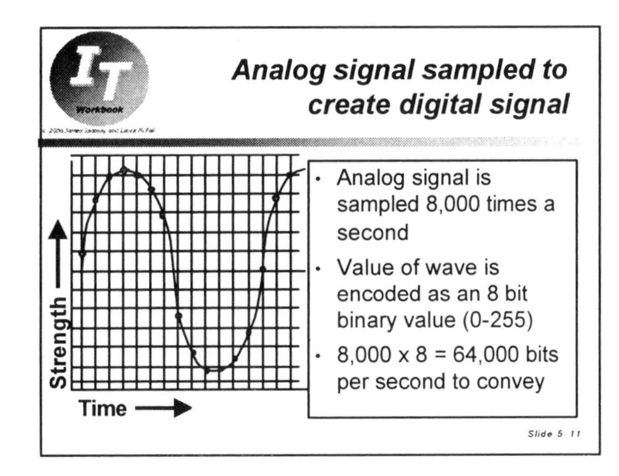
Analog signal sampled to create digital signal
Analog signal is sampled 8,000 times a second
Value of wave is encoded as an 8 bit binary value (0-255)
8,000 x 8 = 64,000 bits per second to convey
Strength
Time
Slide 5-11

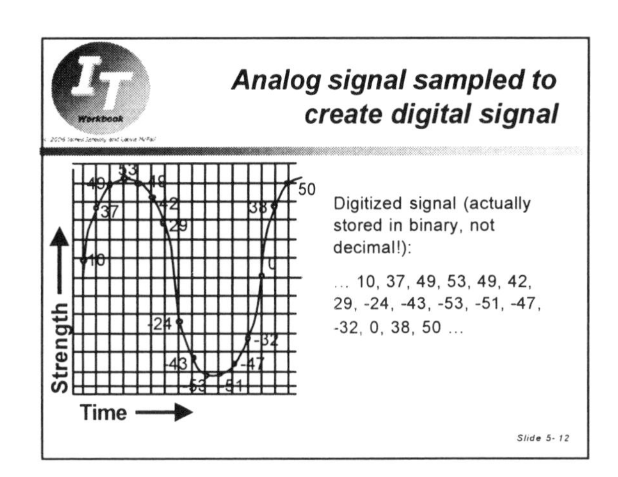
Analog signal sampled to create digital signal
Digitized signal (actually stored in binary, not decimal!):
... 10, 37, 49, 53, 49, 42, 29, -24, -43, -53, -51, -47, -32, 0, 38, 50 ...
Strength
Time
Slide 5-12

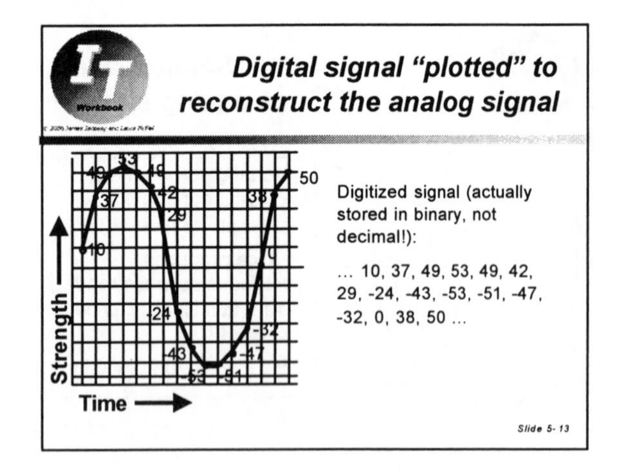

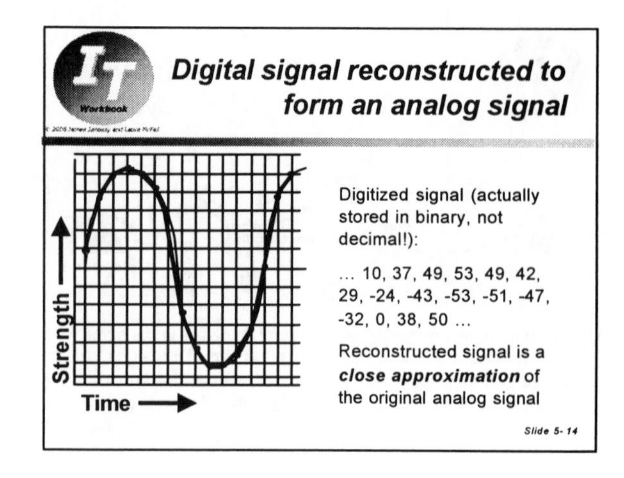

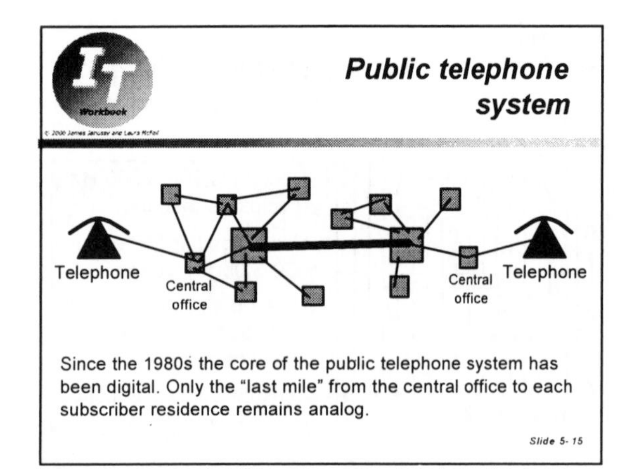

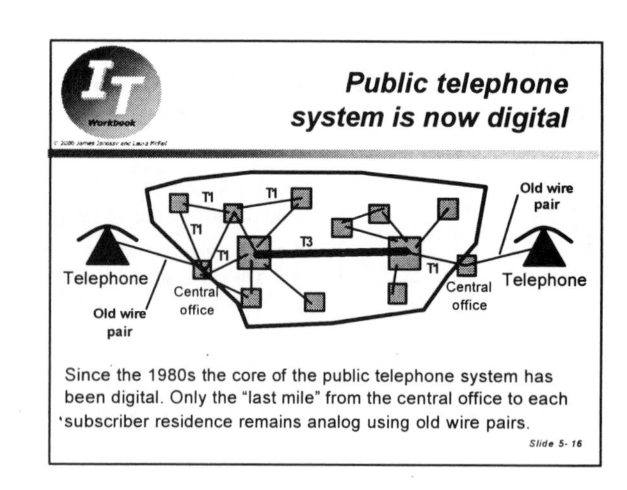

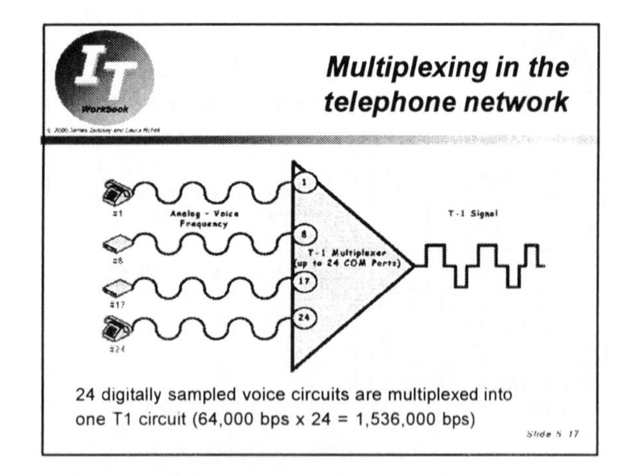

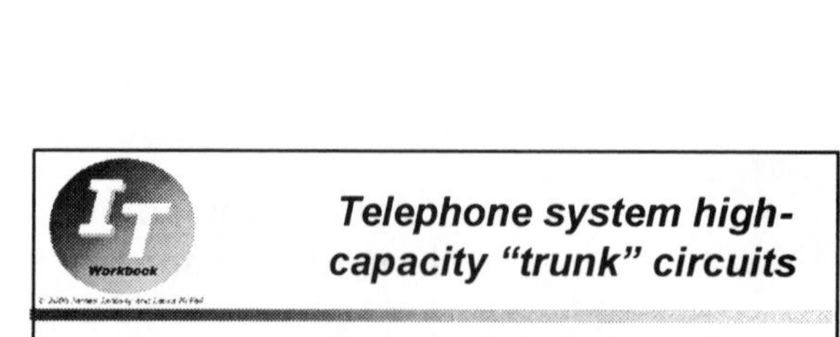

Telephone system high-capacity "trunk" circuits

- **T1** – 1,536,000 bits/second (24 voice)
- **T3** – 45,000,000 bits/second (720 voice)
- **OC3** – 155,000,000 bits/second
- **OC12c** – 620,000,000 bits/second
- **OC48c** – 2,500,000,000 bits/second
- **OC192c** – 10,000,000,000 bits/second!

Slide 5- 18

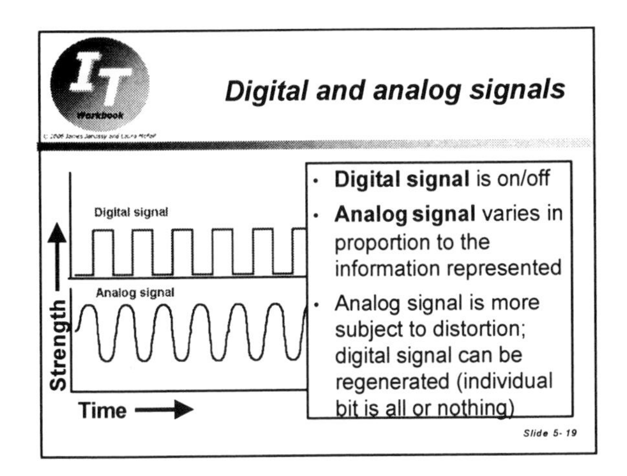
Digital and analog signals
Digital signal
Analog signal
Strength
Time
Digital signal is on/off
Analog signal varies in proportion to the information represented
Analog signal is more subject to distortion; digital signal can be regenerated (individual bit is all or nothing)
Slide 5-19

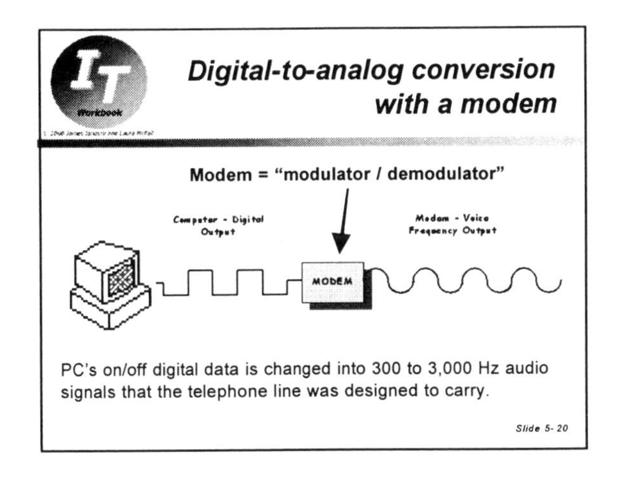
Digital-to-analog conversion with a modem
Modem = "modulator / demodulator"
MODEM
PC's on/off digital data is changed into 300 to 3,000 Hz audio signals that the telephone line was designed to carry.
Slide 5-20

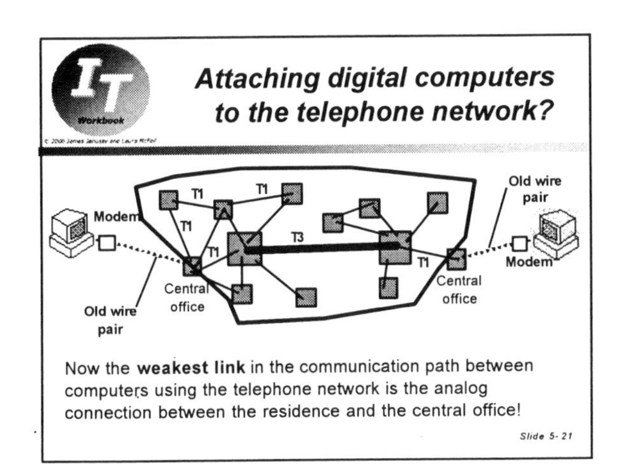
Attaching digital computers to the telephone network?
Modem
Old wire pair
T1
T3
Central office
Now the weakest link in the communication path between computers using the telephone network is the analog connection between the residence and the central office!
Slide 5-21

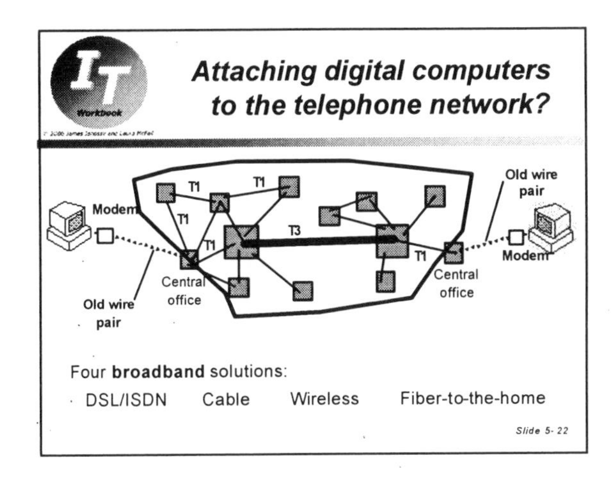
Attaching digital computers to the telephone network?
Modem
Old wire pair
T1
T3
Central office
Four broadband solutions:
DSL/ISDN
Cable
Wireless
Fiber-to-the-home
Slide 5-22

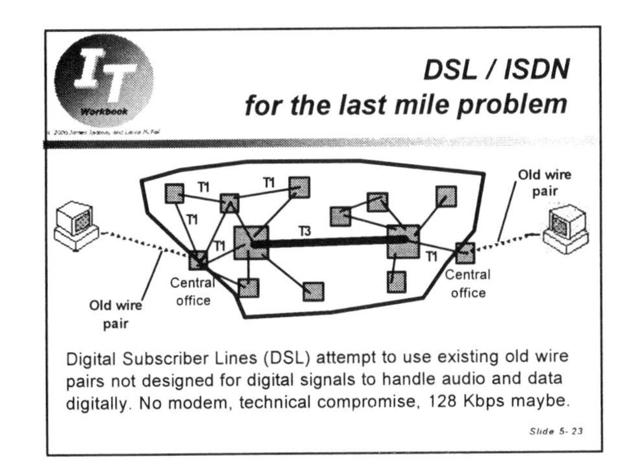
DSL / ISDN for the last mile problem
Old wire pair
T1
T3
Central office
Digital Subscriber Lines (DSL) attempt to use existing old wire pairs not designed for digital signals to handle audio and data digitally. No modem, technical compromise, 128 Kbps maybe.
Slide 5-23

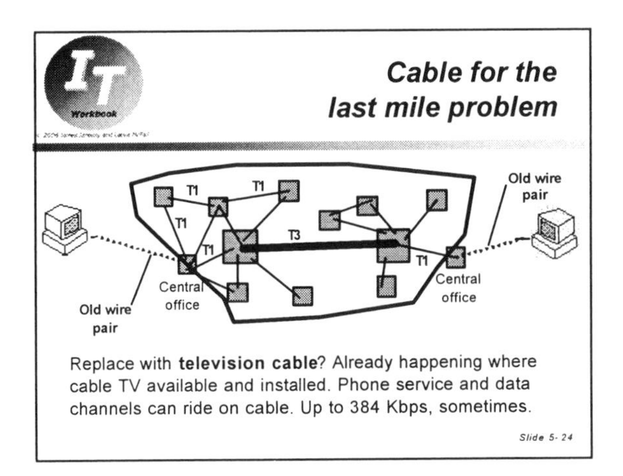
Cable for the last mile problem
Old wire pair
T1
T3
Central office
Replace with television cable? Already happening where cable TV available and installed. Phone service and data channels can ride on cable. Up to 384 Kbps, sometimes.
Slide 5-24

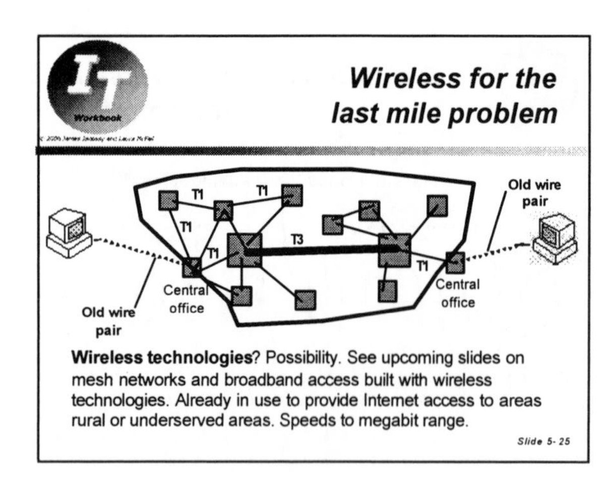
Wireless for the last mile problem
T1
T3
Old wire pair
Central office
Wireless technologies? Possibility. See upcoming slides on mesh networks and broadband access built with wireless technologies. Already in use to provide Internet access to areas rural or underserved areas. Speeds to megabit range.
Slide 5-25

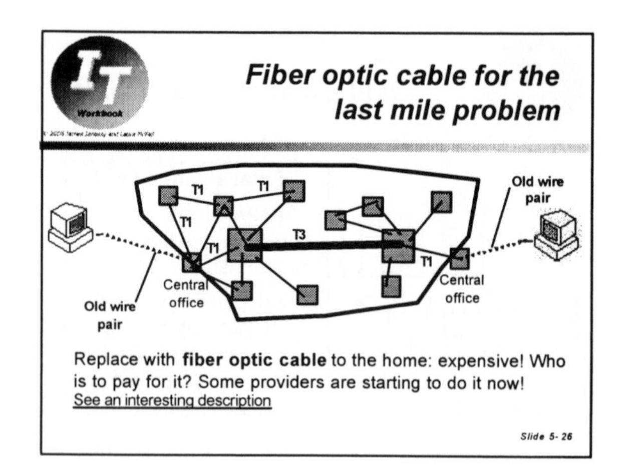
Fiber optic cable for the last mile problem
T1
T3
Old wire pair
Central office
Replace with fiber optic cable to the home: expensive! Who is to pay for it? Some providers are starting to do it now!
See an interesting description
Slide 5-26

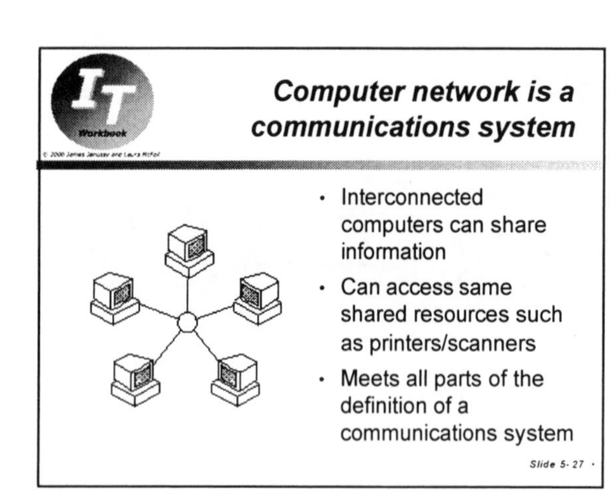
Computer network is a communications system
Interconnected computers can share information
Can access same shared resources such as printers/scanners
Meets all parts of the definition of a communications system
Slide 5-27

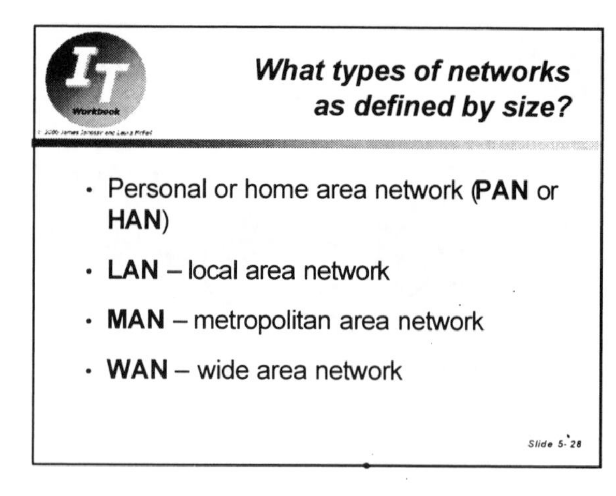
What types of networks as defined by size?
Personal or home area network (PAN or HAN)
LAN – local area network
MAN – metropolitan area network
WAN – wide area network
Slide 5-28

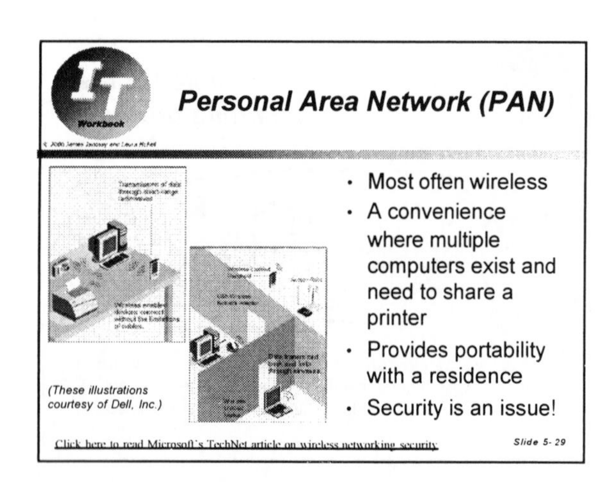
Personal Area Network (PAN)
Most often wireless
A convenience where multiple computers exist and need to share a printer
Provides portability with a residence
Security is an issue!
(These illustrations courtesy of Dell, Inc.)
Click here to read Microsoft's TechNet article on wireless networking security
Slide 5-29

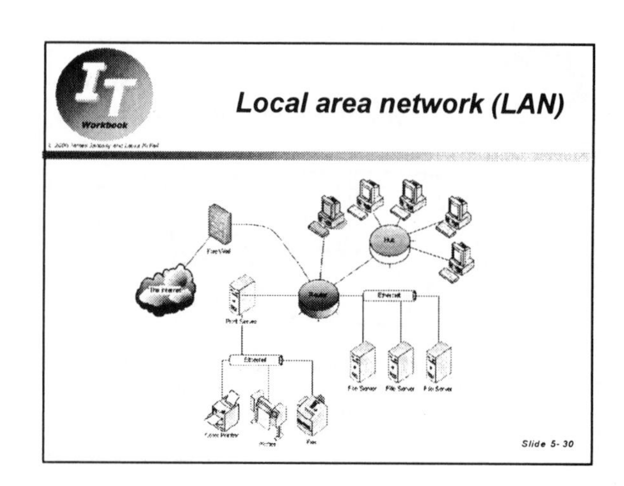
Local area network (LAN)
Slide 5-30

Local area network (LAN)

- Computers are geographically close together; home, office, or small group of buildings such as a college
- Often higher local data rates than WAN
- Do not involve leased telecommunication lines; wire or coaxial cable is physically installed to support them

Slide 5- 31

Metropolitan area network (MAN)

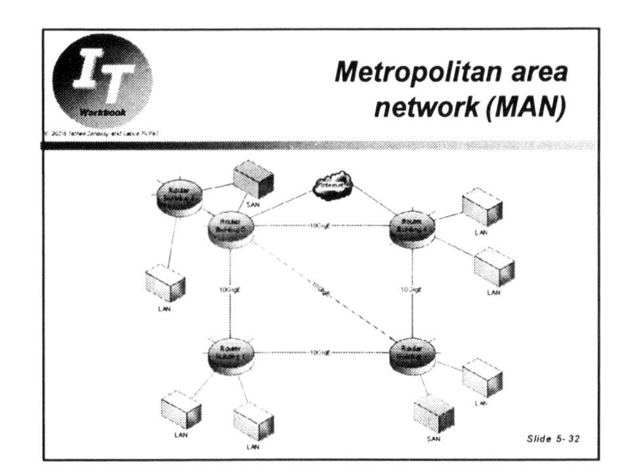

Slide 5- 32

Metropolitan area network (MAN)

- Large computer networks spanning a campus or a city
- Use leased lines, wireless infrastructure, optical fiber connections
- Used to connect local area networks (LANs)
- Many WANs are built for one particular organization and are private
- Best example of an open WAN is the Internet

Slide 5- 33

Network server is just another computer

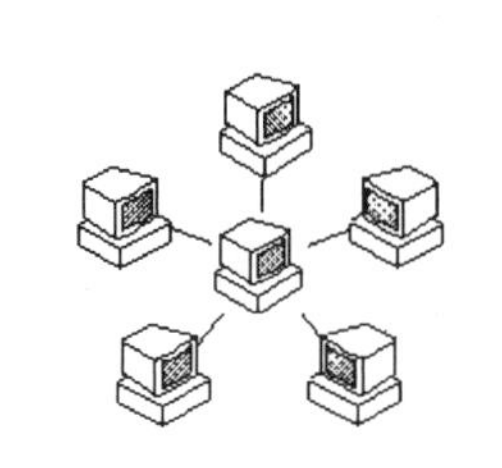

- Server may run a more capable operating system (a network operating system)
- Server might not be used as a regular end-user machine
- How are the server and computers connected?

Slide 5- 34

Three dimensions of the "connection" question

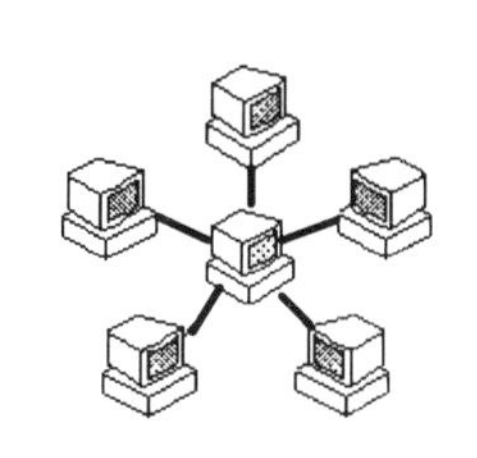

- Server may run a more capable operating system (a network operating system)
- Server might not be used as a regular end-user machine
- How are the server and computers connected?

Slide 5- 35

Three "connection factors" in a computer network

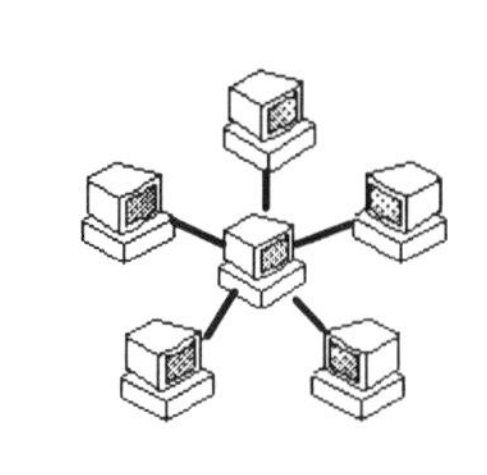

- **Transmission media** (wires, cables, wireless)?
- What **shape** is the connection pattern ("topology")?
- What **signals** are used to communicate ("interoperate")?

Slide 5- 36

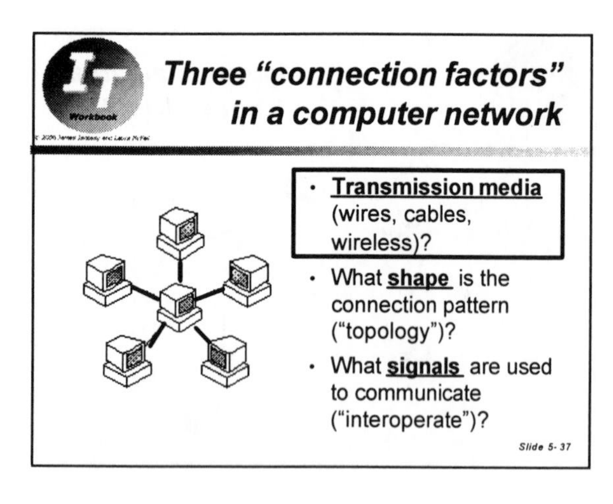
Three "connection factors" in a computer network
Workbook
• Transmission media (wires, cables, wireless)?
• What shape is the connection pattern ("topology")?
• What signals are used to communicate ("interoperate")?
Slide 5- 37

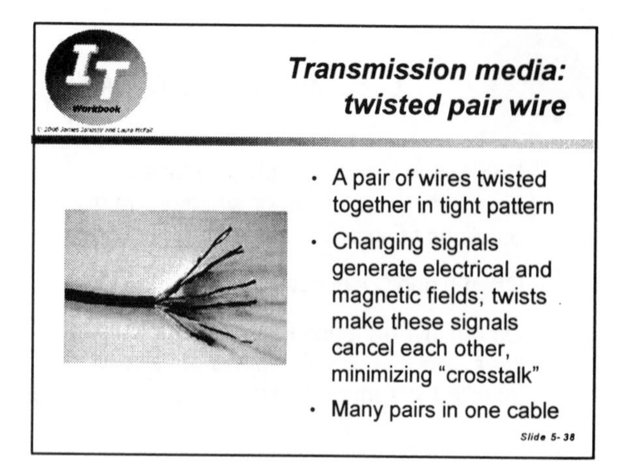
Transmission media: twisted pair wire
Workbook
• A pair of wires twisted together in tight pattern
• Changing signals generate electrical and magnetic fields; twists make these signals cancel each other, minimizing "crosstalk"
• Many pairs in one cable
Slide 5- 38

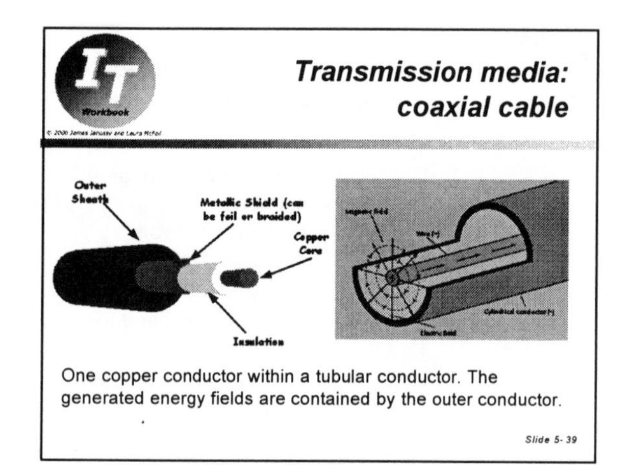
Transmission media: coaxial cable
Workbook
Outer Sheath
Metallic Shield (can be foil or braided)
Copper Core
Insulation
One copper conductor within a tubular conductor. The generated energy fields are contained by the outer conductor.
Slide 5- 39

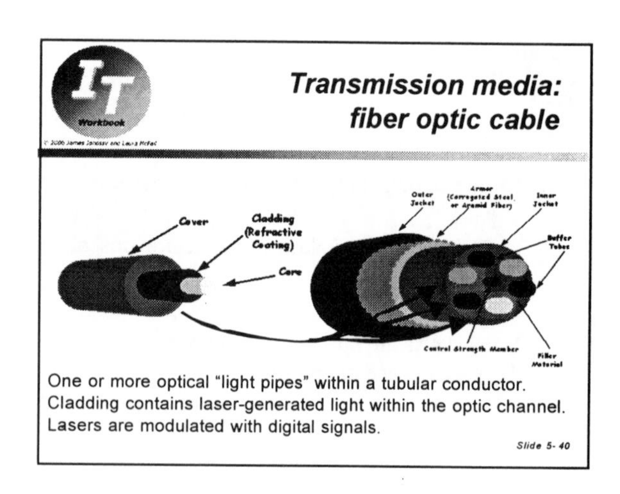
Transmission media: fiber optic cable
Workbook
Cover
Cladding (Refractive Coating)
Core
Outer Jacket
Inner Jacket
Buffer Tubes
Central Strength Member
Filler Material
One or more optical "light pipes" within a tubular conductor. Cladding contains laser-generated light within the optic channel. Lasers are modulated with digital signals.
Slide 5- 40

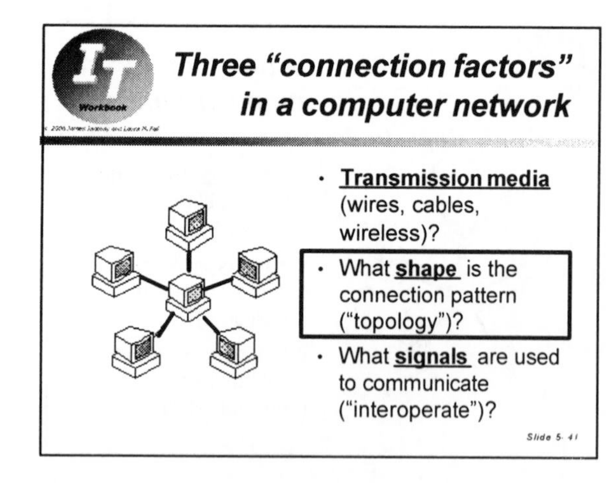
Three "connection factors" in a computer network
Workbook
• Transmission media (wires, cables, wireless)?
• What shape is the connection pattern ("topology")?
• What signals are used to communicate ("interoperate")?
Slide 5- 41

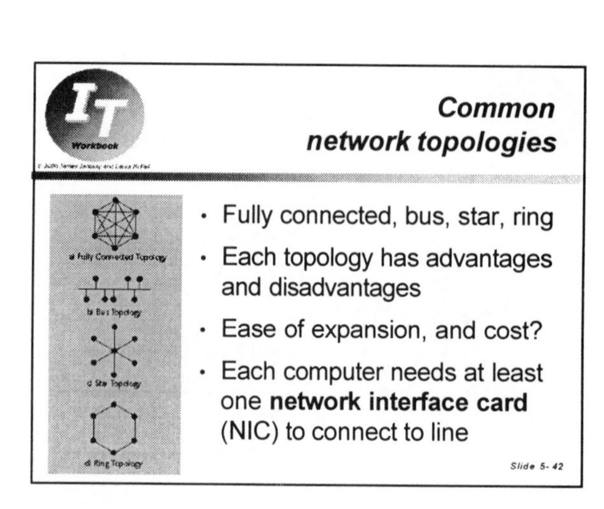
Common network topologies
Workbook
• Fully connected, bus, star, ring
• Each topology has advantages and disadvantages
• Ease of expansion, and cost?
• Each computer needs at least one network interface card (NIC) to connect to line
Slide 5- 42

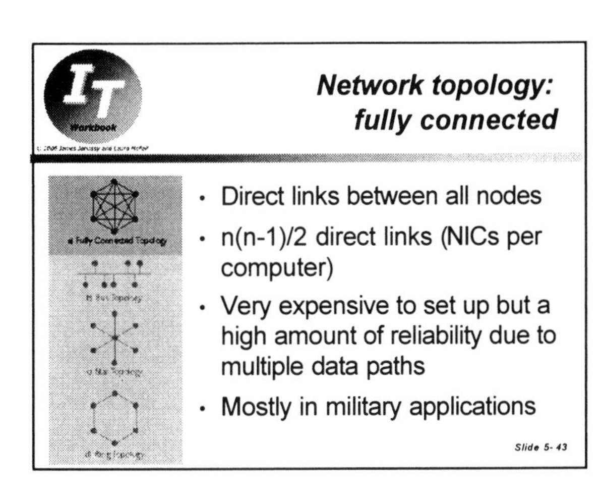
Network topology:
fully connected
Direct links between all nodes
n(n-1)/2 direct links (NICs per computer)
Very expensive to set up but a high amount of reliability due to multiple data paths
Mostly in military applications
Slide 5- 43

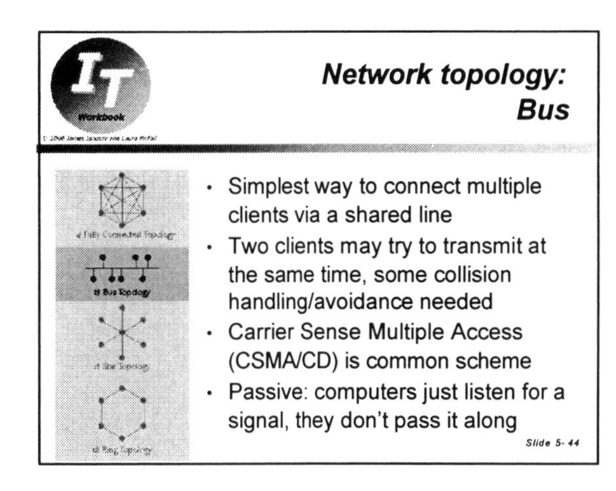
Network topology:
Bus
Simplest way to connect multiple clients via a shared line
Two clients may try to transmit at the same time, some collision handling/avoidance needed
Carrier Sense Multiple Access (CSMA/CD) is common scheme
Passive: computers just listen for a signal, they don't pass it along
Slide 5- 44

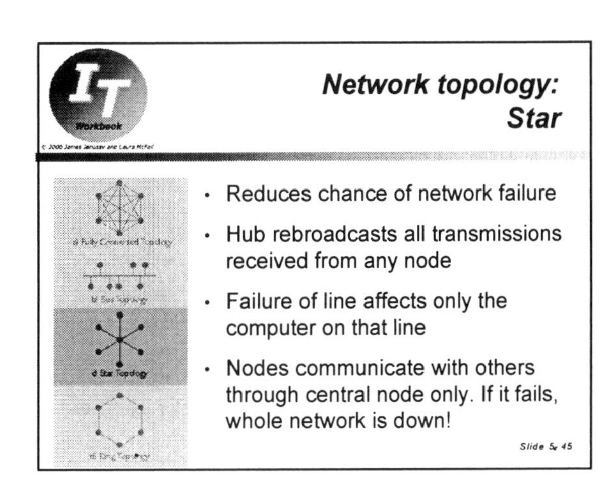
Network topology:
Star
Reduces chance of network failure
Hub rebroadcasts all transmissions received from any node
Failure of line affects only the computer on that line
Nodes communicate with others through central node only. If it fails, whole network is down!
Slide 5- 45

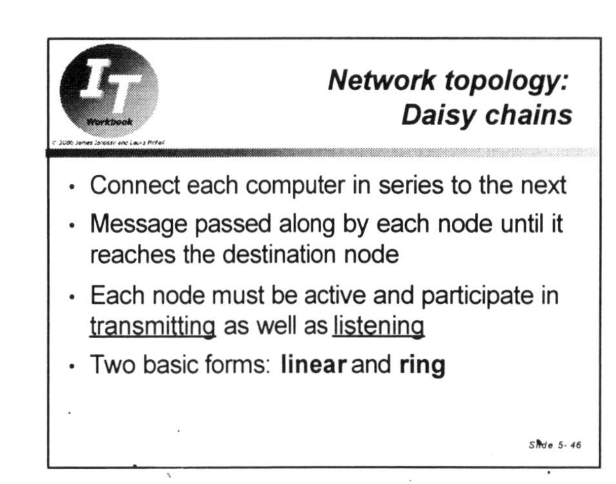
Network topology:
Daisy chains
Connect each computer in series to the next
Message passed along by each node until it reaches the destination node
Each node must be active and participate in transmitting as well as listening
Two basic forms: linear and ring
Slide 5- 46

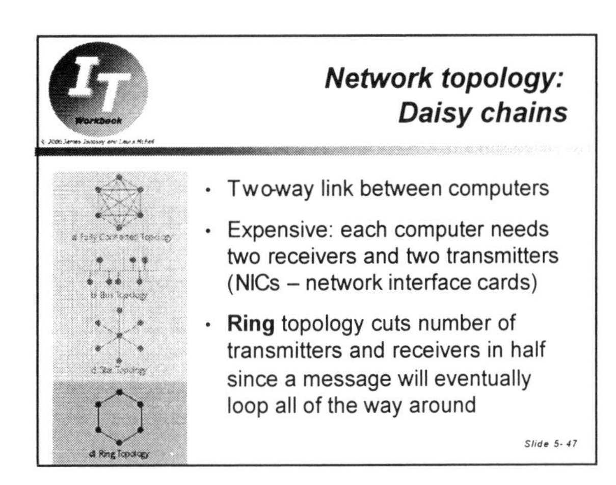
Network topology:
Daisy chains
Two-way link between computers
Expensive: each computer needs two receivers and two transmitters (NICs – network interface cards)
Ring topology cuts number of transmitters and receivers in half since a message will eventually loop all of the way around
Slide 5- 47

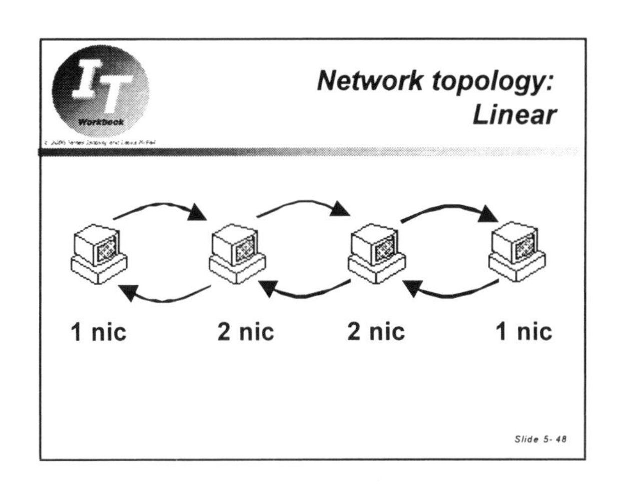
Network topology:
Linear
1 nic
2 nic
2 nic
1 nic
Slide 5- 48

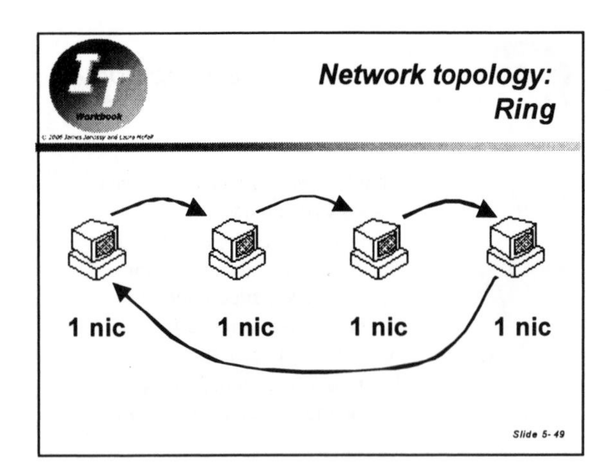
Network topology:
Ring
1 nic
1 nic
1 nic
1 nic
Slide 5- 49

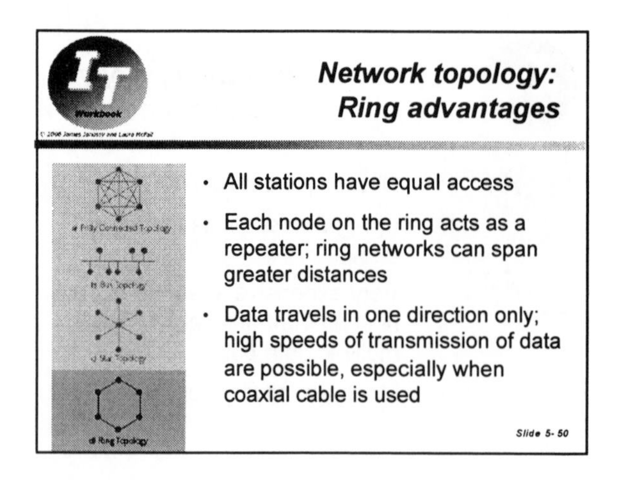
Network topology:
Ring advantages
All stations have equal access
Each node on the ring acts as a repeater; ring networks can span greater distances
Data travels in one direction only; high speeds of transmission of data are possible, especially when coaxial cable is used
Slide 5- 50

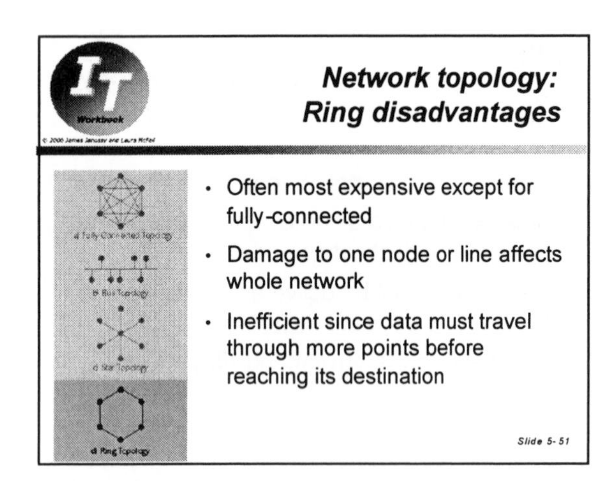
Network topology:
Ring disadvantages
Often most expensive except for fully-connected
Damage to one node or line affects whole network
Inefficient since data must travel through more points before reaching its destination
Slide 5- 51

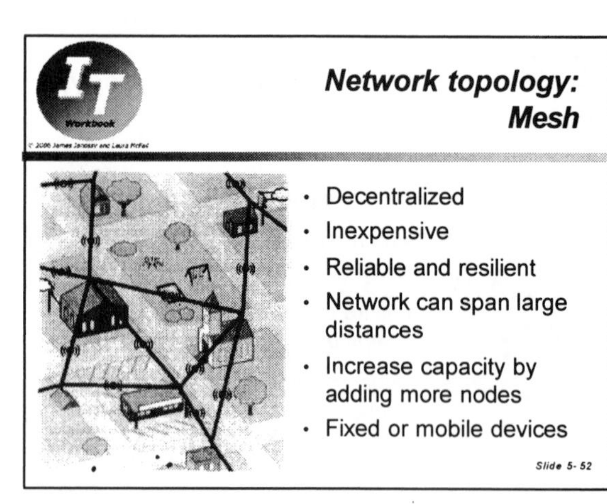
Network topology:
Mesh
Decentralized
Inexpensive
Reliable and resilient
Network can span large distances
Increase capacity by adding more nodes
Fixed or mobile devices
Slide 5- 52

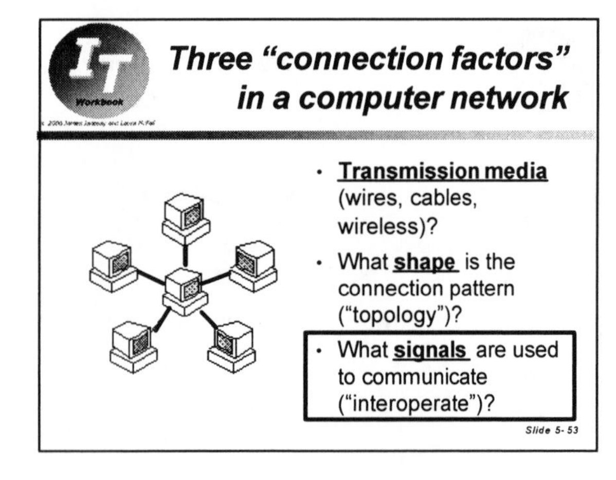
Three "connection factors" in a computer network
Transmission media (wires, cables, wireless)?
What shape is the connection pattern ("topology")?
What signals are used to communicate ("interoperate")?
Slide 5- 53

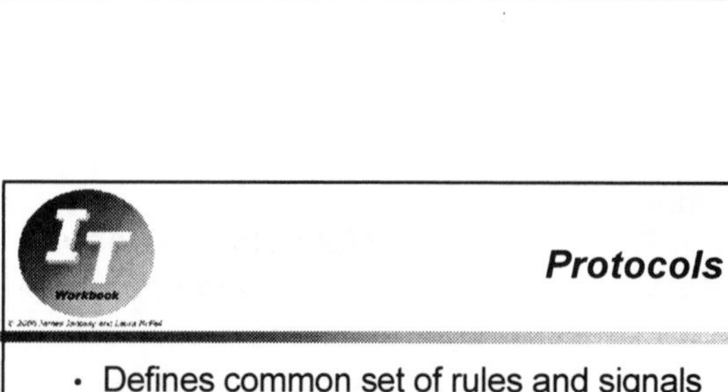
Protocols
Defines common set of rules and signals that computers on a network use
One of the most popular proprietary protocols for LANs is called Ethernet
Another popular proprietary LAN protocol is the IBM token-ring network
TCP/IP is common open Internet protocol
Slide 5- 54

Protocols – more details

- Agreed-upon format for data exchange
- Determines type of error checking, data compression method used
- Defines how sending device will indicate that it has finished sending a message
- Defines how receiving device will indicate that it has received a message

Slide 5- 55

Wireless technologies

- Connect desktop devices without cables
- Connect computers to networks in offices
- Broadband: connect over large areas

Slide 5- 56

Bluetooth

- **Bluetooth:** 10th century king of Denmark who got warring parties to communicate with each other (hence the name by its inventors)
- A convenient way to connect devices like personal digital assistants (PDAs), keyboards, mouse, mobile phones, laptops, PCs, printers, digital cameras without cables at .72 Mbps
- Low -cost, globally available short range radio frequency; intended range up to 30 feet

Slide 5- 57

Wi-Fi

- **Wi-Fi:** a set of product compatibility standards for wireless local area networks (WLAN) based on the IEEE 802.11 specifications, 11 Mb/sec or 54 Mb/sec. ***Not* "wireless fidelity"** (so what *does* it stand for? We can all guess...)
- For mobile devices, LANs, Internet access
- Connect to the Internet when near an access point called a "hot spot" with a laptop
- Offered free by universities and some public places

Slide 5- 58

Wi-Fi advantages

- Uses unlicensed radio spectrum (2.4 GHz)
- Allows LANs to be deployed without cabling
- Inexpensive products are widely available
- Different brands of products are interoperable
- Competition has lowered prices
- Supports roaming from between hot spots
- Supports encryption to protect traffic
- Global standards, works in different countries

Slide 5- 59

Wi-Fi disadvantages

- 1/10 watt limit, prone to interference from other sources such as Bluetooth, microwave ovens, cordless phones (early Wi-Fi only)
- Wired Equivalent Privacy (WEP) encryption is breakable even when correctly configured
- Access point operators might steal personal information transmitted from Wi-Fi users
- Limited range, 150 ft. indoors, 300 ft. outdoors; latches on to wrong hot spot in close quarters

Slide 5- 60

Network components: router / gateway

- Router forwards data packets on the network
- Connected to at least two networks, commonly two LANs or WANs or a LAN and its ISP's network
- Located at gateways, the places where two or more networks connect
- Uses message headers, forwarding tables, and communicated with other routers to determine the best path for forwarding packets

Slide 5-61

Network components: router / gateway

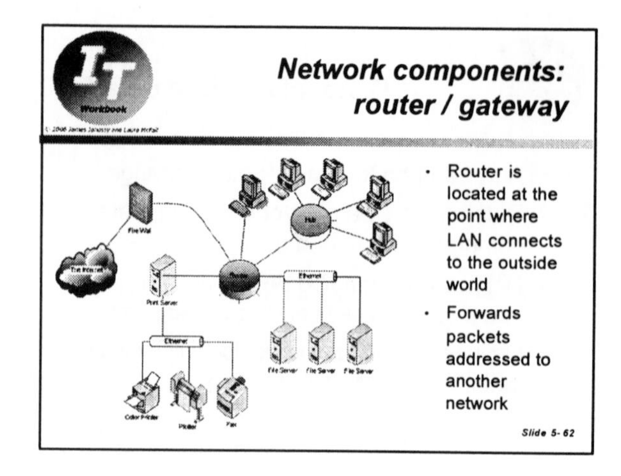

- Router is located at the point where LAN connects to the outside world
- Forwards packets addressed to another network

Slide 5-62

Network components: firewall

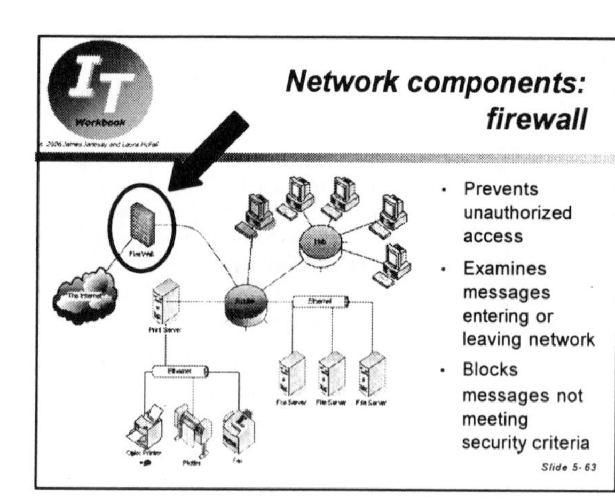

- Prevents unauthorized access
- Examines messages entering or leaving network
- Blocks messages not meeting security criteria

Slide 5-63

Network components: network operating system

- Includes functions for connecting computers and devices into a local-area network (LAN)
- UNIX and the Mac OS have networking functions built in
- Novell Netware, Artisoft LANtastic, Microsoft Windows Server, Windows NT

Slide 5-64

Network components: TCP protocol

- Transmission Control Protocol
- IP protocol deals only with data packets
- TCP enables two hosts to establish a connection and exchange streams of data
- Guarantees delivery of data and that packets will be obtained in the same order in which they were sent

Slide 5-65

Intranet

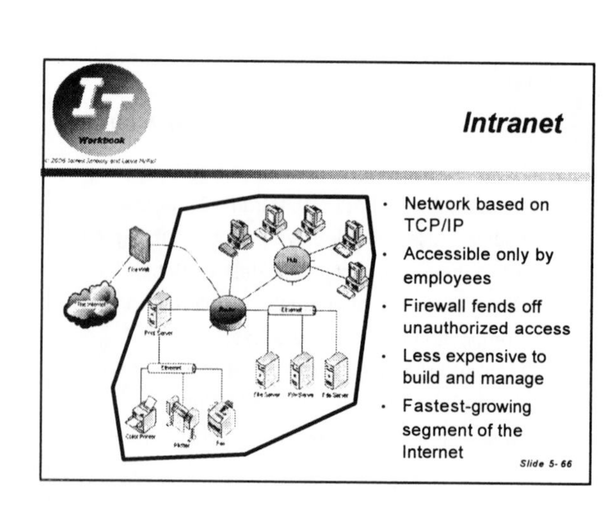

- Network based on TCP/IP
- Accessible only by employees
- Firewall fends off unauthorized access
- Less expensive to build and manage
- Fastest-growing segment of the Internet

Slide 5-66

IT Workbook Assignment 5.1

Student name:

Write your answers to the questions below within the box. In each case, please choose your words carefully to answer the specific questions. Avoid simply copying passages from your readings to these answers.

Explain what communication is, in the general sense, and how we arrange to achieve it over long distances using energy:

1)

Identify the five characteristics of the components of every communication system:

2)

3)

4)

5)

6)

IT Workbook Assignment 5.2

Student name:

Write your answers to the questions below within the box. In each case, please choose your words carefully to answer the specific questions. Avoid simply copying passages from your readings to these answers.

Describe how an electric current is used when an **analog signal** such as a voice is transmitted in an **analog telephone system**:

1)

Describe how an **analog signal** such as a voice is represented and transmitted when the transmission is done **digitally**:

2)

IT Workbook Assignment 5.3

Student name:

Write your answers to the questions below ***within the box****. In each case, please choose your words carefully to answer the specific questions. Avoid simply copying passages from your readings to these answers.*

Describe the landline **public telephone system** in the United States, explaining what parts are **digital**, and what parts are **analog** (if any):

1)

A "T1" telephone line can be leased for dedicated use, and can carry digital signals at the rate of 1,536,000 bits per second. Explain how many voice telephone circuits can be carried on one T1 line through the technique of multiplexing, and how this is achieved:

2) how many voice circuits?

3) how is this multiplexing achieved?

IT Workbook Assignment 5.4

Student name:

Write your answers to the questions below within the box. In each case, please choose your words carefully to answer the specific questions. Avoid simply copying passages from your readings to these answers.

What is the role of a **modem** in achieving communication between two computers located several hundred miles apart, when the telephone system will be used to carry the signals?

1)

Identify and briefly describe the **four different technologies** that are currently available to achieve **broadband communication** between computers located several hundred miles apart:

2)

3)

4)

5)

IT Workbook Assignment 5.5

Student name:

Write your answers to the questions below *<u>within the box</u>*. *In each case, please choose your words carefully to answer the specific questions. Avoid simply copying passages from your readings to these answers.*

Describe, in general terms, what a **computer network** is, and why it is useful:

1)

Identify and briefly describe the **four type of computer networks** commonly defined as to their size (coverage area):

2)

3)

4)

5)

IT Workbook Assignment 5.6

Student name:

Write your answers to the questions below *within the box*. *In each case, please choose your words carefully to answer the specific questions. Avoid simply copying passages from your readings to these answers.*

Identify and describe the three "connection factors" that must be considered when computers are used in a network:

1)

2)

3)

Identify and briefly describe the three common types of **transmission media** currently available for constructing computer networks, and the advantages and disadvantage of each:

4)

5)

6)

IT Workbook Assignment 5.7

Student name:

Write your answers to the questions below within the box. In each case, please choose your words carefully to answer the specific questions. Avoid simply copying passages from your readings to these answers.

Explain what the **topology** of a computer network is, and identify and describe **four of the common wired topologies** currently in use:

1)

2)

3)

4)

5)

Identify what a **protocol** is in connection with computer networking, and identify some common protocols in current use:

6) a protocol is:

7) some protocols in common use in computer networking:

IT Workbook Assignment 5.8

Student name:

Write your answers to the questions below within the box. In each case, please choose your words carefully to answer the specific questions. Avoid simply copying passages from your readings to these answers.

Bluetooth and **WiFi** are two wireless connection technologies in common use. Compare and contrast these in terms of transmission speed, transmission range, and intended use:

1) compare as to transmission speed:

2) compare as to transmission range:

3) compare as to intended use:

Identify what a **router** is, and describe in detail what a router does:

4) a router is:

5) what a router does:

IT Workbook Assignment 5.9

Student name:

Write your answers to the questions below <u>within the box</u>. In each case, please choose your words carefully to answer the specific questions. Avoid simply copying passages from your readings to these answers.

Identify what a **gateway** is, and describe in detail what a gateway does:

1) a gateway is:

2) what a gateway does:

Identify what a **firewall** is, and describe in detail what a firewall does:

3) a firewall is:

4) what a firewall does:

IT Workbook Assignment 5.10

Student name:

Write your answers to the questions below within the box. In each case, please choose your words carefully to answer the specific questions. Avoid simply copying passages from your readings to these answers.

Describe what a mesh network is, and how it is different from any of the four wired computer network technologies in terms of implementation, cost, and robustness.

1) a mesh network is:

2) how a mesh network is different from the four wired computer network technologies in terms of implementation, cost, and robustness:

Discuss how **satellite technology** and **WiMAX** technology are being used to bring Internet connectivity to remote areas that it is not feasible to serve using wireline technologies, including the specific example of Coffman Cove, Alaska:

3)

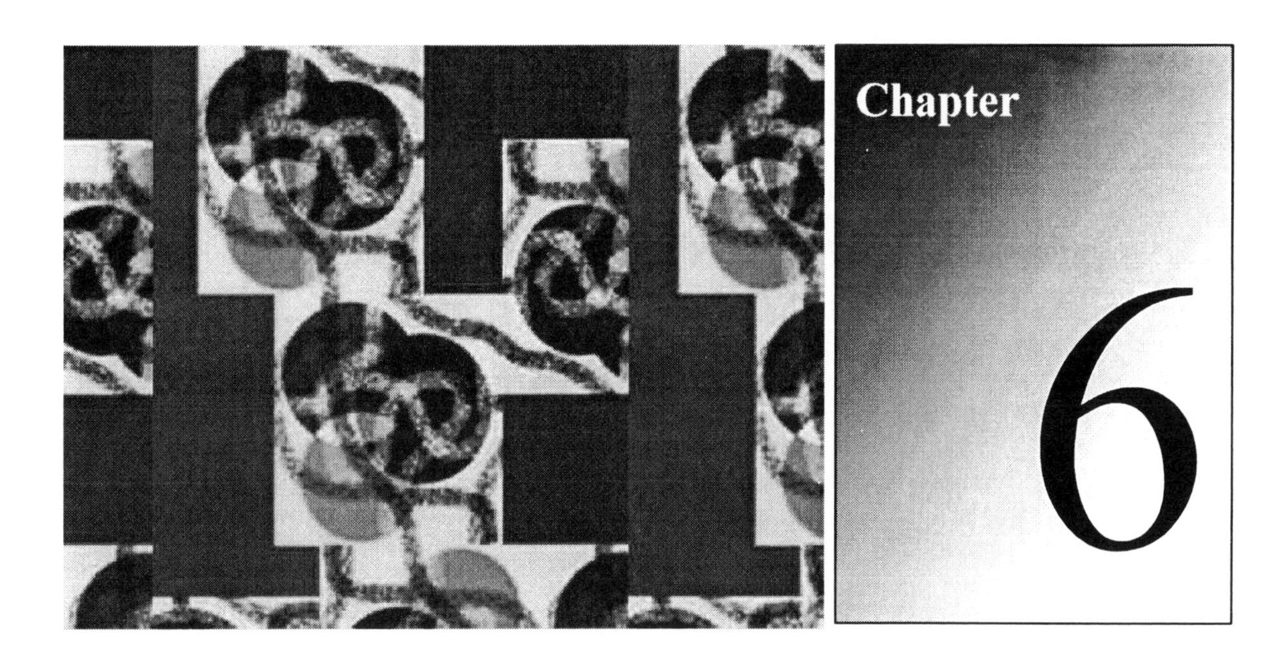

The Internet and World Wide Web

The **Internet** provides flexible connectivity. It is a communication network in which a computer can seek out and connect to another computer anywhere by using the "name" associated with it. In this chapter we examine the history, construction, and operation of the Internet, its components, the traffic of the **World Wide Web** conveyed by the Internet, and some of the programming tools used with it.

Since its commercialization in the early 1990's the internet has completely revolutionized the use of computers and the flow of information. Before the internet computers were connected only using limited local area network arrangements. Now we fully expect that any computer or computing device anywhere can quickly connect to millions of others!

The **signal routers and domain name servers** (DNS) of the internet provide an electronic pipeline. The World Wide Web (WWW) is the major service carried by the internet. It provides **Hypertext Transfer Protocol** (HTTP) which establishes connections between computers. **Hypertext Markup Language** (HTML) is communicated through the internet by HTTP to define how a computer screen should be formatted, while scripting languages such as **JavaScript** and **VBScript** provide interactivity in combination with HTML. Programming languages such as **Java**, **.net**, **C**, **C++**, and **Perl** can interact with the web to provide access to relational databases. The internet also provides **File Transfer Protocol** (FTP) to ship files of information in bulk, and supports a wide variety of other types of communication such as e-mail and instant messaging.

Learning objectives

Chapter 6
The Internet and World Wide Web

This chapter, in combination with the web readings, web links, and podcasts available at www.ambriana.com is focused on providing the knowledge to achieve these learning objectives:

- Understand how the Internet developed, what the Internet is composed of and what functions it performs, and how the World Wide Web relates to the Internet
- Be able to describe the differences between the switched telephone network and packet switching communication, what type of communication the Internet is based on, and the basis for this
- Understand Internet addressing involving URL and IP addresses, how these are handled by the Internet, and how machinery is addressed using MAC addresses
- Be able to trace the route of a transmission to an Internet site, and to describe how the transmission occurs using the results of the trace
- Recognize and be able to distinguish seven specific major Internet applications
- Be able to describe Internet vulnerabilities including denial of service attacks, spoofing, phishing, data sniffing, and keystroke loggers, and how they are perpetrated
- Understand what intellectual property is, and the applicability, specific rights, protections and limitations of copyrights, patents, and trademarks in relationship information technology.

Podcast Scripts – Chapter 6

These readings of the **Information Technology Workbook** are also accessible in audio form as .mp3 files for free download. Visit the workbook web site, www.ambriana.com. You can listen to this material on your own Ipod or .mp3 player or computer, read it, or both listen and read.

Podcast 35
INTERNET STRUCTURE

One of the greatest things about the Internet is that nobody really owns it. The Internet is a global collection of networks, big and small, that connects together in all sorts of different ways to form a single working entity. The name Internet comes from this idea of interconnected networks.

Since its humble beginnings in 1969, the Internet has grown from four host computer systems to tens of millions of host computers. But just because no one owns the Internet doesn't mean that the Internet isn't monitored and maintained. The **Internet Society**, a non-profit group established in 1992, coordinates the formation of proposals for policies that define how we use and interact with the Internet. The World **Wide Web Consortium (W3C)** is an international organization in member groups and various others, including the public, work together to develop standards for the World Wide Web. Led by Tim Berners-Lee, the inventor of the World Wide Web, W3C's mission is to lead the Web to its full potential by developing protocols and guidelines that ensure long term growth. The consortium also engages in education and outreach, develops software, and serves as an open forum for discussion about the Web. Finally, **ICANN**, the Internet Corporation for Assigned Names and Numbers is a non-profit organization that has responsibility for IP addresses, management of the domain name system, governance of domain name registrars, and resolution of domain name disputes. These services were originally performed under U.S. Government contract by the Internet Assigned Numbers Authority (IANA).

Every computer connected to the Internet, including the one in your home, is part of one of the interconnected networks of the Internet. You may use a modem and dial a local number to connect to an Internet Service Provider (ISP) or some type of higher speed access, for a monthly fee. You might be part of a local area network in your workplace, but even so you most likely still connect to the Internet using an ISP that your company has contracted with. When you connect, you become part of their network. The ISP then connects to a larger network and becomes part of their network, and so on.

Large communications companies have their own dedicated high capacity communication lines that connect regions of the world and carry enormous amounts of Internet traffic. In each region it serves, such a company has a Point of Presence (POP). The POP is a place for local users to access the company's network through a local phone number or dedicated line. POPs connect to the high speed backbones at Network Access Points (NAP) of the truly large-scale communications carriers. The amazing thing here is that there is no overall controlling network. The participants in the Internet connect to each other based on interconnection agreements and standardized protocols.

On the Internet, dozens of large Internet providers interconnect at NAPs in various cities, and trillions of bytes of data flow between the individual networks at these points every day. The Internet is a huge collection of networks that agree to all intercommunicate with each other. All of these networks rely on NAPs, backbones and routers to talk to each other. It is incredible that a message can leave one computer and travel halfway across the world through several different networks and yet arrive at another computer in a fraction of a second!

In this process, routers determine where to send information from one computer to another. **Routers** are specialized computers that send your messages and those of every other Internet user speeding to their destinations along thousands of different pathways. A router has two separate but related jobs. It insures that information doesn't go where it's not needed, which is crucial for keeping large volumes of data from clogging the connections of innocent bystanders. And it makes sure that information reliably gets to the intended destination. In performing these jobs, a router is extremely useful and efficient in making the connection between separate computer networks. It also protects each networks from the others, preventing traffic on one network from spilling unnecessarily over to other networks. Routers can do this because their function is precisely defined and remains the same, regardless of the direction and volume of traffic.

Podcast 36
BACKBONES AND ISPs

The Internet is a collection of thousands of individual networks and organizations, each of which is separately operated and managed. Each network cooperates with other networks to direct Internet traffic so that information can pass among them. Together, these networks and organizations make up the wired world that is the Internet.

Large communications providers built the Internet backbones and allow others to access them. The Internet **backbones** are a extremely fast network spanning the world from one major metropolitan area to another. The term refers to the main circuits of the Internet, which consist of a collection of interconnected commercial, government, academic and other high capacity data routes and routers.

Backbones are very high capacity lines that carry enormous amounts of Internet traffic. They are used to transmit data to millions of locations at once. The original Internet backbone, Arpanet, was developed in 1969 and expanded in the 1970s to connect 18 universities and four government agencies. It used communication circuits leased from telephone companies. In 1989 the NSF Net backbone was established, the United States military broke off to use as a separate military network, and ARPANET was shut down. A plan was developed to further expand the NSF Net, then render it obsolete by creating a new network architecture based on decentralized routing through packet switching. The intent was to create a network capable of being operated by non-governmental organizations.

The NSF Net Internet backbone was decommissioned on April 30, 1995, once the privatized Internet was operational. The Internet now consists entirely of various commercial ISPs, private networks, and inter-university networks, all connected together into one vast array .

Today's backbones are physically made up of fiber optic cabling. These **optical carrier (OC) links** use beams of light to transmit data. Confined in fiber optic cabling, this type of communication provides extremely high capacity with no interference. OC connections use a technology called SONET, a complex global standard that that combines many separate signals into one communication stream, a technique known as **multiplexing**. Multiplexing had been used in earlier forms in analog and digital communication, but not on the scale accomplished by optical networks.

Multiple backbone providers exist. These organizations are different from an ISP, which provides user access to the Internet. You might consider an ISP like a retailer that consumers deal with. In this analogy, a backbone provider is a wholesaler. It supplies the ISPs with bulk high-volume access to lines that connect ISPs to each other. Major backbone providers include MCI, Sprint, and UUNET. These communications carriers have large geographic coverage of the United States utilizing a mesh topology. Many of these links run on fiber optic links with speeds capable of up to ten or more gigabytes per second. The fiber links have the capability to provide even greater capacity, but the technology for routing and multiplexing at higher speeds has yet to catch up with that capacity.

Internet Service Providers (ISP) are businesses that offer users access to the Internet and may offer related services, such as domain name registration, web site hosting, dial-up or DSL access, and leased line access. At its most basic level of service, an ISP charges a periodic access fee to its customers. In return, the ISP gives the consumer a user name, password, access phone number, and appropriate connection software. Equipped with a modem, the user can then log on to the Internet, browse the World Wide Web, and send and receive e-mail messages. In addition to serving individuals, ISPs also serve businesses and larger companies, providing a higher-speed connection from the firm's network to the Internet.

Podcast 37
IP ADDRESSES AND URLs

Every machine on the Internet has a unique identifying number, named an Internet Protocol or IP address. An IP address consists of 4 sets of numbers separated by periods. For example, an IP address might be 140.192.22.55. Only one computer on the Internet will have this address (or none might have it).

But computers communicate in binary language, and they must be able to interpret IP addresses. The four numbers in an IP address are called **octets**, because as an arithmetic quantity, they each are expressible as an eight-bit value. So an IP address is actually a 32 bit unsigned binary number. Since each bit can have only two possible states (on or off, 0 or 1) the number of different IP addresses is 2^{32}, which is 4,294,967,296 unique values.

Out of the almost 4.3 billion possible IP addresses, certain values are restricted from common use. For example, the IP address 0.0.0.0 (the lowest address) is reserved for network use, as is IP address 255.255.255.255 (the highest address).

Octets serve a purpose besides simply separating the numbers. They are used to create classes of IP addresses that can be assigned to a particular business, government or other entity, based on size and need.

Let's consider the IP address 101.102.103.104. Octets are split into two sections, Net and Host. The net section always contains the first octet, which is 101 in this example. The net section is used to identify the network a computer belongs to. Host identifies the actual computer on that network. The host section always contains the last octet, which is 104 in this example.

When development of the Internet began, it consisted of a small number of computers hooked together with modems and telephone lines. Users could make connections only by providing the actual IP address of the computer they wanted to establish communication with. This became unwieldy as more and more users came online.

The first solution to the problem of cumbersome Internet addressing was a simple text file maintained by a Network Information Center that mapped names to IP addresses. Soon this text file became so large it was too cumbersome to manage. In 1983, the University of Wisconsin, one of the institutions on the original Internet, created the Domain Name System (DNS), which maps text names to IP addresses automatically using database technology. This way you need only remember the domain name, such as www.ambriana.com , rather than the actual IP address.

The **Domain Name System**, composed of Domain Name Servers, stores information associated with domain names in a distributed database the Internet. The domain name system associates many types of information with domain names, but most importantly, it provides the IP address associated with the domain name. It also lists mail exchange servers accepting e-mail for each domain, and is an essential component of the Internet.

The Domain Name System makes it possible to attach hard-to-remember IP addresses such as 207.142.131.206 to easy-to-remember domain names such as wikipedia.org. This types of a name, which is effectively a synonym for an IP address, is called a Uniform Resource Location (URL). We take advantage of this synonym when we recite URLs and e-mail addresses.

Top level, or first level domain names, include .com, .org, .net, .edu and .gov. Within every top level domain a huge list of second level domains exists. For example, the .com first-level domain contains Google, Yahoo, and Microsoft, just to name just a few. This name is the "dot com" name you make up and hope has not yet been claimed if you envision setting up a web site! Every name in the .com top-level domain must be unique.

The left-most word in a URL, like www, is the host name. It specifies the name of a specific machine, with a specific IP address, in a domain. A given domain could potentially contain millions of host names, as long as they were all unique within that domain.

DNS servers accept requests from programs and other name servers to convert domain names into IP addresses. When a request comes in—such as when you type www.ambriana.com in the address line of your browser—the DNS server can do one of four things with it:

1. It can answer the request with an IP address if it can find the domain name in its local database.
2. It can contact another DNS server and try to find the IP address for the domain name.
3. It can say, "I don't know the IP address is for the domain you requested, but here's the IP address for a DNS server that knows more than I do."
4. It can return an error message because the requested domain name does not exist in any DNS server it can contact.

A URL is a synonym for an Internet address. Here are some simple rules to use when entering or creating a URL:

- a URL never contains spaces
- a URL is usually entered as lower case letters, but is not case sensitive
- a URL always starts with a protocol prefix such as http://, but most browsers will assume those characters if you do not enter them as part of the address
- letters, digits, or hyphens can be used to form the name, but each part of the name between dots has to contain at least one letter
- avoid use of the underscore in a name since it is supported erratically by domain name servers
- Each string of characters between the dots must be 1 and 63 characters long
- a URL is not the same thing as an email address

Podcast 38
INTERNET APPLICATIONS

The world wide web is not the only function of the Internet. A number of applications exist that you probably use on a regular basis, but you may not associate them with the Internet. Among these are e-mail, file transfer protocol, instant messaging, news groups, web logs, or blogs, Internet voice, and distributed computing.

Let's discuss e-mail. **E-mail** is a method of sending and receiving messages over electronic communication systems. The term e-mail applies both to Internet e-mail and to intranet systems that allow users within an organization to send messages to each other. These collaborative systems frequently have some sort of gateway that lets them send and receive Internet e-mail. Some organizations use Internet protocols for internal e-mail service.

What about file transfer, using FTP? **FTP**, which stands for **file transfer protocol**, is commonly used for exchanging files over any network that supports the TCPIP protocol, such as the Internet. Two computers are involved in the transfer, a server and a client. The server runs FTP software and listens on the network for connection requests from other computers. The client, running FTP client software, initiates a connection to the server. Once connected, the client can upload and download files from the server, and rename or delete files on the server.

Any software company or individual programmer can create FTP server or client software because the protocol is an open standard. Every computer platform supports the FTP protocol. This allows any computer connected to a TCPIP based network to manipulate files on another computer on that network, regardless of the operating systems involved. This is a type of **interoperability**.

Let's consider instant messaging. **Instant messaging** differs from e-mail in that text conversations occur in real time. Most Internet Service Providers offer a "presence" information feature that indicates if people on a user's list of contacts are currently online and available to chat. Most instant messaging applications also include the ability to automatically send a status message, like that of a message on an answering machine. Popular instant messaging services on the Internet include MSN Messenger, AOL Instant Messenger, Yahoo Messenger, and Google Talk.

Instant messaging facilitates collaboration over the Internet. In contrast to e-mails or phone calls, the parties know whether the person they want to contact is available. But people are not forced to reply immediately to incoming messages, so communication via instant messaging can be less intrusive than telephone communication.

What about news groups? A **news group** is a repository for messages posted from many users at different locations. Newsreader software is used to read news groups. News group servers are hosted by various organizations and institutions. Most Internet Service Providers host their own news server or rent access to one for their subscribers. Companies exist to sell access to premium news servers. Every host of a news server maintains agreements with other news servers, to regularly synchronize and form a network. When a user posts to one news server, the message is stored locally. That server then shares the message with the servers that are connected to it.

How about web logging? A web log, commonly referred to as a **blog**, is a web based publication consisting of periodic articles, usually in the order of most recent first. Blogs can be hosted by dedicated blog hosting services, or they can be run using blog software on ordinary web hosting services. Some blogs focus on a particular subject, such as politics or local news, while other blogs function as online diaries. A typical blog combines text, images, and links to other blogs, web pages, and media related to its topic. Authoring, maintaining, or adding an article to an existing blog is called **blogging**. A person who posts these entries is called a **blogger**.

A blog has certain attributes that distinguish it from a standard web page. It allows for easy creation of new pages. New data is entered into a simple form and submitted. Automated templates add the article to the home page, creating the new full article page, and add the article to the appropriate archive.

A blogging site allows for easy filtering of content, by date, category, author, or other attributes, and it allows the administrator to invite and add other authors, whose permission and access are easily managed.

Blogging has become a passion with some individuals. It can be addictive, and it can be damaging. For example, if a blogger reveals information about their workplace that is considered confidential, or reveals behavior incompatible with their occupation, employment can be adversely affected. People have been fired for revealing things in this way. If you have an urge to blog, consider that anyone in the universe can read what you write, and that Internet search engines like Google make everything you post instantly accessible to you by name.

Now let's discuss Internet telephone service. Internet Voice, also known as **Voice over Internet Protocol**, or **VoIP**, allows you to make telephone calls using a broad band Internet connection instead of a regular telephone line. Some VoIP services may only allow you to call other people who subscribe to the same service. Other VoIP vendors allow you to call anyone who has a regular telephone number. Some VoIP services only work over your computer or a special phone, while other services let you use a traditional phone through an adapter.

VoIP converts analog voice signals from your telephone into a digital signal that travels over the Internet. The signal is converted back to analog form at the other end of the line. If you make a call using a phone with a VoIP adapter, you'll be able to dial as usual, and the service provider may also provide a dial tone. If your service assigns you a regular phone number, a person can call you from their regular phone without using special equipment.

Let's consider distributed systems. **Distributed computing** refers to any of a variety of computer systems that use more than one computer at different locationg to run an application. Distributed computing usually refers to local area networks that are designed to let a single program run simultaneously at various sites. Distributed processing systems contain sophisticated software that detects idle CPU usage on the network and parcels out programs to utilize this computational horsepower. In this way, tens, hundreds, or even thousands of networked computers can work together on a particularly processing intensive problem.

The growth of distributed processing models has been limited, however, due to a lack of compelling applications, and by bandwidth bottlenecks, combined with significant security, management, and standardization challenges. But a number of vendors now offer products that make use of a distributed computing capability. An innovative worldwide Internet distributed computing project, with the goal of finding intelligent life in the universe, is named "SETI at home." Operated by the Planetary Society, SETI has captured the unused desktop processing cycles of millions of personal computers to analyze radio signal data received from distant space.

RSS is a newer application on the web designed to automatically distribute information. It forms the basis for I-Tunes, Apple's music distribution mechanism that supports its huge population of I-Pod and other sound player users. RSS stands for **Rich Site Summary**, but others refer to it as "Really Simple Syndication." Based on XML, RSS lets a web site "subscribe" to a new, data, or music file feed from another site. Sites that want to distribute information about themselves or data they have gathered can host an RSS feed and make it available to others. Commercial web site operators find this appealing because it provides and automatic way for their site to provide information that draws traffic. Individual subscribers like it because it makes it possible to receive transmissions automatically in off-hours, and have them loaded to their computer or portable device to keep them current.

Podcast 39
WORLD WIDE WEB, HTML, and XML

The Internet is a large voluntary communications network that links connected computers together in flexible ways. It provides most of the electronic traffic on the Internet. Standards and common tools have been developed to build the web and to insure continuity and ease of use around the world.

Although the idea for the web was originated by Tim Berners-Lee in 1989, beginning in the early 1990's the governance and evolution of the web was vested in the World Wide Web Consortium. The World Wide Web Consortium, also known as W3C, is an international organization in which member groups, a full time staff, and the public, work together to develop standards for the web. The consortium also engages in education and outreach, develops software, and serves as an open forum for discussion about the Web and potential improvements. Let's consider several terms and standards the W3C has developed that have helped the web evolve.

Browsers

We'll begin with a web browser, which is a computer program such as Internet Explorer, FireFox, Netscape, or Safari. A web **browser** is a program that lets you display and interact with text, images, and other information from the Internet or computer network. The computer screens displayed by a web browser are usually called web pages. The first browser was the precursor to Netscape, and was developed and released publicly in 1994.

A web page can contain links to other web pages, usually indicated as underlined words, or as symbols or graphic images of clickable buttons. Links on a web page to other web pages are called **hyperlinks**. Although browsers are typically used to access the World Wide Web, they can also be used to access information provided by web servers in private networks or content in file systems. Such a network, internal to an organization, is called an **intranet**.

HTTP

Now, a question. How does the information displayed by the browser on your computer actually get to you? To answer this question, we need to define a few more terms. **HTTP**, which stands for **HyperText Transfer Protocol**, is the method used to convey information on the Web. The development of HTTP was coordinated by the W3C and the Internet Engineering Task Force, culminating in HTTP version 1.1, in use today. HTTP is a protocol, that is, an agreed upon method, for the exchange of information between servers and clients connected by the Internet. An HTTP client initiates a request to the Internet by establishing a Transmission Control Protocol connection to a particular entry point, called a port, on a remote server. When you use Internet Explorer, this happens automatically behind the scenes, and you are not aware of it. Upon receiving the HTTP request, the server sends back a transmission, which can include the requested web page, a requested file, an error message, or some other information.

HTML

At this point, a question has probably arisen in your mind. What, specifically, are web pages? And, how are they formed? Each **web page** is a transmission made up of ordinary letters and numbers. These letters and numbers contain coding known as **HTML**, which stands for **HyperText Markup Language**.

Markup languages predate the Internet. They stem from the type of markings an editor would make on a written document to tell a type setter how a published book was to be formatted for printing. A printer's markup language contains agreed-upon notations, such as a line to cross out a word, a symbol to indicate that a word is to be type set in bold rather than plain letters, and many others. In a similar way, HyperText Markup Language consists of agreed upon special words, codes, and symbols, that a browser such as Internet Explore recognizes. HTML is designed for the creation of web pages that any browser will display in a similar way. In the jargon of the web, when a browser follows the instructions coded in HTML for a web page to display it on a computer screen, we say that the browser ***renders*** the page. In other words, "render" is more or less a synonym for "display" although it actually implies "construct and display." HTML is used to identify the intended formatting for text, denoting headings, paragraphs, lists and so on; HTML specifies the content and intended appearance of the web page. The graphic appearance of the page is not transmitted, but rather, HTML tells the browser how to ***construct*** the page, using the fonts available on the computer, and any transmitted digital images.

Originally defined by Tim Berners Lee, HTML is now an international standard, the specifications for which are maintained by the World Wide Web Consortium. Early versions of HTML were defined with a loose set of syntax rules that aided its adoption by those unfamiliar with web publishing. Web browsers such as earlt versions of Netscape and Internet Explorer commonly made assumptions about the intent of the HTML coding for a web page and proceeded to render pages acceptably even if they deviated from standard HTML coding practices. For example, a web page coded with minor syntax infractions, such as reversed slashes or missing semicolons, appears normally when viewed using Internet Explorer. Modern browsers, however, such as FireFox, often expect HTML to adhere more closely to the official syntax specifications. A page with HTML errors, which appears acceptably using Internet Exploerer, may not be located at all or may appear to have typographical errors when accessed using a newer browser such as FireFox.

XML and XHTML

As an outgrowth of the development of new tools for the web environment, a type of markup language more flexible than HTML was designed, named XML. **XML** stands for **Extensible Markup Language**. XML provides a general way of describing data. In XML, unlike HTML, not all the tags containing formatting or identification words and codes are predefined. Instead, tags can be defined specifically for the needs of a given document. Therefore, an XML file can contain data as well as formatting information. The primary purpose of XML is to facilitate the sharing of data across different systems, particularly systems connected via the Internet. Languages based on XML, such as Geography Markup Language, are defined in a formal way, allowing programs to modify and validate documents in these languages, without prior knowledge of their form.

The tools for Web page coding continue to evolve. **XHTML** is a new version of markup language that may someday replace HTML for web page coding. XHTML incorporates XML features into HTML. At present, XHTML is evolving in parallel with HTML, and HTML is still very heavily used.

Podcast 40
WEB PROGRAMMING TOOLS

It's important to know that web pages are categorized as either static or dynamic. Let's consider the differences between static and dynamic web pages.

The content of a **static web page** is always the same. All of the content and formatting are contained in a single set of HTML code. HTML was designed with the idea that a human would compose the HTML by marking up a document that had fixed, unchanging data content. For example, a web page describing how to contact the operator of the site is usually a static web page. If something, such as a telephone number, needs to be changed in a static web page, a person will manually edit the HTML or the web page to change the information, using a text editor, and will then place the corrected HTML for the web page on the server, to publish it.

A **dynamic web page** displays information that is a combination of static and highly changeable information. For example, a web page that a student accesses to view grades for a term is dynamic. Every student sees similar titling and column labels but the courses and grades displayed are specific to each student. In the case of dynamic web pages, the changeable information usually comes from a database, and the web page is sometimes referred to as data driven.

Data driven web sites cannot be handled by HTML alone, because HTML has no facility to access a database and include data from it in the web page. Early in the Internet era, clever programmers realized that programs could be created, in languages such as Java or Visual Basic, that could access a database to obtain information and then automatically create HTML, combining the data with title, labeling text, and HTML tags.

For dynamic pages, we often say that the HTML for the page is created "on the fly" by a program. Once the HTML created by a program reaches a browser it is processed in exactly the same way that hand-coded HTML is processed. Let's consider programming languages and environments associated with modern computer programming and the web.

Java is a modern full-fledged programming language that can access vast libraries of routines and interfaces designed to deal with the web and with databases accessible via the Internet. Java is object oriented and in many ways the successor to C++ for common programming purposes.

Although its name is similar to Java, **JavaScript** and Java are two entirely different program development tools. JavaScript is a programming language designed to enhance web pages through the use of small programs executed by client computers. JavaScript was originally named "LiveScript" and was developed as an extension of HTML to provide computational ability within a web page. When Java assumed prominence in the web world, LiveScript was renamed JavaScript to capitalize on name recognition. JavaScript is intended to be attached to an HTML file and must be executed by the web browser rendering the web page.

JavaScript cannot deal with databases or files but it provides the means to make many interactive enhancements to a web page. For example, changing the appearance of text on a web page or triggering a popup box when the mouse is used to put the cursor over an area is often implemented in web pages using JavaScript. However, care is needed in the use of JavaScript because different browsers support slightly different versions of the language. If you want to learn JavaScript, you can find many excellent, free tutorials on the web covering it, as well as sites on the web that freely provide already built JavaScript routines for useful functions.

Visual Basic is Microsoft's current implementation of the Basic language. Basic originated in the old timesharing era of minicomputers and was one of the first languages implemented on personal computers in the 1970's. Unlike early versions of Basic, Visual Basic is a full-blown programming language for the Windows GUI environment. It provides the ability to produce formatted screens and deal with databases. In the early days of the web, programmers often used Visual Basic to access data from databases and form HTML dynamically to create some of the first dynamic web sites. It now supports object-oriented concepts, and it is still possible to use it to form web pages on the fly.

.net ("dot net") is Microsoft's current large-scale program development toolset. This includes a collection of programming languages within a comprehensive development and production environment. All languages in the environment can intercommunicate, since all of these languages—J#, C#, and Visual Basic, and several others—actually produce the same intermediate code, which is executed in a runtime environment. J# ("J-sharp") is Microsoft's equivalent alternative to Java, and C# is its equivalent alternative to C++. Microsoft has provided tools for program development, for production library operation, and to address maintainability and security issues in the .net environment. This environment is aimed squarely at competing with Java in application and web development.

Active Server Pages (ASP) is another of Microsoft's server-based software environments for the creation of dynamically-generated web pages. ASP web pages are coded in VBScript or JScript, which are languages similar to JavaScript. ASP includes objects that provide support for many functions commonly needed in this work, such as preserving variables from one web page to another, and for testing dynamic web sites.

Web Services is a recently-coined term for software systems and components designed to "support interoperable machine-to-machine interaction over a network" (to quote the W3C definition). Under Web Services concepts, messages from an application requesting data or services are expressed in XML and passed to other software applications over a network using HTTP. Software applications can be written in different programming languages and can be run on different types of computers, but they can smoothly communicate, exchange data, and support one another over computer networks, since open standards are followed for the interfaces. Under the Web Services concept, vendors and software developers are encouraged to provide interfaces in an open standard format. The W3C is currently evolving standards to facilitate Web Services.

Podcast 41
INTELLECTUAL PROPERTY RIGHTS

Intellectual property laws confer certain exclusive rights to the specific form in which information is expressed, but not to the ideas or concepts themselves. It's important to understand that the term intellectual property denotes the specific legal rights that authors, inventors and other holders of intellectual property may hold and exercise.

Intellectual property laws are designed to protect different forms of intangible subject matter, such as pictures and ideas. Let's get familiar with some associated terms.

A **copyright** may exist in creative and artistic works, such as books, movies, music, photographs, and software, giving a copyright holder the exclusive right to control reproduction or adaptation of such works for a certain period of time such as 50 to 70 years. Copyright laws vary from country to country.

A copyright not only protects the creator's economic interests but also the integrity of a work, by authenticating its originality. No one can copyright a work that has prior copyright protection. But someone might challenge the validity of an original copyright if that person can prove that a work was stolen, plagiarized, or adopted from an existing work and simply modified in minor ways.

A **patent** may be granted in relation to an invention that is new and useful, and not simply an obvious advancement over what existed at the time. A patent grants the holder an exclusive right to commercially exploit the invention for a certain period of time, usually 20 years from the filing date of a patent application.

A **trademark** is a distinctive sign used to distinguish the products or services of one business from those of another business. Trademarks can be valuable business assets if they are associated with a brand or a well-recognized product that has built up a level of public recognition and trust.

An **industrial design right** protects the form of appearance, style or design of an industrial object, such as parts, furniture or textiles.

A **trade secret** is an item of confidential information concerning the commercial practices or proprietary knowledge of a business.

Patents, trademarks and designs all fall into a subset of intellectual property known as industrial property. Like other forms of property, the exclusive rights to intellectual property can be transferred, with or without payment, or licensed to third parties. In some cases, it may even be possible to use intellectual property as security for a loan.

The basic public policy rationale for the protection of intellectual property is that intellectual property laws encourage the pursuit and eventual disclosure of innovation into the public domain, to be used for the common good, by granting authors and inventors exclusive rights to exploit their works and invention for a period of time.

Let's consider copyright law in our digital era. When you think about the rights and restrictions that apply to online text, think about how easy it is to distribute a document or entertainment in digital form. If someone posts a document to a mailing list, for example, that document might be accessible through an archive for years to come. Copies of it might be made available at countless different web sites, or redistributed repeatedly via e-mail. If the original author wants to correct an error in the original document, or revise it with important updates, recalling all of the copies of the original version would be impossible. The author no longer controls the document in the same way that a publisher can control print editions of a book.

When text was restricted to physical print, controlling distribution of that text was fairly easy. Copy and fax machines eroded that control and forced people to rely more heavily on compliance with and enforcement of copyright laws. The rapidly growing body of digital products has once again upset that delicate balance. Now, more than ever, people must assume ever greater responsibility for voluntary compliance with existing laws, and should be no surprise that copyright protection is on the minds of publishers, authors, and entertainers.

You have now completed all of the podcasts for Chapter 6 of the Information Technology Workbook. Make sure you review the material at the web links for Chapter 6 provided at the workbook web site, at www.ambriana.com.

Lecture Slides 6

Introduction to Information Technology

The Internet and World Wide Web

Course resources

PowerPoint presentations, podcasts, and web links for readings are available at www.ambriana.com > IT Workbook

Print slides at 6 slides per page

Homework, quizzes and final exam are based on slides, lectures, readings and podcasts

Slide 6-2

Main topics

- Internet components
- World Wide Web components
- The Internet as a pipeline
- Packet vs. traditional switching
- Connecting to the Internet
- Internet traffic and vulnerabilities

Slide 6-3

What is the Internet?

- Interconnected network
- Connects thousands of LANS
- Connects millions of computers
- 500,000,000 people around the world
- 160,000,000 people in the United States (more than half the U.S. population)
- Audience growing by millions annually

Slide 6-4

What is the Internet?

- Internet pipeline that connects computers
- Physically composed of **telephone lines**, **fiber optic lines**, and other forms of **high speed communication links**
- Based on signal routing technology and communication protocols (standards) different than telephone network

Slide 6-5

What is the World Wide Web?

- The most popular service supported by the Internet
- Represents just one form of communication handled by the Internet
- Relies on standard "page definition" language (HTML) and images (.JPG, .GIF, .PNG)
- Relies on standard browsers for viewing

Slide 6-6

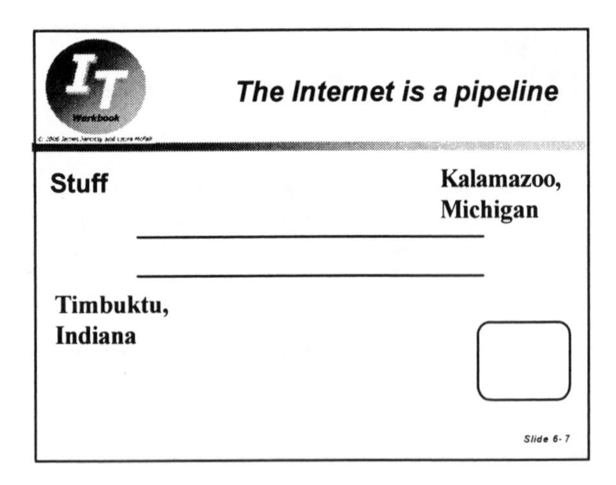

The Internet is a pipeline
Stuff
Kalamazoo, Michigan
Timbuktu, Indiana
Slide 6-7

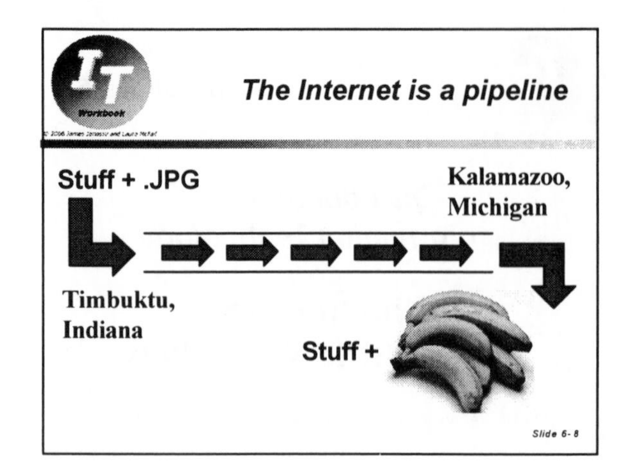

The Internet is a pipeline
Stuff + .JPG
Kalamazoo, Michigan
Timbuktu, Indiana
Stuff +
Slide 6-8

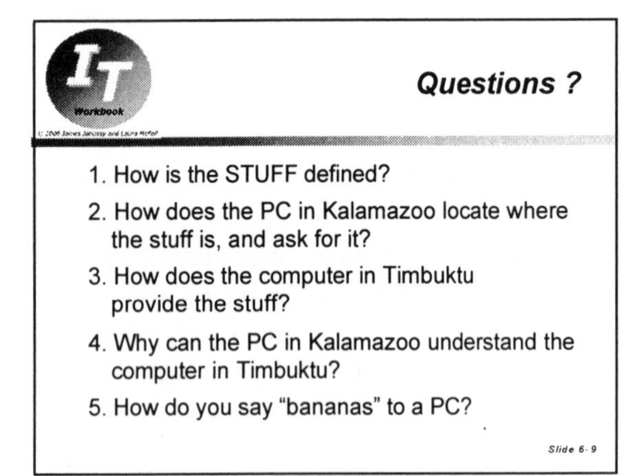

Questions ?
1. How is the STUFF defined?
2. How does the PC in Kalamazoo locate where the stuff is, and ask for it?
3. How does the computer in Timbuktu provide the stuff?
4. Why can the PC in Kalamazoo understand the computer in Timbuktu?
5. How do you say "bananas" to a PC?
Slide 6-9

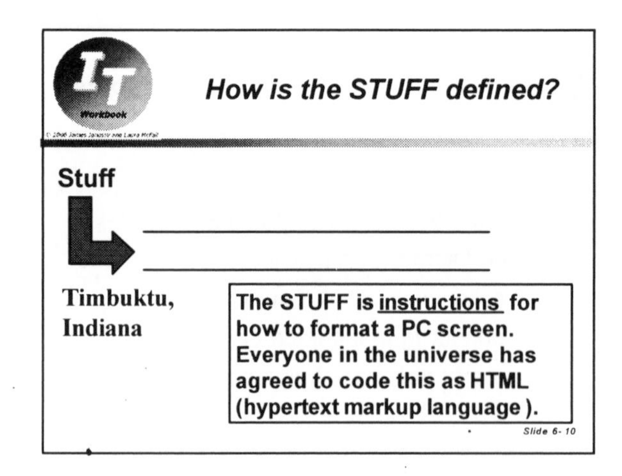

How is the STUFF defined?
Stuff
Timbuktu, Indiana
The STUFF is instructions for how to format a PC screen. Everyone in the universe has agreed to code this as HTML (hypertext markup language).
Slide 6-10

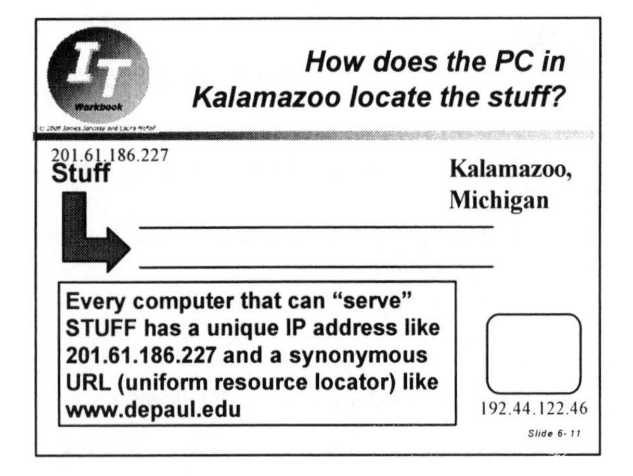

How does the PC in Kalamazoo locate the stuff?
201.61.186.227
Stuff
Kalamazoo, Michigan
Every computer that can "serve" STUFF has a unique IP address like 201.61.186.227 and a synonymous URL (uniform resource locator) like www.depaul.edu
192.44.122.46
Slide 6-11

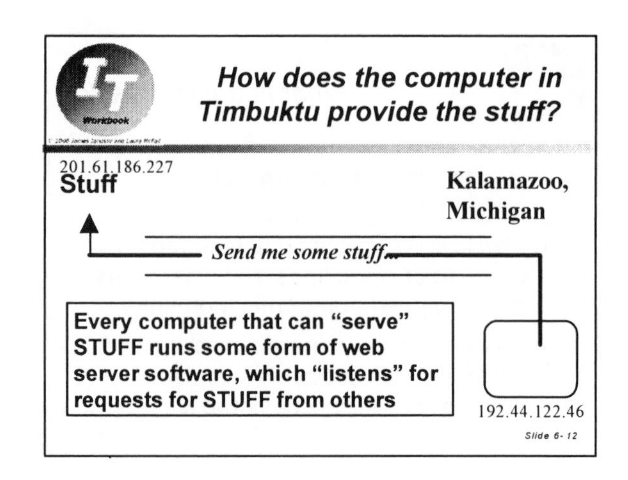

How does the computer in Timbuktu provide the stuff?
201.61.186.227
Stuff
Kalamazoo, Michigan
Send me some stuff...
Every computer that can "serve" STUFF runs some form of web server software, which "listens" for requests for STUFF from others
192.44.122.46
Slide 6-12

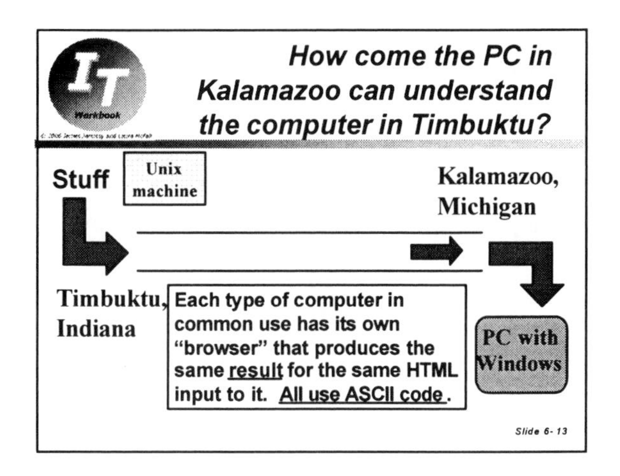

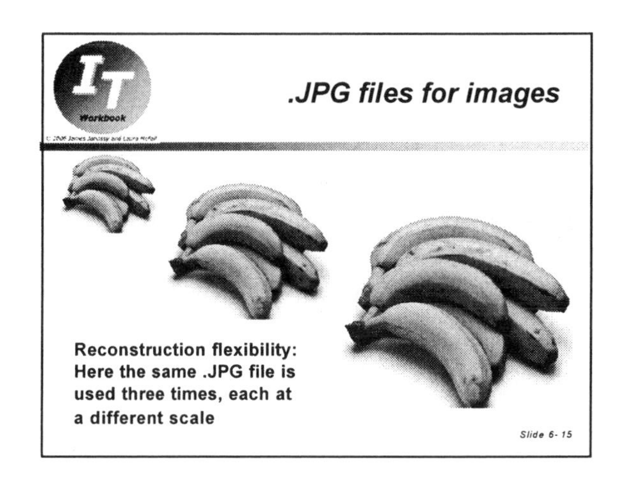

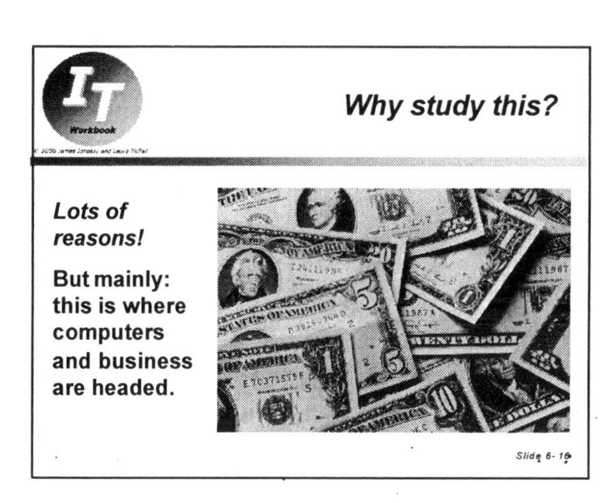

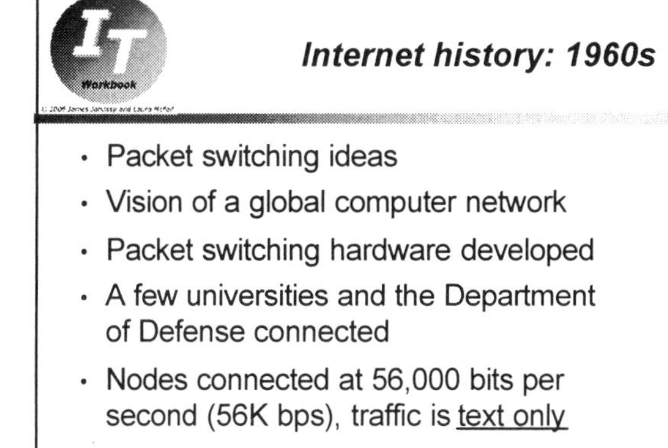

Packet switching

- Digitized text is broken into groups of bytes called "packets"
- Each packet is labeled and separately sent through the network
- Packets in the same message can take different routes through the network
- Packets are reassembled into original order at the receiving end by TCP/IP

Slide 6- 18

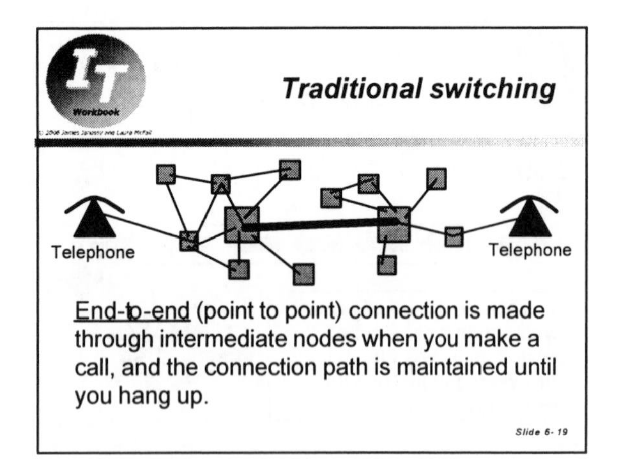
Traditional switching
Telephone
Telephone
End-to-end (point to point) connection is made through intermediate nodes when you make a call, and the connection path is maintained until you hang up.
Slide 6- 19

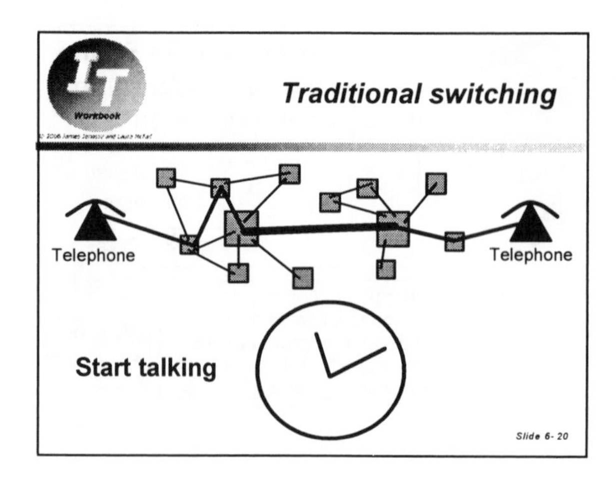
Traditional switching
Telephone
Telephone
Start talking
Slide 6- 20

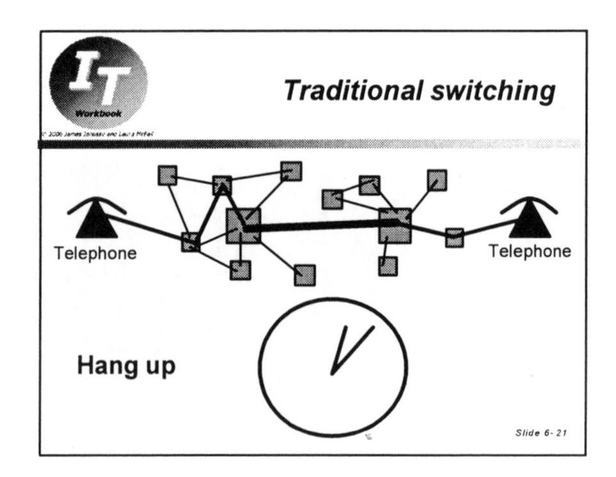
Traditional switching
Telephone
Telephone
Hang up
Slide 6- 21

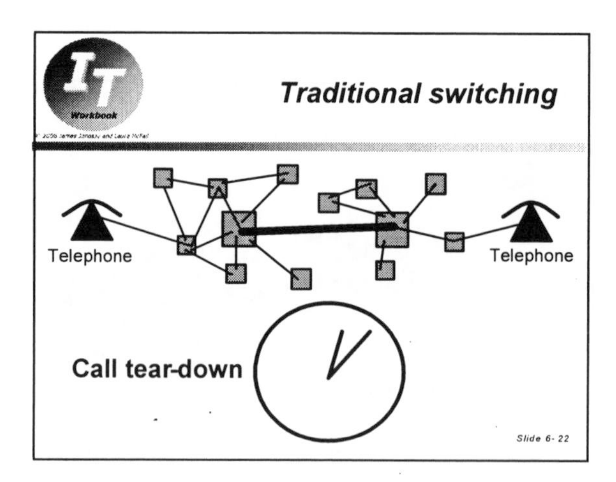
Traditional switching
Telephone
Telephone
Call tear-down
Slide 6- 22

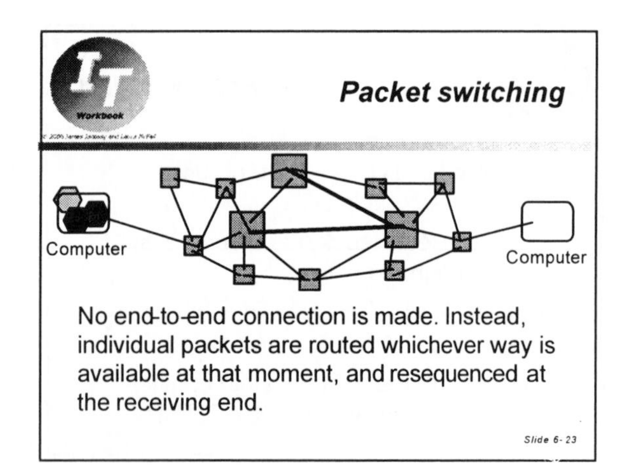
Packet switching
Computer
Computer
No end-to-end connection is made. Instead, individual packets are routed whichever way is available at that moment, and resequenced at the receiving end.
Slide 6- 23

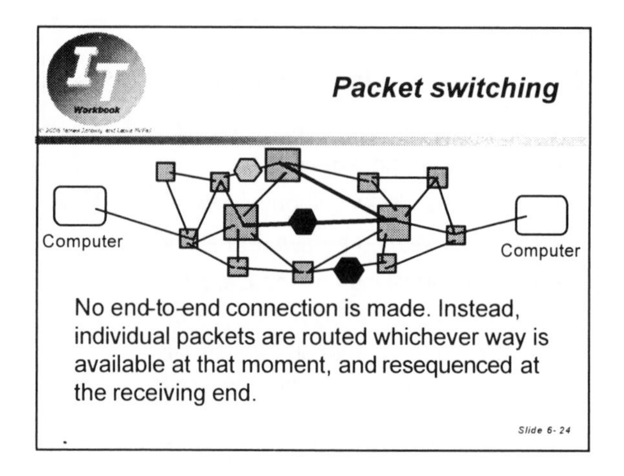
Packet switching
Computer
Computer
No end-to-end connection is made. Instead, individual packets are routed whichever way is available at that moment, and resequenced at the receiving end.
Slide 6- 24

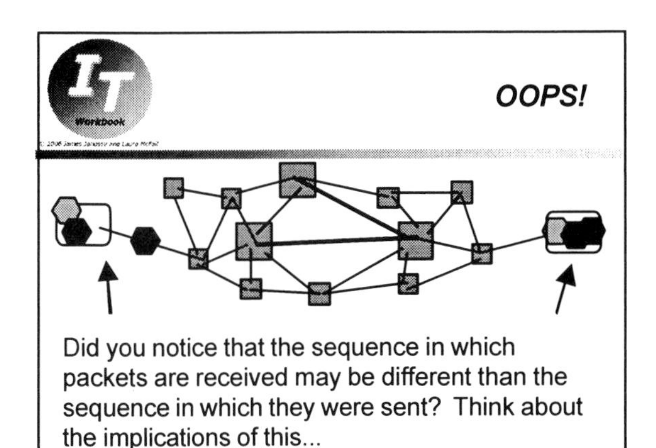

OOPS!

Did you notice that the sequence in which packets are received may be different than the sequence in which they were sent? Think about the implications of this...

Slide 6-25

Internet history: 1970s

- E-mail invented
- Ethernet communication between computers
- First local area networks (LANs)
- TCP/IP invented (the communication protocol of the Internet)
- All of this occurs <u>before</u> PCs are invented

Slide 6-26

Internet history: 1980s

- PCs invented
- ARPANET connects several universities
- Telnet and ftp invented
- First NSF-funded backbone
- Internet carries text documents only
- 1989: Berners-Lee proposes WWW based on HTML (http invented)

Slide 6-27

Internet history: 1990s

- NSFNET takes over from ARPANET
- Now 21 nodes
- Connected at 45 megabits per second
- 50,000 networks attached worldwide (29,000 in U.S.)
- 1994: First graphical browser MOSAIC, Netscape, first commercialization

Slide 6-28

Internet history: 1998-2000s

- Investment boom (venture funding) to "dot.com" firms reaches extraordinarily high levels
- Lots of foolhardy ventures
- Retrenchment ("dot.com bubble bursts") about 2000, shakes out marketplace
- E-commerce starts to settle down

Slide 6-29

Internet history: Late 1990s, 2000s

- Internet first era begins in 1994
- U.S. government privatizes the Internet
- 100 million users by 1997
- 400 million users by 2001
- Estimated 800 million users by 2003
- Highly extensible: meaning it can continue to grow

Slide 6-30

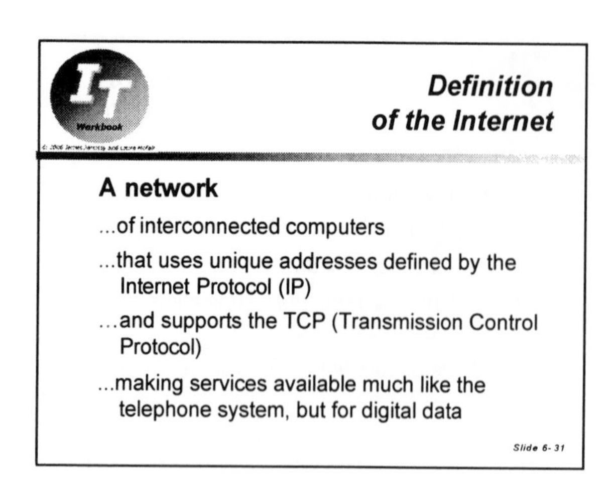
IT
Workbook
Definition
of the Internet
A network
...of interconnected computers
...that uses unique addresses defined by the Internet Protocol (IP)
...and supports the TCP (Transmission Control Protocol)
...making services available much like the telephone system, but for digital data
Slide 6-31

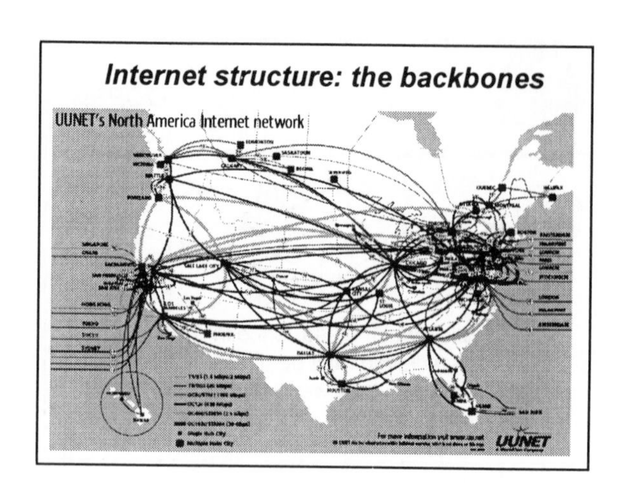
Internet structure: the backbones
UUNET's North America Internet network
UUNET

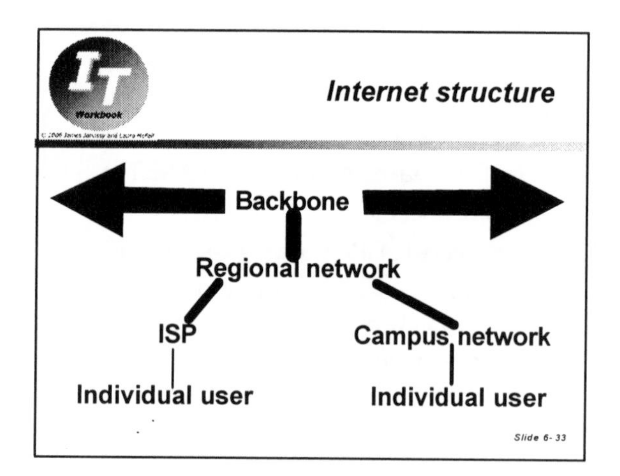
IT
Workbook
Internet structure
Backbone
Regional network
ISP
Campus network
Individual user
Individual user
Slide 6-33

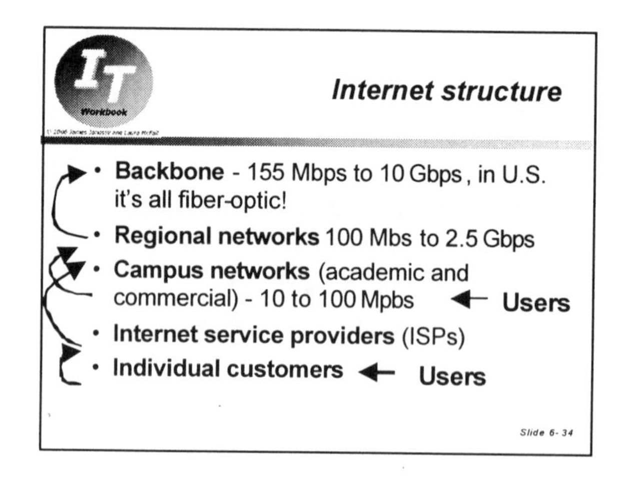
IT
Workbook
Internet structure
• Backbone - 155 Mbps to 10 Gbps, in U.S. it's all fiber-optic!
• Regional networks 100 Mbs to 2.5 Gbps
• Campus networks (academic and commercial) - 10 to 100 Mpbs
Users
• Internet service providers (ISPs)
• Individual customers
Users
Slide 6-34

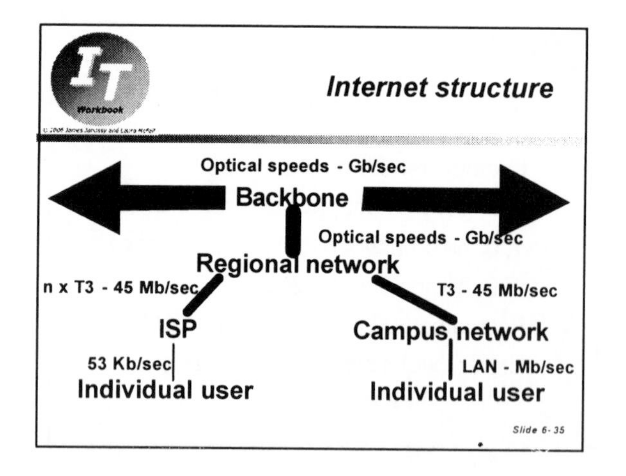
IT
Workbook
Internet structure
Optical speeds - Gb/sec
Backbone
Optical speeds - Gb/sec
Regional network
n x T3 - 45 Mb/sec
T3 - 45 Mb/sec
ISP
Campus network
53 Kb/sec
LAN - Mb/sec
Individual user
Individual user
Slide 6-35

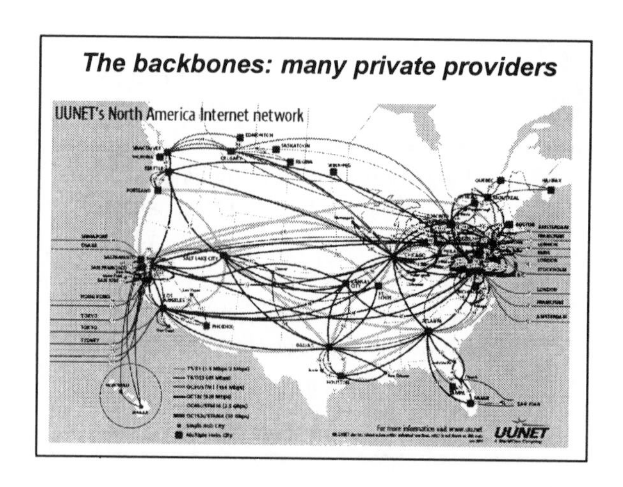
The backbones: many private providers
UUNET's North America Internet network
UUNET

ISPs

- **Internet Service Providers** attach to the Internet at the regional network level
- Individual customer can subscribe to Internet service with ISP such as America Online, MSN network, AT&T, or any of 5,000 other smaller ISPs
- **Individual users** connect to their ISP via dialup, DSL, or cable modem

Slide 6- 37

Wire connection to ISP

- **Modem** uses ordinary telephone line and is typically 30 Kbps to 56 Kbps bits per second (slow!)
- **Digital subscriber line** (DSL) is dedicated telephone line enhanced to handle 150 Kbps to about 1 Mbps
- **Cable modem** uses cable TV coaxial cable for 350 Kbps to 1 Mbps rates

Slide 6- 38

Wi-Fi wireless connection to ISP?

- Low-power 2.4 GHz radio connection to a wireless hub, but the hub is connected to ISP by wire line
- 11 Mb/sec or 54 Mb/sec "802.11b/g" standard; faster than connection of the local wireless hub and ISP
- Capacity shared between multiple users at the location

Slide 6- 39

Commercial Internet connections

- **T1** is dedicated, always-connected telephone circuit, 1.54 Mbps, costs $300 to $400 per month
- **T3** is dedicated, always-connected telephone circuit, 45 Mbps, costs $5,000 to $10,000 per month
- Even greater capacity optical connections exist at higher costs

Slide 6- 40

Types of dedicated circuits

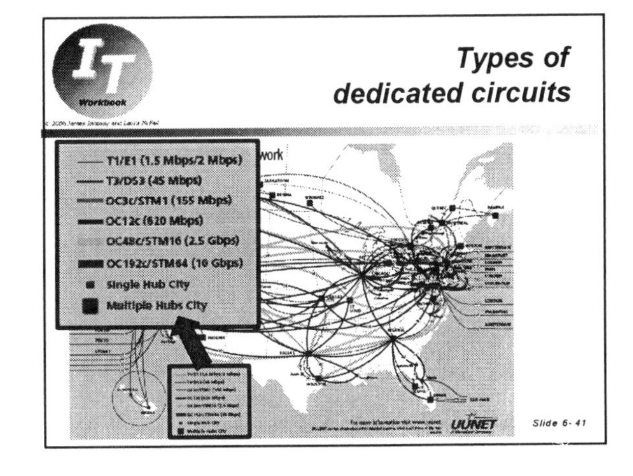

Slide 6- 41

Intranet

- **TCP/IP** network used internally to connect computers
- Uses same technology as Internet
- Replaces "proprietary" local area network such as Novell and Windows/NT
- Usually less expensive, more flexible

Slide 6- 42

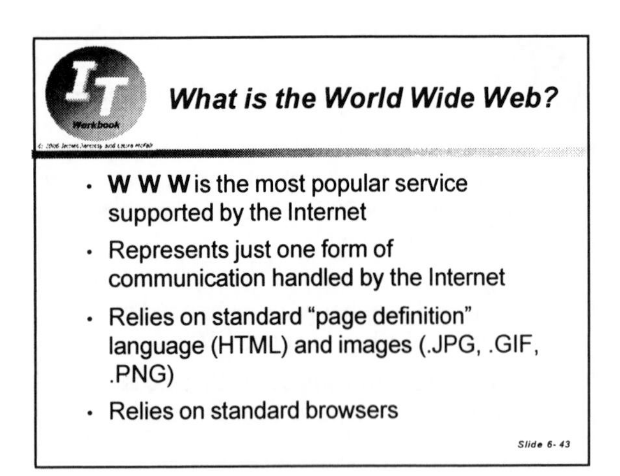
What is the World Wide Web?
• W W W is the most popular service supported by the Internet
• Represents just one form of communication handled by the Internet
• Relies on standard "page definition" language (HTML) and images (.JPG, .GIF, .PNG)
• Relies on standard browsers
Slide 6- 43

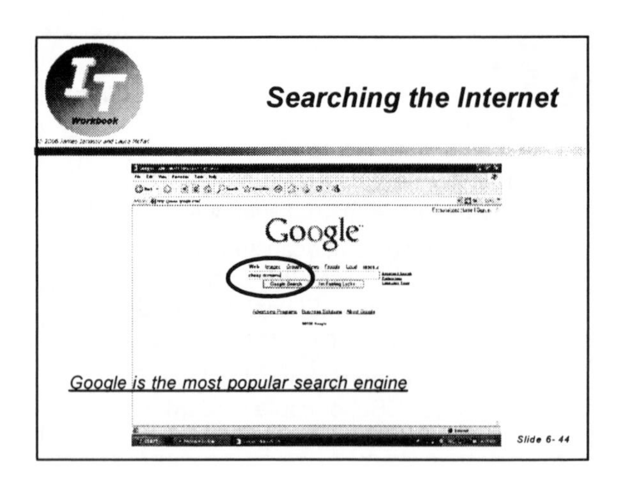
Searching the Internet
Google
Google is the most popular search engine
Slide 6- 44

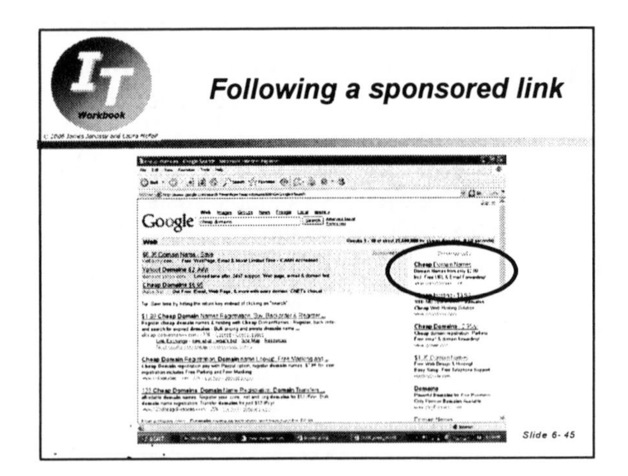
Following a sponsored link
Google
Slide 6- 45

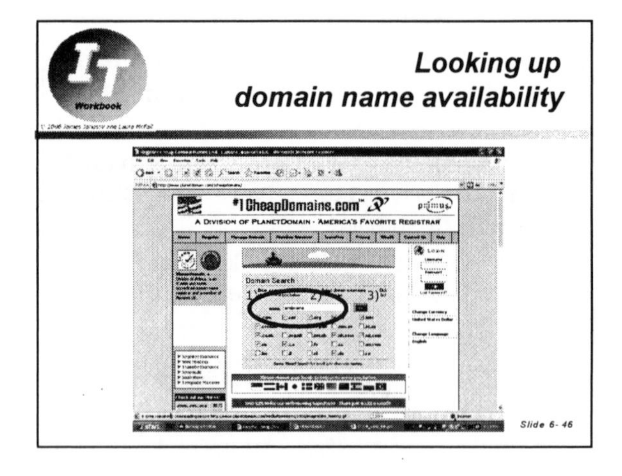
Looking up domain name availability
#1CheapDomains.com
Slide 6- 46

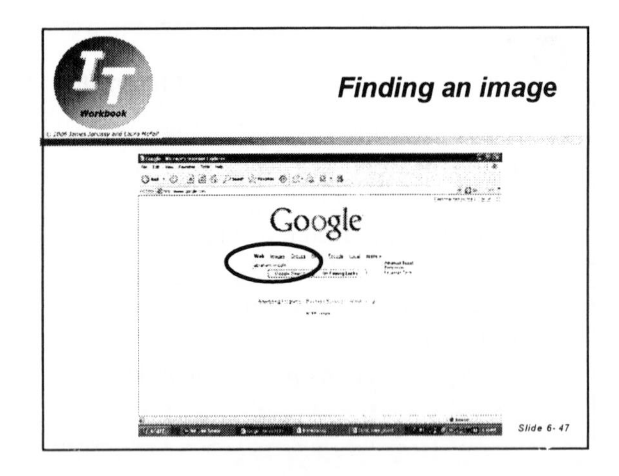
Finding an image
Google
Slide 6- 47

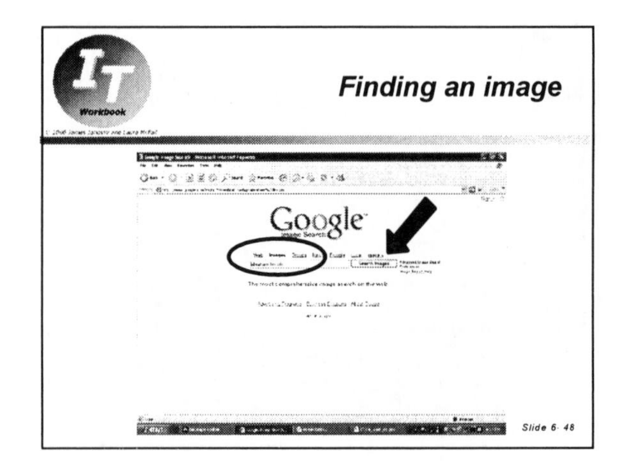
Finding an image
Google
Slide 6- 48

Finding an image

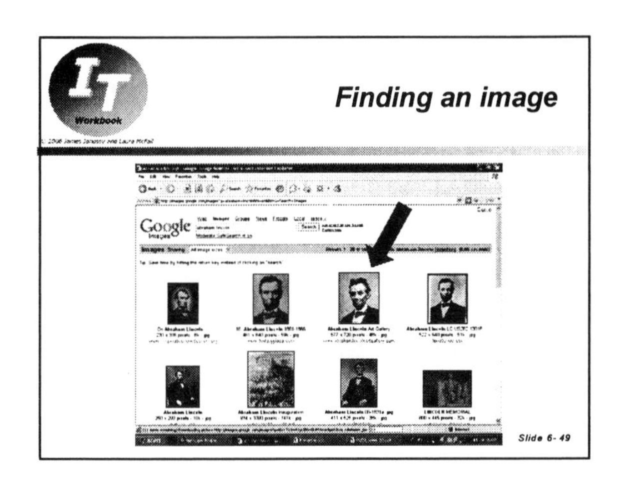

Slide 6- 49

Finding an image

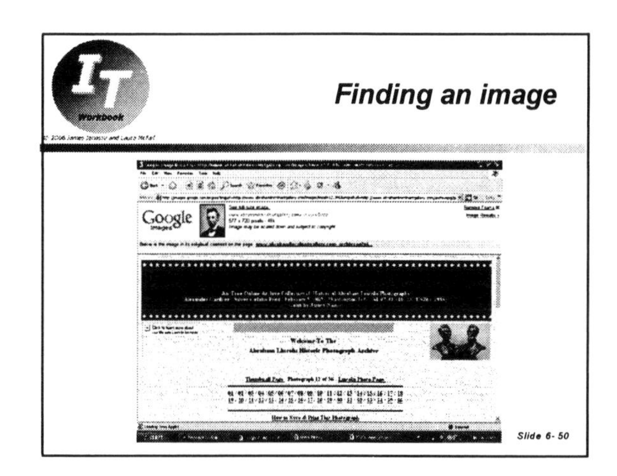

Slide 6- 50

Other Internet traffic?

- **W W W** is the most popular service supported by the Internet
- **FTP:** File transfer protocol; transfer whole files
- **RSS:** XML-based feeder / subscription service; I-Tunes is an example
- **VoIP:** "voice over Internet Protocol"

Slide 6- 51

Internet vulnerabilities

- **Denial of service attacks** 
- Viruses and worms
- Spoofing
- Phishing
- Data sniffer
- Keystroke logger

Flooding a server with a request for a response; from many "invaded" computers

Slide 6- 52

Internet vulnerabilities

- Denial of service attacks
- **Viruses and worms**
- Spoofing
- Phishing
- Data sniffer
- Keystroke logger

Virus plants harmful program on computer; worm "proliferates" on network

Slide 6- 53

Internet vulnerabilities

- Denial of service attacks
- Viruses and worms
- **Spoofing** 
- Phishing
- Data sniffer
- Keystroke logger

Tamper with e-mail packet header and "from" to fool recipient into providing information

Slide 6- 54

Internet vulnerabilities

- Denial of service attacks
- Viruses and worms
- Spoofing
- **Phishing**
- Data sniffer
- Keystroke logger

Send e-mail looking communication from a bank, ask for "confirmation" of personal info

Slide 6- 55

Internet vulnerabilities

- Denial of service attacks
- Viruses and worms
- Spoofing
- Phishing
- **Data sniffer**
- Keystroke logger

"Inside job"... technician uses instrument to observe data streams on network

Slide 6- 56

Internet vulnerabilities

- Denial of service attacks
- Viruses and worms
- Spoofing
- Phishing
- Data sniffer
- **Keystroke logger**

Software records keyboard actions, send to a website with visible indication

Slide 6- 57

Intellectual property

- Writing, artistic creation, invention, software
- Laws protect the right of the originator to sole control or exploitation of it for a period of time
- Copyright or patent?

Slide 6- 58

Copyright

- Copyright holder has exclusive right to control reproduction or adaptation
- For a period of time, 50 to 70 years
- Laws vary by country
- **Intention:** encourage development of works that eventually reach the public domain, benefiting everyone

Slide 6- 59

Patent

- Inventor holder has exclusive right to exploit the invention (new device or process)
- For a period of time, 20 years
- **Intention:** encourage development of works that eventually reach the public domain, benefiting everyone

Slide 6- 60

IT Workbook Assignment 6.1

Student name:

Write your answers to the questions below <u>within the box</u>. In each case, please choose your words carefully to answer the specific questions. Avoid simply copying passages from your readings to these answers.

Explain in simple terms what the Internet is, what it is composed of, and what functions it performs:

1) what the Internet is:

2) what the Internet of composed of:

3) what functions the Internet performs:

Explain what the World Wide Web (www) is, and how it relates to the Internet:

4) what the World Wide Web is:

5) how the World Wide Web relates to the Internet:

IT Workbook Assignment 6.2

Student name:

Write your answers to the questions below <u>*within the box*</u>*. In each case, please choose your words carefully to answer the specific questions. Avoid simply copying passages from your readings to these answers.*

Explain in detail how the **plain old telephone system** (POTS) establishes communication between a person making a call, and the person who is being called:

1)

Explain in detail how the Internet's **packet switching** method of communication works, how it this differs from that of the ordinary switched telephone system, and why this method of communication was chosen by the Internet's inventors:

2) how packet switching works:

3) how packet switching differs from the ordinary switched telephone system:

4) why the inventors of the Internet chose to use packet switching:

IT Workbook Assignment 6.3

Student name:

Write your answers to the questions below within the box. In each case, please choose your words carefully to answer the specific questions. Avoid simply copying passages from your readings to these answers.

Describe **major events** in the history of the **Internet and World Wide Web** from the 1970s through the present time:

1)

Describe the **backbone** of the Internet, its **communication capacity**, and what type of **transmission medium** it relies on:

2) the Internet "backbone" is:

3) backbone communications capacity is:

4) the backbone is primarily what type of transmission medium?

IT Workbook Assignment 6.4

Student name:

Write your answers to the questions below within the box. In each case, please choose your words carefully to answer the specific questions. Avoid simply copying passages from your readings to these answers.

Explain what a **router** used in the Internet is, how it is different from a router used in a small local area network, and the two jobs it must perform in making the Internet work:

1) what a router used in the Internet is:

2) how an Internet router is different from a router used in a small local area network:

3) one job an Internet router must perform:

4) another job an Internet router must perform:

Explain what a **packet** is, what's in it, and its approximate size:

5) what a packet is:

6) what's in a packet:

7) the size of a packet:

IT Workbook Assignment 6.5

Student name:

Write your answers to the questions below within the box. In each case, please choose your words carefully to answer the specific questions. Avoid simply copying passages from your readings to these answers.

Explain what a **MAC address** is, how it is assigned, where it is stored, and what it does:

1) what a MAC address is:

2) how a MAC address is assigned:

3) where a MAC address is stored:

4) what a MAC address does:

Explain what a **logical address** is, how it is assigned, where it is stored, and what it does:

1) what a logical address is:

2) how a logical address is assigned:

3) where a logical address is stored:

4) what a logical address does:

IT Workbook Assignment 6.6

Student name:

Write your answers to the questions below within the box. In each case, please choose your words carefully to answer the specific questions. Avoid simply copying passages from your readings to these answers.

The **Traceroute** program on a Microsoft Windows personal computer can be used to trace the route a packet takes from your computer to an internet site. This shows you how many routers handle the packet between its origin and the site it is addressed to. Use Traceroute, named **tracert**, as described at the http://computer.howstuffwork.com/router10.htm web site, to trace the route from your computer to www.ambriana.com, which is hosted in Santa Monica, California. Write here the number of routers involved in this transmission:

1)

Then trace the route from your computer to this web site, which is located in Rome, Italy:

http://www.vatican.va

Below is a sample of the output you will see. Compare the number of routers involved in this to the number of routers in your tracert to www.ambriana.com, and write the number of routers here:

2)

(If your request times out, as in this example, write the number of lines of output you receive, such as 20 here.)

```
Tracing route to www.vatican.va [212.77.1.246]
over a maximum of 30 hops:

  1   177 ms   179 ms   173 ms  ipt-dtcd13.dial.aol.com [205.188.70.109]
  2   176 ms   173 ms   173 ms  iptfarmd-dr3-sw0-vlan3.conc.aol.com [205.188.120
.126]
  3   170 ms   173 ms   167 ms  172.18.30.60
  4   171 ms   173 ms   173 ms  172.18.52.69
  5   171 ms   173 ms   173 ms  pop1-vie-P4-0.atdn.net [66.185.141.113]
  6   171 ms   173 ms   173 ms  Level3.atdn.net [66.185.139.86]
  7   171 ms   179 ms   179 ms  ae-1-55.bbr1.Washington1.Level3.net [4.68.121.12
9]
  8   262 ms   263 ms   263 ms  as-2-0.bbr2.Frankfurt1.Level3.net [4.68.128.169]

  9   289 ms   269 ms   281 ms  as-1-0.mp2.Milan1.Level3.net [4.68.128.186]
 10   273 ms   275 ms   281 ms  ge-1-1.car2.Milan1.Level3.net [4.68.125.254]
 11   275 ms   281 ms   281 ms  213.242.65.134
 12   273 ms   281 ms   275 ms  81-208-50-50.ip.fastwebnet.it [81.208.50.50]
 13   292 ms   299 ms   293 ms  83-103-100-169.ip.fastwebnet.it [83.103.100.169]

 14   297 ms   299 ms   293 ms  83-103-100-218.ip.fastwebnet.it [83.103.100.218]

 15     *        *        *     Request timed out.
 16   298 ms   287 ms   299 ms  85-18-194-198.ip.fastwebnet.it [85.18.194.198]
 17   285 ms   275 ms   287 ms  83-103-94-182.ip.fastwebnet.it [83.103.94.182]
 18     *        *        *     Request timed out.
 19     *        *        *     Request timed out.
 20     *        *     ^C
```

IT Workbook Assignment 6.7

Student name:

Write your answers to the questions below within the box. In each case, please choose your words carefully to answer the specific questions. Avoid simply copying passages from your readings to these answers.

Explain what a **denial of service** attack is, and how some hackers managed to accomplish one in 2000, before more sophisticated measures were implemented to thwart them:

1) what a denial of service attack is:

2) how hackers have been able to accomplish a denial of service attack:

Explain why, with the current IP addressing method in which an IP address takes the form 172.18.30.60, there are a possible **4,294,967,296** unique Internet addresses possible.

3)

Explain what a **Uniform Resource Locator** (URL) is and how it is associated with an IP address on the Internet:

4)

IT Workbook Assignment 6.8

Student name:

Write your answers to the questions below within the box. In each case, please choose your words carefully to answer the specific questions. Avoid simply copying passages from your readings to these answers.

Identify and concisely describe **seven major internet applications** :

1)

2)

3)

4)

5)

6)

7)

IT Workbook Assignment 6.9

Student name:

Write your answers to the questions below within the box. In each case, please choose your words carefully to answer the specific questions. Avoid simply copying passages from your readings to these answers.

Describe what the term **intellectual property** identifies:

1)

Does intellectual property mean a **creation** such as a work of literature or a computer program, or does it mean the **legal rights** that authors or inventors might hold in relationship to the creation? Discuss this as it applies to software in general:

2)

Describe how **copyrights**, **patents**, and **trademarks** each operate in connection with intellectual property:

3) copyrights:

4) patents:

5) trademarks:

IT Workbook Assignment 6.10

Student name:

Write your answers to the questions below within the box. In each case, please choose your words carefully to answer the specific questions. Avoid simply copying passages from your readings to these answers.

What types of creations can be **copyright**?

1)

Can an **idea** be copyright? Discuss this in connection with software that implements a calculation for which the formula is generally known and available to all:

2)

Identify and briefly describe the **five exclusive rights** that the holder of a copyright possesses:

3)

4)

5)

6)

7)

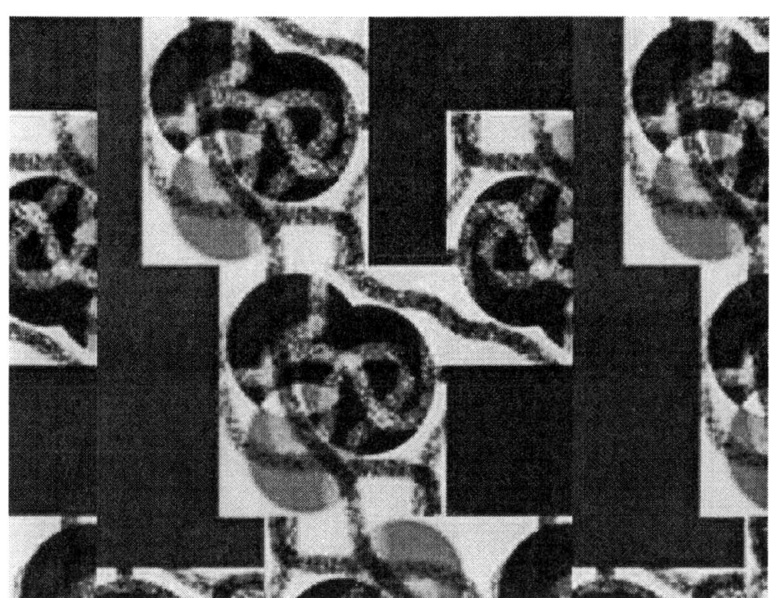

E-Business

In this chapter we delve further into the effects of the Internet on society by examining the phenomenon known as e-business. **E-business** is defined as **commerce conducted using the Internet**.

A business such as a retail store that has no Internet presence is called a **"bricks and mortar"** business. A bricks and mortar business relies on a physical building that customers must visit. In contrast, such a business that also has an "online storefront" web site to take orders for physical products is known as a **"bricks and clicks"** operation. This type of business often existed before the Internet, but has adopted the additional channel provided by the Internet for distribution of information about its goods and usually now also sells good via the Internet.

Completely digital businesses are entirely electronic, using the internet to take orders and to distribute their fully-electronic products such as music, documents, images, or software. These are sometimes called **"pure player"** enterprises.

The Internet has radically redefined commerce. It has made it possible for some vendors to eliminate the "middlemen" such as distributors and retailers from their "value chain" to deal directly with customers, a process known as **disintermediation**. Even more importantly, the Internet and the World Wide Web have leveled the consumer playing field, making it possible for almost everyone to become knowledgeable of prices for products across a huge number of vendors. Three major internet business models have emerged. Business-to-consumer (**B2C**) encompasses online retail storefronts and digital businesses. Business-to-business (**B2B**) includes information exchanges and industry-specific "hubs" serving commercial customers. Consumer-to-consumer (**C2C**) web sites support online auctions and the development of online communities.

Learning objectives

Chapter 7
E-Business

This chapter, in combination with the web readings, web links, and podcasts available at www.ambriana.com is focused on providing the knowledge to achieve these learning objectives:

- Understand and be able to describe what e-business is
- Be able to describe what the value chain is, and what it is comprised of for a manufacturing or sales organization
- Understand the role and functions of electronic data interchange (EDI) and electronic funds transfer (EFT) and how they predated the Internet and e-commerce, but are benefited by it
- Be able to distinguish between "bricks and mortar," "bricks and clicks," and "pure play" business organizations
- Understand the characteristics of information asymmetry, how the World Wide Web has affected it, what the phenomenon of disintermediation is, and to what extent it has occurred due to the Web
- Recognize and be able to describe how e-commerce has altered marketing and selling, and the several defined e-commerce business models
- Understand how the Internet has provided important advantages for both consumers and vendors, and what these specific advantages are
- Understand what public key encryption is, what it is based on, and how it is used in e-commerce
- Recognize what the "digital divide" is, important facts about it, and why it is essential to close this divide.

Podcast Scripts – Chapter 7

These readings of the **Information Technology Workbook** are also accessible in audio form as .mp3 files for free download. Visit the workbook web site, www.ambriana.com. You can listen to this material on your own Ipod or .mp3 player or computer, read it, or both listen and read.

Podcast 42
THE DIGITAL WORLD

The world is going digital. As Michael Rappa has noted, "Entire industries have been transformed by the new digital reality. Whether it is in the field of music, photography, publishing, journalism, banking, finance, manufacturing, health care, education, or entertainment—no segment of industry or government is untouched." [1]

What does it mean to say that the world is going digital? It means that the visual, auditory, and textual communications by which the interactions of business, government, and civilization itself take place are increasingly represented electronically, in the digital form that computers can store, manipulate, and transmit. The processes by which goods are manufactured, marketed, bought and sold, and delivered are becoming digital as well, with computers aiding in design and manufacturing activities. And since the Internet is a gigantic computer network that routes and circulates digital information around the world in a split second, it is becoming the circulatory system of the digital world.

More than a billion people now use the Internet every day. They use it to communicate with each other using e-mail, to locate information, to send business documents, and to do myriad other things, many of which were not possible to do before the age of the electron. In this chapter of the Information Technology Workbook we examine in particular what e-business consists of, as facilitated by the Internet and the World Wide Web. How does e-business work? How it is organized, and how does it generate revenue?

We'll do this by first examining the four major types of e-commerce:

- business to consumer (B2C)
- business to business (B2B)
- consumer to consumer (C2C)
- peer to peer (P2P).

In keeping with the trendy naming tendencies of the technical world, the abbreviations for these four types of Internet business use the number "2" in place of the word "to".

We'll first take a look at the perspective of each of these types of Internet business. We'll then look at the initial rise and collapse of Internet e-commerce expectations in what has been called Internet Era I, from the point at which the world became aware of the World Wide Web in 1994, to the "dot com" economic collapse of the year 2000. In that short span of six years, many notions of what would succeed on the Internet as e-commerce were tested. Some concepts succeeded, and some were proven to be based on incorrect assumptions or just plain wrong. We'll dig into what was learned in that era and distill our current understanding of the business models and marketing methods that are succeeding in the electronic environment.

We'll explore innovative payment methods that have evolved to facilitate e-commerce, and ways that business and personal communications are made secure through the use of encryption.

Finally, we'll consider a potentially disruptive effect of the digitization of the world and e-commerce on human society. We'll explore the "digital divide" that is widening between nations and societies that are benefiting from the new forms of interaction and communication, and those that are not. Experience is proving that simply providing electronic access is not enough for a given society to benefit. What is designed for conveyance by the new means of access is even more important to the provision of benefit to societies confronting modernization and industrialization.

[1] See http://digitalenterprise.org/introduction/intro.html

Podcast 43
EDI and EFT

E-business, or e-commerce as some people refer to it, is any business that is facilitated or conducted using electronic communication. This broad definition of e-business and e-commerce actually encompasses the computer-to-computer exchange of business information named **Electronic Data Interchange (EDI)**, which predates the Internet. The development of EDI was a thrust of large business organizations in the 1980's. With the EDI effort, large businesses agreed to standardize the documents by which they purchased raw materials, shipped them, billed for them, and conducted most of their other interactions of commerce. The EDI standards that grew out of these efforts prescribe the formats and data elements that wholesale sellers and buyers, raw materials producers and manufacturers, distributors and retailers, and other business partners agreed to use in day-to-day goods and value exchanges.

EDI documents cover hundreds of different types of business-oriented communications and are entirely electronic. They were originally conveyed using proprietary computer networks, but when the Internet became available, EDI documents began to be transmitted using it. Huge manufacturers such as General Motors were among the most ardent contributors to the development of EDI. They eventually insisted that all of their many subcontractors equip themselves to be able to transaction business using EDI. It presented quite an investment for a smaller firm to adapt its bidding, shipping, and billing systems to create EDI documents and to receive communications in this format. It also presented quite a burden to smaller firms to equip themselves to tie into the proprietary communications networks that predated the Internet. Now, with the ease of communication provided by the Internet, the cost of participating in e-commerce as a supplier to large firms has been reduced, helping to level the playing field.

Electronic Funds Transfer (EFT) is another electronic facilitator of commerce and qualifies as an early form of e-commerce. EFT is a system for transferring money from one bank account directly to another without the use of a paper document such as a check. Direct Deposit is a widely used EFT program that began in the late 1980's. Using Direct Deposit, employers output a data transmission from their payroll systems rather than printing paper paychecks. The data transmission is sent to a bank that routes the transactions to the Automated Clearinghouse (ACH) Network, which debits the employer's bank account and credits each employee's bank account.

EFT is being extended now to allow consumers to pay most of their bills online, eliminating the expense, time delays, and potential for theft posed by paper statements and bills, and the expense to companies of processing paper check payments. The economies and savings of electronic payment mechanisms are huge since it has been estimated that the cost of processing a paper payment document such as a check is now over $3!

Podcast 44
INFORMATION ASYMMETRY and DISINTERMEDIATION

The Internet brings three significant advantages to consumers, and several advantages to sellers of product and services. It changes the factors in the "discovery" element of purchasing, and it offers new alternatives to marketing that make it possible even for small enterprises to project a large image. In this podcast we examine how the Internet has changed business from the consumer's side of the transaction.

A consumer interested in a specific product can now search for the product and information about it online, quickly, and in private. The consumer can become much more aware of competing products and competing vendors offering the same product, and also see reviews of products and vendors submitted by other consumers. Search engines and shopping "bots" like www.bizrate.com and www.mysimon.com not only make competitors and their offerings highly accessible and visible, many serve as bulletin boards at which happy or unhappy customers can post their praise or complaints—about the vendors as well as the products! A slogan of the Yellow Pages phone directory in bygone years was "let your fingers do the walking," which encouraged people to look up the phone number and location of stores, and (possibly) to make a telephone call to check on items and prices. People are using their fingers these days, but not so much to make telephone calls to check on products, but to tickle a keyboard and use the Internet. Smart shoppers these days perform their "due diligence" by comparing vendor web sites, seeking out pure-player sites that fly under the radar of sales tax collectors, and learning what past consumers say about specific products and vendors.

After using the Internet and these capabilities to become more informed consumers, many people cannot envision commerce in a world where sellers held a monopoly on pricing and product availablity information. Such one-sided access to information about costs, pricing, and availability is called **information asymmetry**. Information asymmetry worked to the advantage of sellers in the past in the same way that the "company store," which held a monopoly on goods in a small mining town, had the upper hand—buyers had little choice and not nearly as much knowledge as the sellers. Used car vendors formerly had such an advantage, even though several might exist in a town. But who would now buy a used car as all people once did, by shopping physically at one or two used car lots? Only, sad to say, the uninformed. After using an online site such as Carmax to choose from a huge variety of used vehicles, it's apparent that anyone now has access to a much larger inventory of used cars to compare features, condition, and prices, without even leaving their own home. Unfortunately, the people who must still pay the higher price for relatively uninformed purchasing are those least able to afford it, the segment of society that cannot yet afford Internet access, or don't know how to use it effectively. This form of disenfranchisement is one aspect of fairness that the modern communication revolution must still address. But as the Digital Divide organization points out, it's not only a matter of access, but a matter of education. Even a person who cannot afford a computer has access to public libraries, nearly all of which provide free access to computers and the Internet. But there's a learning curve to this new environment, and it presents a distinct hurdle that must be surmounted.

Podcast 45
INTERNET EFFECTS ON MARKETING

The Internet brings advantages to consumers and several advantages to sellers of product and services. In this podcast we examine how the Internet has changed business from the seller's side of the transaction. Sellers of products are the recipients of at least five major benefits of the Internet.

First, it's now possible for sellers to **electronically publish much larger catalogs** than they were able to with paper promotional materials. This is how even a small business enterprise can have a presence on the web as great as a much larger business. In the past, the printing expense of a large catalog made it something only a large, high-volume business could afford to produce and distribute.

Secondly, sellers can now **reach consumers at all hours of the day, across a geography that is essentially unlimited** Every business now potentially has a global reach. The overhead associated with keeping the doors of a physical store open (lighting, cooling or heating, and personnel costs) are reduced to the cost of hosting a web site. And creating awareness and achieving a presence in the marketplace no longer relies solely on expensive advertising or postal mailings. Knowledge of a vendor's offerings is now made available by the exposure provided by search engines such as Google. A form of advertising on search engines to give prominence to "sponsored links" follows a model based on "click-throughs" where the major cost of the advertising is based on how many people actually click on the link after seeing it.

Third, sellers can **adjust prices instantly** to reflect a changing economic situation or in response to a change in availability. In former times, since advertising was expensive, a vendor's flexibility was more tightly constrained. Since print advertising "lives" longer and constitutes an offer, it elicits a customer expectation of the advertised price.

Fourth, manufacturers may be able to **eliminate some or all intermediaries**, that is, middlemen such as distributors and retailers, and deal directly with consumers, allowing more competitive pricing. In fact, the process of eliminating middlemen in the distribution chain, called **disintermediation**, has not occurred as fully as some people expected in the early days of the Internet. It has definitely occurred in those instances such as book selling, where the commodity can be satisfactorily examined online (as to content, in any case) by a prospective purchaser, the product needs no "installation," it's conveniently shipped, has a shelf life of months or years, and is not perishable. But many types of products are not so readily sold without the added value provided by in-person inspection and selection in a physical place of business, configuration, customization such as fitting or adaptation, installation, and after-sales support and service. For these reasons, while customers are now able to use the Internet to become more price and competitor aware and can locate multiple vendors offering products they want, few people buy clothes, groceries, major appliances, cars, hardware, personal grooming products, or expecially bulky or heavy items online. On the other hand, a great many people now buy airline tickets and book hotels online, without the use of a travel agent. Disintermediation has occurred, but not across the board as some people earlier thought was inevitable.

And fifth, the Internet is not only a communication pipeline, but represents a whole **new factor in the marketing of goods and services**. It provides two-way communication that allows sellers to track customer preferences and behaviors, and to **customize product offerings**. This has been developed into a science by service providers that can assist a vendor to acquire consumer demographic and Internet activity information, allowing innovative web retailers to maximize customer convenience and product appeal.

Podcast 46
THE VALUE CHAIN

The World Wide Web and the electronic communications network of the Internet that supports it developed in a world in which a business concept known as the value chain already existed. In this podcast we'll examine what the value chain is and how it helps refine and streamline business processes for competitive advantage.

The **value chain** concept was articulated by Michael Porter in the 1980's. The concept was described in a book he wrote entitled ***Competitive Advantage: Creating and Sustaining Superior Performance***. Porter's concept identifies where and how value is added to goods in the process of manufacturing, and helps strategists shape improvements in those processes. The connection between Porter's value chain concept and the Internet is quite direct. The Internet appeared as a communications revolution at precisely the moment that business strategists were seeking to improve the integration of the elements of the manufacturing process. Since communication is a key element of process integration, the communication benefits of the Internet have become enormously important to business process improvements based on value chain analysis.

The value chain consists of five categories of activity. These activities comprise what a manufacturing organization does as it acquires raw materials, transforms them into products, markets and sells them to customers, fulfills orders for the products and delivers them to customers, and maintains its relationship with its customers. Porter identifies these activities in this way:

1. **Acquiring raw materials**. This activity is known as **inbound logistics**. The raw materials are the items that the business regards as its production inputs. In the case of a manufacturing industry such as the automobile industry, the raw materials include assemblies such as carburetors, tires, and lamps which are the products of other vendor's value chains. The timing of acquisition activities is an important consideration in inbound logistics. Interruptions in raw materials arrival can upset and delay manufacturing processes, but the premature arrival of raw materials can also be disruptive since the materials then have to be stored awaiting use. It's important to regard human labor and the organization of it as a raw material to be acquired in sufficient quantity for the business process.

2. **Transforming raw materials into products**. This activity is known as **production**. This is the series of steps by which raw materials are transformed into goods of interest to other businesses or consumers.

3. **Outbound logistics** is the activity that delivers manufactured goods to the places where they can be distributed to the sales, marketing, and maintenance functions. This may involve the stockpiling of manufactured products in warehouses as volumes of products are pumped through the distribution pipeline to wholesale distributors and onto retail store shelves or as inputs to the inbound logistics activities of other businesses.

4. **Marketing and selling products**. Marketing consists of actions that make potential customers aware of the goods manufactured by production activity. In order to be effective, marketing must deal with goods for which potential customers actually have or perceive a need, offered at a price that they consider reasonable. This requires an understanding of the competitive environment and the availability of the same or similar goods from others. It also involves advertising, which in many cases can promote the growth of the perception of need for a product on the part of potential customers, and the perception that the product offered is in some way superior to products that compete for the customer's interest. Selling is an activity that includes the involvement of some form of contact with the customer, including possibly product demonstration, leading up to the critical moment when the customer commits to the purchase of the product and places an order for it.

5. Post-sale activities are those that follow the placement of an order by a customer. These are called **maintenance activities**. These include the fulfillment of the customer's order by delivery of goods, billing for and collecting payment for the goods, and very importantly, maintaining the good will of the customer so that the customer will be as likely as possible to place additional orders in the future.

The first four processes in the value chain, namely, inbound logistics, production, outbound logistics, and marketing and selling products are the focus of enterprise resource planning systems, known as ERP systems. While ERP systems were designed prior to the advent of the Internet, in the modern world their communication requirements are greatly facilitated by the Internet. Many actions of the fifth value chain component, maintenance activities, are supported by Customer Relationship Management (CRM) systems. CRM systems assemble information about the contacts between an organization and its customers and make this accessible to customer service, marketing, and product development personnel. CRM systems make heavy use of e-mail as a means of maintaining contact with customers. Most ERP systems now offer an integrated CRM component.

Podcast 47
WEB GROWTH and TYPES OF E-COMMERCE

Two major types of e-commerce exist. One of these is business-to-consumer, and the other is business-to-business. Two other types of e-commerce exist but are of much less scale in terms of the number of web sites involved: consumer-to-consumer, and peer-to-peer. In this podcast we will discuss the major types of web e-commerce, commonly referred to as B2C and B2B.

Business-to-consumer (B2C) e-commerce is the type of e-commerce most of us are familiar with. It consists of any type of business transaction that occurs on the web where an ordinary consumer is involved. In other words, this is retail trade on the web. In this context a transaction is defined as the exchange of money for something of value: physical goods or a service.

Business-to-business (B2B) e-commerce consists of transactions between businesses, not involving an ordinary consumer. B2B occurs, for example, when a manufacturing business buys a large quantity of raw materials in bulk form through a broker or intermediary on the Web. The easy way to think about this is to characterize it as wholesale trade. This type of commerce occurs behind the scenes and out of view of ordinary consumers. The vendors involved are typically not those with which consumers would deal, or even be aware of. This is the type of commerce that EDI was originally designed to facilitate.

You might have an inkling that B2C e-commerce—the e-commerce retail trade many of us participate in—is a large business. It is projected to grow from $60 billion per year in 2000 to over $250 billion in 2006. However, B2B e-commerce, businesses trading with businesses and facilitated by electronic means using the Internet, was already over $700 billion in 2000, and will grow to over $5 trillion dollars in 2006. As far as the web itself goes, more than four billion pages were already posted on the web in 2003, and at that time the web was growing by 7 million pages every day! This is growth on an astronomical scale and it has been measured and characterized in other ways too.

In comparison to the acceptance of other new technologies, the Internet and the World Wide Web have grown at a faster rather than did radio or television. Commercial broadcast radio, for example, was introduced in 1920, and it took 38 years to achieve 30% penetration of American households, reaching this level in 1958. Television, which was introduced in 1946, took 17 years to achieve this level of penetration in 1963. But the Internet and the web took only seven years, from 1994 to 2001, to achieve 30% penetration of America households. Why did the Internet and web move so quickly to become an accepted part of the landscape? Most likely this was because much of the infrastructure to support it was already in place with personal computers and telephone communication to the home. A vast number of homes were already equipped personal computers by 1994. The only thing a person who already had a computer needed to acquire to begin accessing the Internet was a modem to attach the computer to the telephone system, and a browser for the computer. Modems typically cost less then $100, and the Mosaic browser, forerunner of Netscape, was distributed free.

It took only a few years after the release of the Mosaic browser in 1994 for people to begin characterizing businesses in one of three ways. Established businesses without a web presence are called "bricks and mortar" businesses. They continue to deal in retail trade at a physical store. This type of business is increasingly rare. Many existing businesses saw the value of establishing a website to either make potential customers aware of their store or to open a mail order sales channel. For example, BestBuy, a major appliance and consumer electronic equipment chain, established a web site to showcase its wares, make it possible for customers to order them, and to pick up their purchases at a store or have them shipped. Thus it changed from a "bricks and mortar" business to a "bricks and clicks" business—a business that has a physical presence and also an online web presence. Even restaurants, the quintessential "bricks and mortar" business, often establish a web site to make their menus available to potential customers, and become to some extent "bricks and clicks" businesses.

Some businesses were formed and exist only on the web. Amazon.com is such a business. There was no Amazon before the .com era. The business was created only after the Internet became available, and its entire business plan is based on the web as its electronic storefront. While Amazon has expanded into selling goods other than books, books still accounted for 70% of its sales in 2005, attesting to the soundness of its original business plan. Businesses like Amazon are sometimes called "pure play" e-commerce businesses. It has clicks, but no bricks. Some pure play e-commerce is entirely digital, and never even ships a physical product, but instead delivers digitized content via download over the Internet. I-Tunes is such a business, and so is photos.com, which provides royalty-free visual images in digital form for a subscripion fee.

The Internet and the web have changed commerce, and the change continues. The entire world is modernizing as a result of the globalization of communication and commerce. Part of the turmoil we see in the world today results from the changes this inevitably makes in cultures and civilizations that are confronting modernity. They are now having the same issues with modernization that began to affect Western Europe hundreds of years ago with the expansion of exploration, trade, and the confrontation with new ideas that challenged established ways.

Podcast 48
E-COMMERCE REVENUE MODELS

Revenue models describe the mechanisms by which money is exchanged for something of value. In one categorizations of models, five basic revenue models describe the essential ways that business-to-consumer e-commerce generates revenue. In other categorizations, up to ten or more distinct variations of revenue models have been identified. In this podcast we'll consider the five best recognized e-commerce revenue models.

The **advertising** model applies when a site is constructed for the purpose of garnering traffic in the form of visits, so that users see something they are drawn to—a search engine, news about world events, news about celebrities, reviews of products, and so forth—and also can't avoid seeing advertising around the edges of the screen, as pop-ups, as pop-unders, or as parts of the screen that overlay content and remain present for a period of time. Just as with print, radio, and television advertising, the site has more value to advertisers the more people see it. A heavily trafficked site can charge advertisers more than a site that generates few visits. This is one reason that RSS feeds, an application that automatically feeds information from news gathering and other sites to subscribing sites, are very popular. It allows sites that profit from advertising to build up readership by offering fresh and frequently-changing content.

The **subscription** model charges for the privilege of viewing and possibly downloading content. E-magazines, referred to as e-zines, use this model. So do online encyclopedias such as Encyclopedia Britannica and pure play companies that allow users to download royalty-free stock photography for a periodic fee. Typically a user subscribes and supplies a credit card that is billed every month, and the user is provided with a password-controlled account.

The **transaction fee model** applies to any web business that charges for specific services performed. Perhaps the single best example of this is PayPal, which charges sellers a fee for a payment handled by PayPal from a buyer. E-bay falls into this category too, by assessing a fee per transaction for "arranging" a meeting place for buyers and sellers. Airline and hotel booking services implement this revenue model as well, since they charge either the buyer or seller or both a small fee for handling each transaction.

The **sales model** is the revenue model of typical "bricks and clicks" retailers. They generate revenue by selling products via electronic storefronts on the web. The customer pays for each product. How the sales model is implemented varies between businesses. The traditional method is for the customer to pay for products using a credit card at the time of purchase. But for small payments, under $15, credit cards are often more expensive than desirable from the merchant's point of view, due to the fee credit card companies assess for each transaction. Sales-model companies have two ways of dealing with this. A micropayment handler such as PayPal typically charges the seller less per transaction than a credit card company, and is ideal for small payments. Some sales-model companies, however, have a special case if what they deal in results in recurring purchases. These companies implement the sales model by establishing an "account" for customers and having each customer place an amount in the account, so that each purchase simply deducts from the balance of the account. For example, a vendor processing digital photos, uploaded by the customer, into print form for 19 cents each can establish an account for each customer and require a $15 minimum amount to be paid at the start. Each print ordered deducts 19 cents from the balance, and when the balance declines to $5 or less, the customer is notified to pay more money into the account. This method benefits the seller in multiple ways. It reduces the amount of credit card processing cost, and, since the customer has prepaid for future purchases, he or she is more likely to make those additional purchases.

The **affiliate model** is based on directing traffic from one site to another. Under this model, a site offers something of interest to viewers, and includes links to other web sites. Each "click through" from the original web site to the affiliate site is counted, and the affiliate site pays the site directing traffic to it a fee based on the number of click-throughs. Search engines such as Google derive significant revenue from click-throughs to sites at "sponsored links." The "sponsors" are the web sites of the retailers who pay for the prominence at search pages related to their wares.

To review: the five major web revenue models are the advertising model, the subscription model, the transaction fee model, the sales model, and the affiliate model. You should now be able to explain how each of these models operates and be able to provide an example of a web site for each model.

Podcast 49
VIRAL MARKETING

One of the most interesting methods of marketing developed with the World Wide Web is named viral marketing. Viral marketing takes it names from viruses, but not in the negative sense of malicious software unleashed on computer networks. Rather, the term viral marketing refers to the way actual viruses act, which is to convert a host organism to produce and pass on copies of the virus. **Viral marketing** consists of creating an item of software or digital entertainment that in some way refers to a product to be marketed, or stories about a such a product, that are so interesting and entertaining that people who see it pass it on to their friends. In this way, viral marketing seeks, as Wikipedia notes, "to exploit pre-existing social networks to produce exponential increases in brand awareness…"[2]

The term viral marketing originated when Steve Jurvetson described Hotmail using this term in 1997. Hotmail offers a free e-mail account to people, but attaches advertising for itself at the end of every message sent. The message encouraged others to establish hotmail accounts, so in effect, every existing user of Hotmail advertises it simply by using it.

Several types of viral marketing exist:

- **Pass-along:** this is simply a message that encourages recipients to forward it to friends. It often includes a clever and entertaining joke or story, or a sound or video clip.
- **Incentivised:** some type of reward is offered in exchange for providing someone else's e-mail address. This type of marketing is

[2] See http://en.wikipedia.org/wiki/Viral_marketing

more successful if the second party also has to participate in order for the first party to receive an added benefit.

- **Undercover:** some form of allusion is made to a product in an entertaining transmission, but the product being advertised is not obvious. the product reference may remain veiled, or it may be a clue that the reader discovers.
- **Edgy gossip:** similar to the way Hollywood sometimes uses scandal or illicit love affairs of stars to generate interest just as movies in which they appear are about to be released. Edgy gossip may be created by advertising or message that create controversy. Blog sites are a popular place to plant these types of stories, which then form the basis of conversations between people, who then "tune in" to learn more for themselves.
- **Anonymous matching:** a fad on dating web sites, this involves people interested in establishing relationships with others forming lists of friends and acquaintances with whom they may wish to become romantically involved. The web site accomplishes a match between people when someone on one user's list also lists that person on their list. This type of web site often allows the person forming a list to send anonymous e-mails to people on their list telling them that someone has a crush on them. A significant percentage of people receiving these e-mails respond and set up accounts and lists of their own, expanding the population of users. In effect, users are driven by the possibility of liaisons provided by the service, and the traffic provides the basis for advertising and affiliate revenue streams.

Viral marketing differs from spam in a very significant way. Spam is trash e-mail, sent by sellers who hope to have a tiny percentage of a vast number of e-mails generate responses. Viral marketing campaigns start with an e-mail campaign that aims at generating a high percentage of active participants who continue the marketing, forming an ongoing presence. Whereas nearly all people will forget spam immediately after deleting it, viral marketing depends on recipients actively passing it on or passing on information about it to others they know.

Podcast 50
SPAM vs. PERMISSION MARKETING

Spam is a common term for "junk" e-mail. It's the electronic analogy to the advertising postal mail that many people receive and mostly toss away. Spam is a real problem because so much of it is generated that it often clogs electronic mail boxes. Spam filters in e-mail systems are often used to scan each incoming item and assess the likelihood of it being spam, giving each item a numeric score based on content and format. If an item receives a score above a certain threshold, the spam filter can automatically tag the item or direct it into a "quarantine" folder to eliminate it from visibility. Spam filters also function in the same way to identify potentially harmful e-mail attachments such as viruses and executable programs.

How is spam generated? Several address list vendors exist on the web and provide thousands to millions of postal and e-mail addresses for a fee, typically in the range of a few cents to a dollar per address. The largest of these vendors culls telephone directories, advertising, and online sources to compile large databases of people and business organizations, and classifies them by location, income level, industry, or other characteristics. Vendors who want to launch a marketing campaign often attempt to judge what the characteristics of likely customers are, and secure a mailing list of people or businesses with those characteristics. The result can be either a postal mailing campaign, or e-mail sent out in bulk to each person on the list.

In the United States, it is legal to direct unsolicited e-mail to people. In Britain, a law that took effect in 2003 prohibits this and allows e-mail marketing to be directed only to people who have previously given consent to receive it. The consent given to this type of e-mail is called **opt-in**. That is, people opt (choose) to receive it.

Opt-in marketing is not popular in the United States, but some marketers do engage in it. It is also called **permission marketing**. Typically, a vendor from whom you are purchasing something will ask if you want to be placed on their mailing list to receive information about other products or sales.

Some vendors feel that permission marketing is much more beneficial than simply broadcasting spam to large numbers of people since the recipients of permission marketing have already expressed a potential interest in the vendor's products. While this is true, it has been estimated that only 10 to 15% of people across the board opt to receive advertising.

The opposite of opt-in is **opt-out**. In opt-out marketing, unsolicited advertising is directed to people, and a button or process is provided to let recipients indicate that they do not wish to receive any more of it from this source. There's little difference between opt-out advertising and ordinary spam. In fact, it's advisable not to respond to messages such as this in spam, because quite often unscrupulous spammers don't actually respond to it in the way they imply. Instead of actually eliminating you from further e-mailings, your response is actually a confirmation that e-mail directed to this address is read. By responding to the apparent opt-out message you may be making yourself a target for even more e-mailings from the spammer.

Especially unscrupulous spammers often use software to **spoof** the sender's name on an e-mail, plugging in names at random. The purpose of this is to fool the recipient into actually opening the e-mail rather than deleting it based on the sender and content of the subject line. Spammers engaging in this typically generate and dump millions of e-mails into the Internet on a daily basis. With this quantity of e-mails, even a very small percentage of responses to this advertising for watches, drugs, mortgage financing and stock or investment advice—the largest categories of items advertised with spam—can generate large enough revenues to be worthwhile to the perpetrators, at the expense of much wasted time and irritation to millions of people.

Podcast 51
RSS MARKETING

RSS, which stands for Rich Site Summary, has emerged as an adjunct to web sites that generate revenue from advertising or affiliate links. Advertising and affiliate revenue models depend on attracting large numbers of viewers to a site. Many ISP entry sites are designed to serve as portals, in order to become destinations, not just sites that a user passes through to get to e-mail. Such sites build traffic by receiving RSS feeds.

RSS is an application pioneered on the web in 1997 by Userland and Netscape to share content for pages that can attract large numbers of users. It has caught on and become a very popular method by which sites that have news, articles of common interest, or other frequently changing information can distribute it to sites that subscribe to the feed. RSS provides the way for a site with news to distribute it, and for sites that want to attract traffic to give themselves varying content of potential interest to people. The larger sites that provide RSS feeds such as Slashdot, Motley Fool, and Wired News have a vested interest in making a constant stream of headlines and attention-grabbing news clips available. Many people seeing the summary at a site subscribing to the feed will track back to the originator's site to read the article, increasing traffic there. RSS is also the mechanism by which podcasts can be distributed as they become available.

RSS has begin to replace several other methods for one site to obtain content from another site. Alternative methods include taking the HTML source from a site and analyzing it ("parsing" it) to pick out the desired content, or using an application program interface (API) provided by a site to share data. These methods require custom programming that is specific to each site offering content. RSS, which is based on XML, makes the process much simpler and standardizes it. A site offering RSS content identifies the content of its offerings in XML. Once a site establishes the software to receive an RSS feed, obtainable as shareware, the site can receive an RSS feed from any site sharing content in this way. As Jonathan Eisenzopf noted in an article at www.newarchitectmag.com in February 2000, "The RSS format is a marriage made in heaven for extending readership... Users read the headline and click on the link to read the story, and both sites get their page views." Aggregator sites build on the capabilities of RSS feeds by making multiple such feeds available at one site (they perform "aggregation") and may also provide tools making it easier for a subscribing site to obtain content.

Podcast 52
CREDIT CARDS AND E-COMMERCE

Credit cards are the most common payment vehicle for Internet business-to-consumer (B2C) e-commerce. In formal terms a credit card is the plastic card, usually conforming to a standard named ISO 7810, that is part of a "retail transaction settlement and credit system."

Two types of cards similar to a credit card exist, debit cards and charge cards. A **debit card** is simply an electronic check; you can only spend money using a debit card that you have in a checking account. While you might use a debit card in an online transaction, it is unwise to do so. Debit cards do not enjoy the same protection under Federal regulations that credit cards do. For example, if your credit card is stolen and used by a thief, under Federal law your liability is limited to $50. But if your debit card is used by someone else, no such protection exists, and your entire bank account may be stolen, with no recourse by you to reclaim it.

A **charge card** represents a loan of money to the purchaser using it, but the money borrowed to make a purchase using a charge card must be paid in full each month. Thus a charge card actually arranges a very short-term loan from the issuer to the card holder.

A **credit card** provides a "revolving" arrangement whereby the person using it is loaned money to make purchases, up to a credit limit established for each card holder, and the money can be repaid in small installments over a variable and potentially long period of time. Credit card issuers very much prefer card holders to make the minimum payment each billing period, and to carry as large a continuing balance as possible. The legislative lobbies funded by credit card issuers have been very successful in the recent decade in securing relaxation of Federal regulations concerning the interest rates and fees issuers are permitted to impose on card holders. Interest rates on credit card debt are several times higher than for bank loans, making credit cards by far the most expensive way to borrow money. Late fees and over-credit-limit penalties of up to $29 per occurrence are another ways that credit card issuers are now permitted to extract money from card holders.

Credit cards are a major convenience, however, if the purchaser actually pays the full balance at each billing period. In such a case, no interest is charged. If however, only part of the balance is paid, interest is charged on the full amount of the balance before the payment is applied! And, of course, credit cards are the primary way that purchases can be made on the Internet.

Credit card issuers charge merchants for each transaction in which payment is made by credit card, so they earn revenue both from the seller and potentially from the buyer.

A major concern exists that people whose limited assets or credit history prevent them from obtaining credit cards are shut out of the economies that can exist in e-commerce retail trade. Typically, most credit card issuers will not extend credit to a person who earns less than $15,000 per year. To some extent, alternative payment mechanisms such as PayPal partially compensate for this, but only if the consumer actually has the cash on hand to make purchases; in this way, PayPal is similar to a debit card.

It's important for people approaching adulthood to consider the factors that go into the computation of creditworthiness and to work toward building a credit rating that allows full participation in the credit-based commerce of modern society. The usual steps to doing this are earning money through part-time work for which income taxes are deducted and paid, avoiding problems with the law, and securing some limited form of credit, such as a department store credit card, using it moderately, and making payments on time, to demonstrate responsibility in handling credit.

Another common method to build a reputation of creditworthiness is to buy a new or used car and placing a down payment amount that is sufficient to persuade a bank or auto credit organization to grant a loan for which no co-signer is required, and making all required payments on time. These actions are recorded in the credit reports that major services compile for all consumers, which credit-issuing organizations look at in determining your creditworthiness.

Ultimately your creditworthiness becomes a major factor in your ability to secure a mortgage on the purchase of real estate, such as a house. Purchasing a house is, for most working people, the largest and most beneficial investment they will ever make. Your ability to do this depends a great deal on how you have demonstrated your ability to handle credit, especially credit cards.

Podcast 53
PUBLIC KEY / PRIVATE KEY ENCRYPTION

You should think of messages, documents, and files that are to transmitted via the Internet not as sealed envelopes, but as if they were written on postcards. Postcards have an address on one side and a written message on the other side. Anyone handling the mail can read postcards. Would you write your private information, including bank account numbers, balances, and passwords on a postcard and mail this information to someone in a distant city? In fact, legislation exists that make it a Federal crime to tamper with and read someone else's ordinary envelope-sealed postal mail, but practically speaking the same legal protection does not exist for postcards and e-mail. Yet a secure means of electronic communication is absolutely essential for e-commerce to take place, since we must be able to securely send credit card numbers and other sensitive data through the Internet!

Sensitive information is conveyed electronically by encrypting it, that is, encoding it, scrambling it in a way that makes it difficult for anyone but the intended recipient to decode it. The most secure way of doing this is to use what is known as "public key / private key" encryption. Various forms of this exist, but all result in the generation of pairs of key values that are related. When one of the pair of values is used in a formula with the data to be encrypted, the data is transformed into a coded form. The coded value can only be decoded using the other key of the pair—even the key used to encrypt the data cannot be used to decrypt it.

When a key pair is generated and issued to a user, the user regards one of the values as his or her "private" key and never reveals it to anyone else. The other key in the pair is declared the user's "public" key, and the user makes it known to everyone else, usually by publishing it on the web. Any message encrypted with the public key can only be decoded by the holder of the corresponding private key. Anyone can now send a confidential message to the holder of the private key, by scrambling the message with that person's public key. The beauty of this is that the message has become secret and can only be read by the intended recipient, but no key had to be securely provided to the sender by the intended receiver. Prior to public key / private key encryption, in single-key encryption, the same key used to encrypt a message was used to decrypt it. This meant that a key such as a code book or password had to be conveyed from the sender to the receiver, posing a potential for unintended people to obtain the decoding key.

Private key / public key encryption also provides the way to authenticate a message or transmission. If a person encrypts a message using his or her private key, only the matching public key can decrypt the message. Since everyone knows everyone else's public keys, or can locate them on the web, anyone can determine if the purported sender is really the sender, simply by attempting to decode the message. If it decodes with a person's public key, that's the person who sent it.

A form of encryption is used automatically between web sites when the key symbol appears on the browser page. You will notice that this symbol appears when you are buying an item online from a reputable web site, and you are asked to enter your credit card number.

If you need to make a secure data transmission involving an e-mail attachment, you need to encrypt the attachment yourself. Software exists on the web to accomplish this. One such program is **PGP**, which stands for "pretty good privacy". PGP offers a number of options including private key / public key generation and traditional single-key encryption. Encryption will most likely be built into future operating systems such as Windows, because the same protection it provides for Internet transmissions would be beneficial to safeguarding ordinary data storage.

Podcast 54
ROLE OF THE HASH VALUE

A quantity known as a **hash value** plays a role in the secure transmission of electronic messages and files, separate and distinct from encryption. Whereas public key / private key encryption deals with keeping things secret, and authenticating the sender of a transmission, a hash value is used to insure that the contents of the transmission have not been altered, either by accident or deliberately.

The technique of using hash values has existed and been used long before sophisticated methods of encryption were developed. As hash value is simply a value determined from the content of a message or transmission, using a formula known to both the sender and the recipient of the transmission. The sender uses the formula to compute the hash value, and sends the hash value either along with the transmission, or separate from it. The recipient of the transmission applies the same formula to compute a hash value from the transmission as received, and compares it to the hash value sent by the sender. If the locally computed hash value is the same as the one sent by the sender, the transmission is known to have been received unaltered. If the locally computer hash value is different from the hash value sent by the sender, the message has been altered, cannot be trusted, and must be resent.

Various formulas can be used to compute the hash value. A very simple formula might be to take the ASCII value of each byte of the transmission as a numeric quantity expressed in binary (a value in the range of 0 through 255) and add these up. While this hash value could catch some forms of alteration of the transmission, the hash value would not differ if the sequence of characters were simply changed. A more sophisticated method could be to take the numeric value of each byte, as in the first method, but then multiply each byte by a different value, so that the position of the byte in the transmission affected the amount the byte contributed to the sum. For example, the value of the first byte might be multiplied by 1, the value of the second byte multiplied by 2, and so forth. After the multiplier reach a value such as 20, we might start again with 1 as a multiplier. This way, some types of transpositions of bytes in the transmission would produce a different hash value.

Much more sophisticated hashing algorithms are used in actual practice. MD4 and MD5 have been popular hashing algorithms but have been proven to be insufficient since it is possible to compute the same hash value with them from two very different transmissions. Other standard algorithms exist with the names WHIRLPOOL, SHA-1 and RIPEMD-160. These algorithms are complex but are implemented in libraries associated with modern programming languages.

Podcast 55
THE DIGITAL DIVIDE

The world is rapidly dividing into two camps: those people who can benefit from digital technology and those who can't. This is actually about who is served by and helped by the technology, not necessarily who has direct access to it. It's important to understand this distinction, because the question is not just a matter of connecting people to the Internet, but about delivering digital services that make it possible for people to become informed members of a world that is modernizing.

Closing the gap between digital "haves" and "have nots" is important for several reasons, but three of the most important are:

- it is a **precondition for reducing poverty**, just as the provision of adequate infrastructures such as water systems, health care, and education;
- It is a **precondition for reducing terrorism**, because the communication it fosters helps break down barriers of ignorance, and aids in overcoming conceptions based on hearsay and stereotypes.
- It is a **precondition for sustaining global economic growth**, since much of the modern economy is based on supply chains that reach around the world.

Asia is now the largest market for the installation of high capacity broadband networks. Rather than the cybercafés of Asian cities of the 1990s, which often attracted young residents simply to play video games, the new networks are expanding access to the information resources and e-commerce of the Internet. Modernization is a painful process for societies just encountering it, just as it was (and in some ways continues to be) painful for western societies that began to be affected by the industrial and scientific revolutions in the 1700's. But widespread communication helps to level the playing field and economic opportunity as the world economy inevitably changes. The disenfranchised of the third world will be more impoverished the more they are unable to communicate with, understand, and meaningfully participate in the changes taking place in the world.

But indications are that the digital divide is widening, not shrinking. The provision of high speed communication needs to be integrated with efforts to improve the infrastructures of developing nations, rather than to simply provide the same types of electronic entertainment prominent in developed countries. Unfortunately, the transfer of technology must be geared not to what products currently sell well in developed countries and produce the most profit, but the products that will actually enhance the lives of residents in the environment in which they live. This takes a coordinative thrust of private industry and governments capable of organizing multi-year, multi-disciplinary efforts. The result would be end-to-end solutions for societies much different from those of the developed western nations, whose cultures and economies do not revolve around the same priorities or assumptions as do those of the currently industrialized nations.

The digital divide exists even in developed nations, in which the people too poor or too poorly educated to benefit from modern e-commerce are effectively denied its benefits. These benefits are not limited simply to making more economical online purchases, but also result from the greater amount of awareness and information that digital communication makes available. Just as poorer areas are often those with the least market competition and the most limited access to modern financial services, the poorest segments of society are those most constrained from benefiting from the dissolving of information asymmetry that the Web provides.

Life will inevitably change for people as a result of the digitizing of information and communication. Whether these changes can be channeled in positive directions for more of the world's population, or will result in greater disparities of economic well-being and quality of life, depends on how well the benefits of the information age and modern societal infrastructures are shared by populations emerging from primitive and highly localized economies.

Lecture Slides 7

Introduction to Information Technology

E-Business

(C) 2006 Jim Janossy and Laura McFall, Information Technology Workbook

Course resources

PowerPoint presentations, podcasts, and web links for readings are available at
www.ambriana.com > IT Workbook

Print slides at 6 slides per page

Homework, quizzes and final exam are based on slides, lectures, readings and podcasts

Slide 7-2

Topics

- Definitions: e-business, value chain, B2B, B2C
- Internet era I, collapse, era II
- Marketing selling before the web, now
- Seven unique features of e-commerce
- E-commerce business models
- Internet irritations and dangers

Slide 7-3

Definitions

- **E-business:** the use of the internet and web to transact business *(limited definition)*
- E-business: any business process empowered by an information system *(broader definition)*
- Business processes along the whole value chain

Slide 7-4

Definitions

- **Value chain:** the generic value-adding activities of an organization.
- Manufacturing: purchasing, production processes, packaging, sales and marketing, order processing, customer service, maintenance
- Internet supports multiple parts of the value chain

Slide 7-5

Types of E-commerce

Distinct types of E-commerce

B2C
B2B
C2C
P2P

Business to consumer is web sales to retail consumers. Largest market in terms of quantity of customers, but only 10% of e-commerce revenue.

Slide 7-6

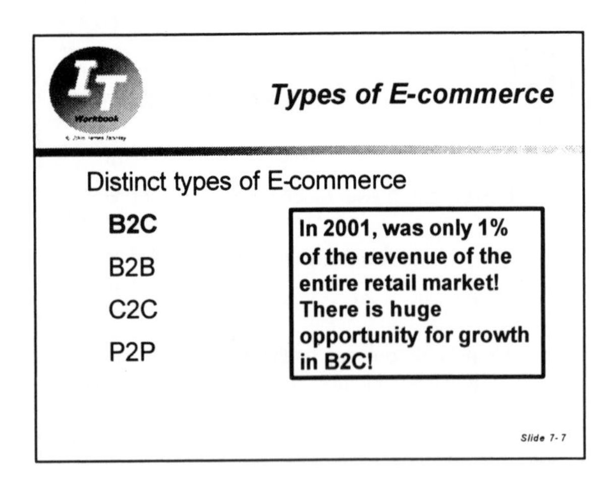
Types of E-commerce
Distinct types of E-commerce
B2C
B2B
C2C
P2P
In 2001, was only 1% of the revenue of the entire retail market! There is huge opportunity for growth in B2C!
Slide 7-7

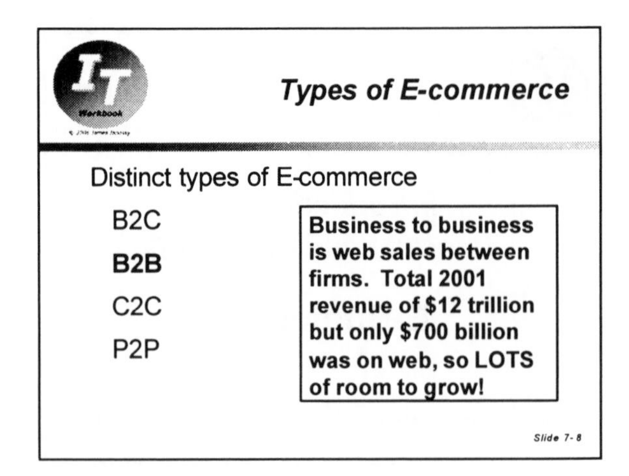
Types of E-commerce
Distinct types of E-commerce
B2C
B2B
C2C
P2P
Business to business is web sales between firms. Total 2001 revenue of $12 trillion but only $700 billion was on web, so LOTS of room to grow!
Slide 7-8

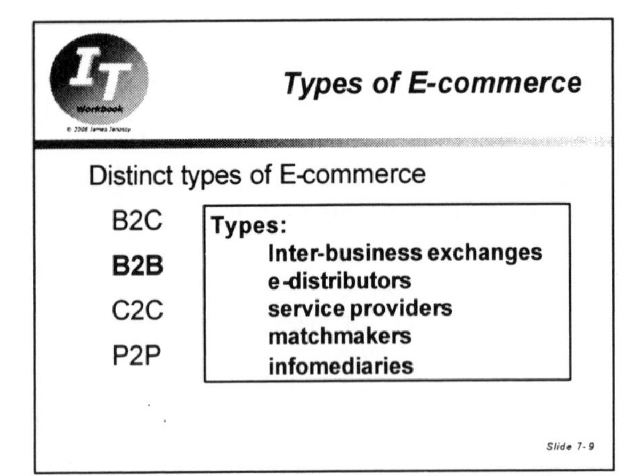
Types of E-commerce
Distinct types of E-commerce
B2C
B2B
C2C
P2P
Types:
Inter-business exchanges
e-distributors
service providers
matchmakers
infomediaries
Slide 7-9

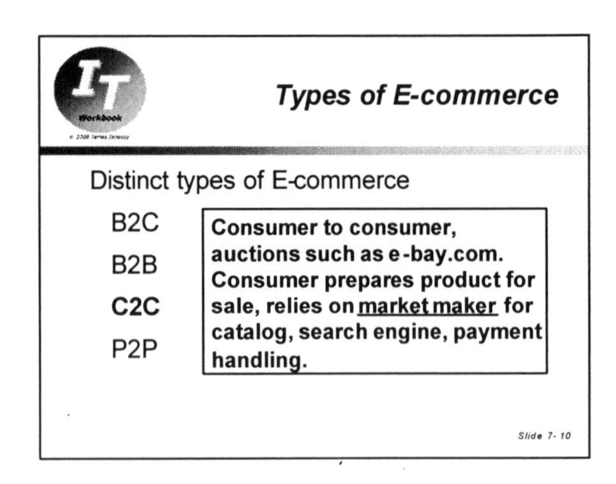
Types of E-commerce
Distinct types of E-commerce
B2C
B2B
C2C
P2P
Consumer to consumer, auctions such as e-bay.com. Consumer prepares product for sale, relies on market maker for catalog, search engine, payment handling.
Slide 7-10

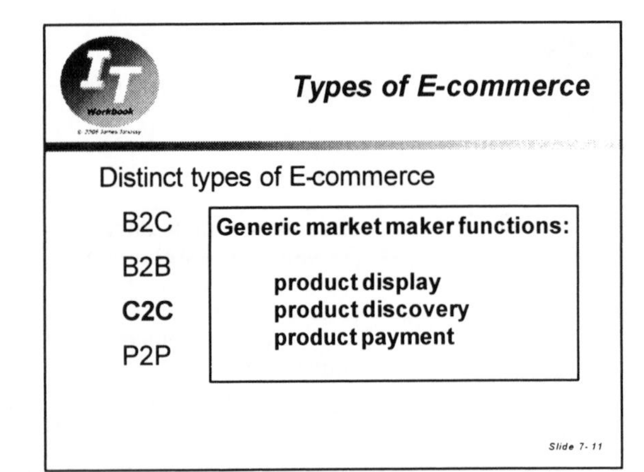
Types of E-commerce
Distinct types of E-commerce
B2C
B2B
C2C
P2P
Generic market maker functions:
product display
product discovery
product payment
Slide 7-11

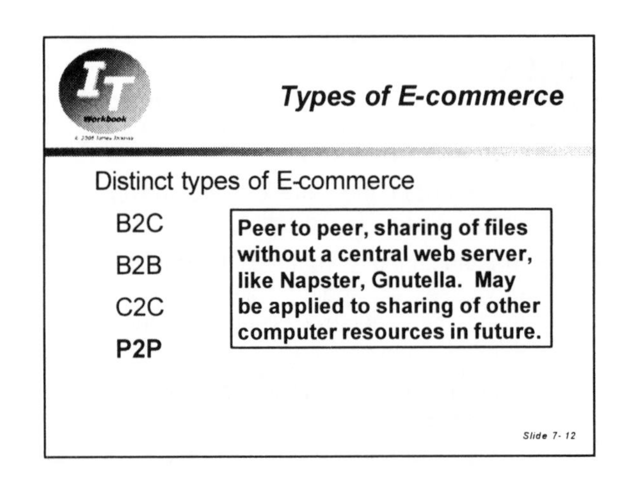
Types of E-commerce
Distinct types of E-commerce
B2C
B2B
C2C
P2P
Peer to peer, sharing of files without a central web server, like Napster, Gnutella. May be applied to sharing of other computer resources in future.
Slide 7-12

B2C vs. B2B

- **B2C**: business-to-consumer 10%
- **B2B**: business-to-business 90%
- **B2C** is most visible to the majority of the population but it is actually dwarfed by business-to-business transactions
- **B2B** via the internet/web is overtaking EDI (electronic data interchange)

Slide 7-13

E-commerce

- 1994: $ 0 2000: 60 billion B2C
- 700 billion B2B
- 1994: Internet use growing 2300% / yr
- Enormous changes in firms, markets, consumer behavior
- Fastest growing type of commerce

Slide 7-14

E-commerce growth

- 1994: $ 0 2000: 60 billion B2C
- **+420%** 700 billion B2B
- 2006: 250 billion B2C
- 5,400 billion B2C
- 4,000,000,000 web pages exist 2003
- 7,000,000 web pages added daily

Slide 7-15

E-commerce growth

- 1994: $ 0 2000: 60 billion B2C
- 700 billion B2B
- 2006: 250 billion B2C
- **+770%** 5,400 billion B2C
- 4,000,000,000 web pages now exist
- 7,000,000 web pages added daily

Slide 7-16

Growth compared to other technologies

B2C e-commerce:

Radio took **38 years** to achieve 30% household penetration (1920-58)

Television took **17 years** to achieve 30% household penetration (1946-63)

Web took only **7 years** to achieve 30% household penetration (1994-2001)

Slide 7-17

Internet Era I 1995 to 2000

- Pieces of underlying technology were all in place by 1994; browsers were last piece; Mosaic triggered explosion in users
- Overly-confident "cowboy" dot.coms
- Lots of swashbuckling investors
- Much dot.com activity in B2C area

Slide 7-18

Internet Era I 1995 to 2000 - Why?

Pattern of technological revolutions, as with electricity, telephone, radio, TV, cars

1. **Explosion** of entrepreneurial activity paves the way ("first mover" start-ups)
2. **Retrenchment**: weaker, less organized players exit while stronger take over
3. **Continued exploitation** by established firms

Slide 7-19

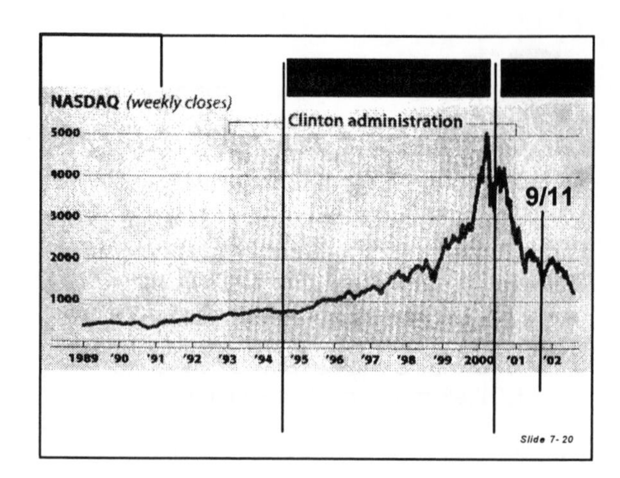

Slide 7-20

Internet Era I collapse factors

1. Many tech companies profited from Y2K efforts, suffered when clients were Y2K'd
2. Telecomm industry overbuilt capacity
3. Christmas 1999 had less sales growth than expected, shows high tech is hard!
4. Valuations of dot.com's too high, 400 x earnings (typical companies 10-15 x); many never showed **ANY** profit!

Slide 7-21

Fallacy of the "first mover" notion

Common idea in Era I - "**first movers**" can gain the market, will lose money first, then dominate; but it doesn't work that way!

Reality: being first isn't enough. You need to have a good business plan, act on it, and need much greater financial strength to develop mature markets

Slide 7-22

Plus... some ideas just aren't so good!

Some innovative ideas sounded good but just weren't viable (wishful thinking)

There are some things that people feel comfortable buying at a distance, and some things people buy in person

Some services are handy, some are too much trouble or inconvenient online

Slide 7-23

B2C E-commerce now...

is alive and well:

"...has moved into the mainstream life of ***established business concerns*** *that have the market brands and financial muscle required for long-term deployment of e-commerce technologies and methods."*

Slide 7-24

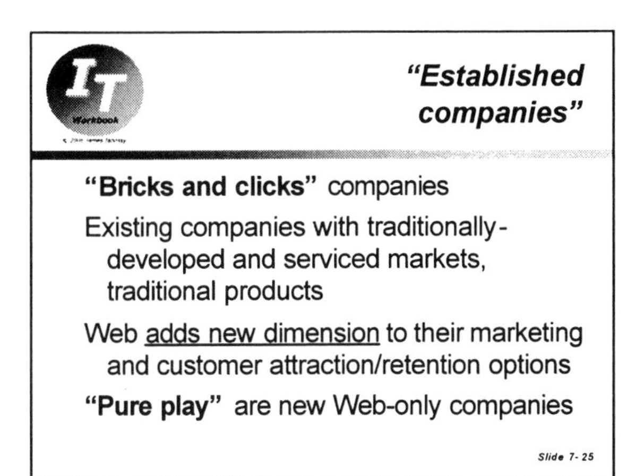
IT
Workbook
"Established companies"
"Bricks and clicks" companies
Existing companies with traditionally-developed and serviced markets, traditional products
Web adds new dimension to their marketing and customer attraction/retention options
"Pure play" are new Web-only companies
Slide 7-25

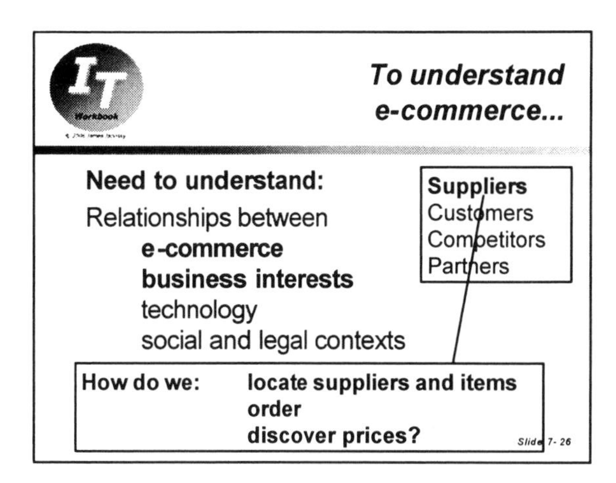
IT
Workbook
To understand e-commerce...
Need to understand:
Relationships between
e-commerce
business interests
technology
social and legal contexts
Suppliers
Customers
Competitors
Partners
How do we: locate suppliers and items
order
discover prices?
Slide 7-26

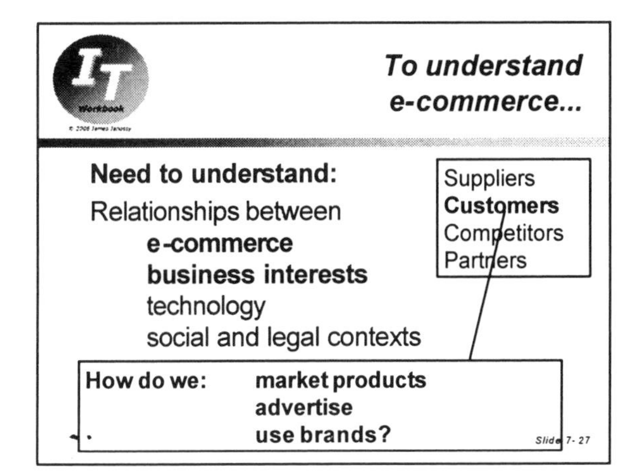
IT
Workbook
To understand e-commerce...
Need to understand:
Relationships between
e-commerce
business interests
technology
social and legal contexts
Suppliers
Customers
Competitors
Partners
How do we: market products
advertise
use brands?
Slide 7-27

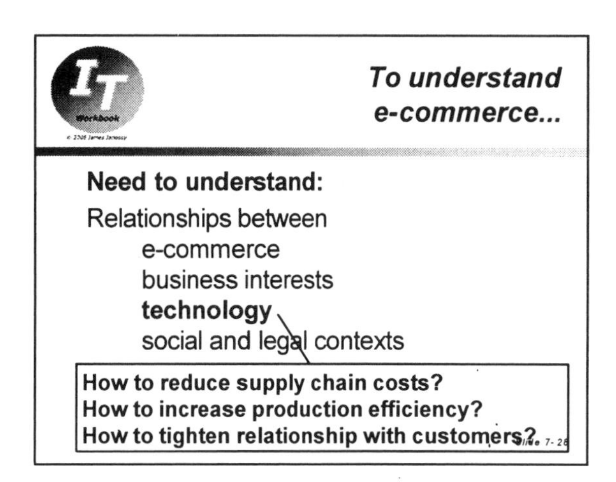
IT
Workbook
To understand e-commerce...
Need to understand:
Relationships between
e-commerce
business interests
technology
social and legal contexts
How to reduce supply chain costs?
How to increase production efficiency?
How to tighten relationship with customers?

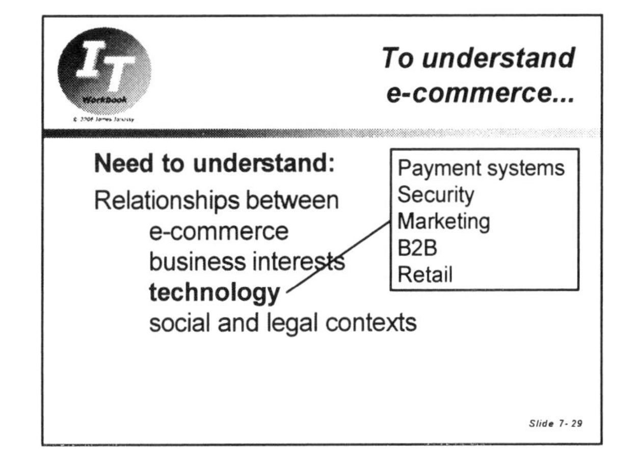
IT
Workbook
To understand e-commerce...
Need to understand:
Relationships between
e-commerce
business interests
technology
social and legal contexts
Payment systems
Security
Marketing
B2B
Retail
Slide 7-29

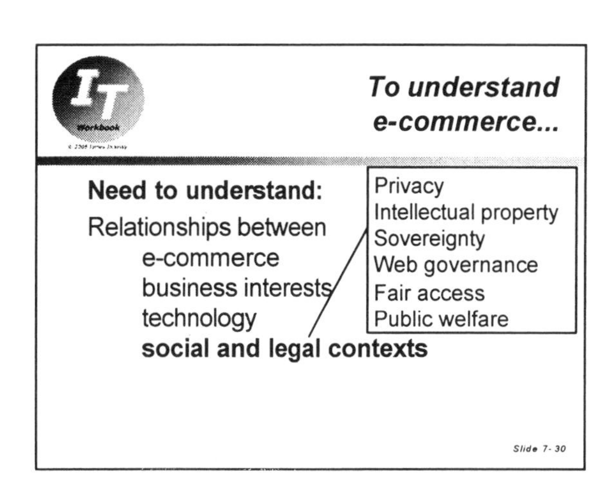
IT
Workbook
To understand e-commerce...
Need to understand:
Relationships between
e-commerce
business interests
technology
social and legal contexts
Privacy
Intellectual property
Sovereignty
Web governance
Fair access
Public welfare
Slide 7-30

Amazon.com

Founding ideas:

- Audience expanding (Web growth)
- Less need to touch and feel books to buy them than many other items
- Large source of supply (2,500 publishers)
- Largest stores had only 12% of market
- Major distributors stock books; no need for local inventory

Slide 7-31

Amazon.com

Founding determinations:

- Market exists
- Books can be sold at a distance
- No one else owns the source of supply
- Competition is not unified
- Distributors hold inventory; few premises needed, few employees: lower cost

Slide 7-32

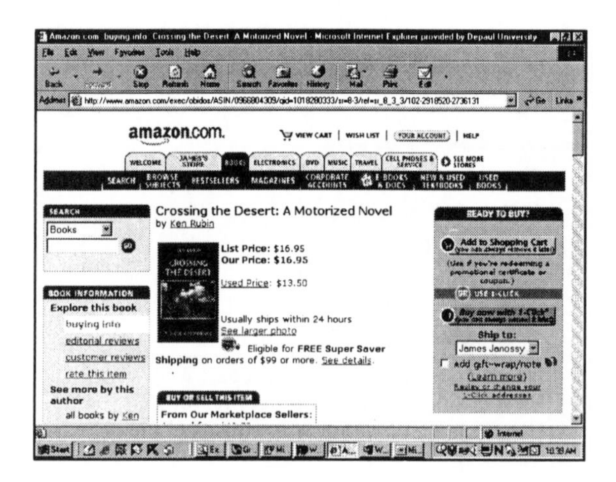

Amazon.com

Compelling factors for customers:

Selection: Million titles (books, CDs, DVDs)

Convenience: anytime, anywhere,simplified ordering ("1-click")

Price: discounts from regular retail price

Service: order confirmation e-mails, notifications of out of stock situations, affiliate (used book) vendors

Slide 7-34

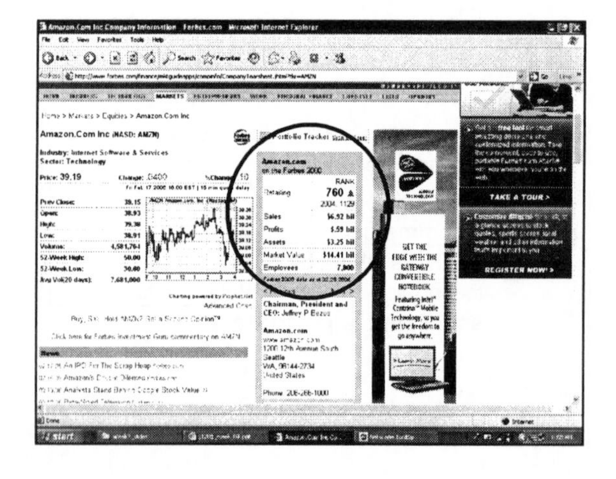

Amazon.com

Yet despite it's diversification into other product lines, in 2005 sales of books, CDs and DVDs still accounted for 70% of Amazon sales!

Slide 7-36

Information asymmetry is. . .

Any disparity in relevant market information among the parties in a transaction.

Slide 7-37

Marketing and selling before the web

Mass marketing

Salesforce driven

Consumers seen as **passive targets**

Campaigns and brands aimed to influence consumers product perceptions and purchasing behavior

Consumers trapped by geographical and social boundaries

Information asymmetry

Slide 7-38

Information asymmetry

Consumers unable to search widely for best price and quality

Information about...

- **prices**
- **costs**
- **fees**

...could be hidden from consumer!

Slide 7-39

Marketing and selling - now

Altered by 7 unique features of e-commerce

- Ubiquity
- Global reach
- Universal standards
- Richness
- Interactivity
- Information density
- Personalization / customization

Slide 7-40

Marketing and selling - now

Altered by 7 unique features of e-commerce

- **Ubiquity**
- Global reach
- Universal standards
- Richness
- Interactivity
- Information density
- Personalization / customization

"Marketspace" extends everywhere, including mobile. Shopping is 24x7 and shopper costs are reduced.

Slide 7-41

Marketing and selling - now

Altered by 7 unique features of e-commerce

- Ubiquity
- **Global reach**
- Universal standards
- Richness
- Interactivity
- Information density
- Personalization / customization

Commerce enabled across borders without modification

Slide 7-42

Marketing and selling - now

Altered by 7 unique features of e-commerce

- Ubiquity
- Global reach
- **Universal standards**
- Richness
- Interactivity
- Information density
- Personalization / customization

One set of communication technology, namely, Internet TCP/IP, HTML, browsers

Slide 7-43

Marketing and selling - now

Altered by 7 unique features of e-commerce

- Ubiquity
- Global reach
- Universal standards
- **Richness**
- Interactivity
- Information density
- Personalization / customization

Message is not limited to text or audio; video, audio, and text all possible, with visual cues

Slide 7-44

Marketing and selling - now

Altered by 7 unique features of e-commerce

- Ubiquity
- Global reach
- Universal standards
- Richness
- **Interactivity**
- Information density
- Personalization / customization

Consumer is engaged in a dialog, as a co-participant in discovering goods

Slide 7-45

Marketing and selling - now

Altered by 7 unique features of e-commerce

- Ubiquity
- Global reach
- Universal standards
- Richness
- Interactivity
- **Information density**
- Personalization / customization

Currency, timeliness, accuracy of information increases; price transparency

Slide 7-46

Marketing and selling - now

Altered by 7 unique features of e-commerce

- Ubiquity
- Global reach
- Universal standards
- Richness
- Interactivity
- Information density
- **Personalization / customization**

Messages possible to individuals not just groups; dialog can be tailored to appeal to individuals

Slide 7-47

Concurrent Era I visions

Thinking of many was:

- **Universal access**
- Info asymmetry reduced
- Middlemen disappear
- Extraordinary profits
- Easy to segment market
- Profit from efficiencies
- Deconstruct traditional distribution

Everyone would have a computer, web access, quickly

Slide 7-48

Concurrent Era I visions

Thinking of many was:
- Universal access
- **Info asymmetry reduced**
- Middlemen disappear
- Extraordinary profits
- Easy to segment market
- Profit from efficiencies
- Deconstruct traditional distribution

Friction-free commerce, huge number of suppliers

Slide 7-49

Concurrent Era I visions

Thinking of many was:
- Universal access
- Info asymmetry reduced
- **Middlemen disappear**
- Extraordinary profits
- Easy to segment market
- Profit from efficiencies
- Deconstruct traditional distribution

Disintermediation; manufacturers deal directly with consumers

Slide 7-50

Concurrent Era I visions

Thinking of many was:
- Universal access
- Info asymmetry reduced
- Middlemen disappear
- **Extraordinary profits**
- Easy to segment market
- Profit from efficiencies
- Deconstruct traditional distribution

Lots of ways to profit from large new markets and marketing strategies

Slide 7-51

Concurrent Era I visions

Thinking of many was:
- Universal access
- Info asymmetry reduced
- Middlemen disappear
- Extraordinary profits
- **Easy to segment market**
- Profit from efficiencies
- Deconstruct traditional distribution

Identify groups with different needs and price sensitivities

Slide 7-52

Concurrent Era I visions

Thinking of many was:
- Universal access
- Info asymmetry reduced
- Middlemen disappear
- Extraordinary profits
- Easy to segment market
- **Profit from efficiencies**
- Deconstruct traditional distribution

Price very low to grab market share, enable low pricing through new efficiencies

Slide 7-53

Concurrent Era I visions

Thinking of many was:
- Universal access
- Info asymmetry reduced
- Middlemen disappear
- Extraordinary profits
- Easy to segment market
- Profit from efficiencies
- **Deconstruct traditional distribution**

Gain visibility fast, "first movers" seek to replace traditional distribution channels

Slide 7-54

Internet Era II

- 2001 and onward
- New technological capabilities
- E-commerce learns from failures and successes of Era 1
- Predictions for the future?
- The impact of wireless (WiFi) access?

Slide 7-55

Understandings needed

- Nature of electronic markets
- E-commerce **business models**
- Firm and industry value chains
- Consumer behavior in e-markets
- Privacy, regulation, taxation issues

Slide 7-56

Business model

- Set of **planned activities** designed to make a profit
- Business models apply to all types of business, not just e-commerce
- **8 ingredients** of all business models
- **Business plan:** document describing the business model

Slide 7-57

Business plan

1. Value proposition
2. Revenue model
3. Market opportunity
4. Competitive environment
5. Competitive advantage
6. Market strategy
7. Organizational development
8. Management team

All factors of the **business model** are important to document in a **business plan**

Slide 7-58

Business plan

1. **Value proposition**
2. Revenue model
3. Market opportunity
4. Competitive environment
5. Competitive advantage
6. Market strategy
7. Organizational development
8. Management team

Heart of model. **How the product or service fulfills the needs of customers.** What is unique? Why use us?

Slide 7-59

Business plan

1. Value proposition
2. **Revenue model**
3. Market opportunity
4. Competitive environment
5. Competitive advantage
6. Market strategy
7. Organizational development
8. Management team

How will the business earn revenue, generate profit, produce a return?

Slide 7-60

Five major E-commerce revenue models

Advertising model
Subscription model
Transaction fee model
Sales model
Affliliate model

Site content and services draw hits. **Sell advertising, banners, links.** User retention is called "stickiness". One of earliest models, now copied and diluted.

Slide 7-61

Five major E-commerce revenue models

Advertising model
Subscription model
Transaction fee model
Sales model
Affliliate model

Subscription charge for content or service. **Content must have high added value**, not readily available elsewhere, no easy substitutes.

Slide 7-62

Five major E-commerce revenue models

Advertising model
Subscription model
Transaction fee model
Sales model
Affliliate model

Fee for enabling or executing a transaction. Like Ebay.com for auction, or E-trade.com for a stock buy or sell.

Slide 7-63

Five major E-commerce revenue models

Advertising model
Subscription model
Transaction fee model
Sales model
Affliliate model

Sell goods.
Physical: office supplies, books, crafts, foods, anything that could be sold mail order.
Electronic: product is downloaded.
Services: software usage is "rented"

Slide 7-64

Five major E-commerce revenue models

Advertising model
Subscription model
Transaction fee model
Sales model
Affiliate model

Referral fee or % commission for steering hits to a site. Maybe give "points" incentive to customers to shop via the site links instead of directly to sites.

Slide 7-65

Business plan

1. Value proposition
2. Revenue model
3. **Market opportunity**
4. Competitive environment
5. Competitive advantage
6. Market strategy
7. Organizational development
8. Management team

Intended marketspace. Realistic opportunity defined by revenue potential of each **niche.**

Slide 7-66

Business plan

1. Value proposition
2. Revenue model
3. Market opportunity
4. **Competitive environment**
5. Competitive advantage
6. Market strategy
7. Organizational development
8. Management team

How many competitors? How large? Market share? How profitable? Their price?

Slide 7-67

Business plan

1. Value proposition
2. Revenue model
3. Market opportunity
4. Competitive environment
5. **Competitive advantage**
6. Market strategy
7. Organizational development
8. Management team

Having a: superior product, lower price, wider market, branding.

Slide 7-68

Business plan

1. Value proposition
2. Revenue model
3. Market opportunity
4. Competitive environment
5. Competitive advantage
6. **Market strategy**
7. Organizational development
8. Management team

Plan that shows how you intend to enter a new market and attract customers. **Partnering? Advertising? Samples?**

Slide 7-69

Business plan

1. Value proposition
2. Revenue model
3. Market opportunity
4. Competitive environment
5. Competitive advantage
6. Market strategy
7. **Organizational development**
8. **Management team**

Plan that shows how the business is going to be staffed as it grows, and how management leads.

Slide 7-70

Let's look at these Internet business models

- **B2C - business to consumer**
- **B2B -business to business**
- C2C - consumer to consumer
- P2P - peer to peer

Slide 7-71

_B2C_

1. **Portal**
2. E-tailer
3. Content provider
4. Transaction broker
5. Market creator
6. Service provider
7. Community provider

Integrated package of content and services, also ISP. E-mail, news, chat rooms, personals, shopping. No longer a "**gateway**" but a **destination**.

Slide 7-72

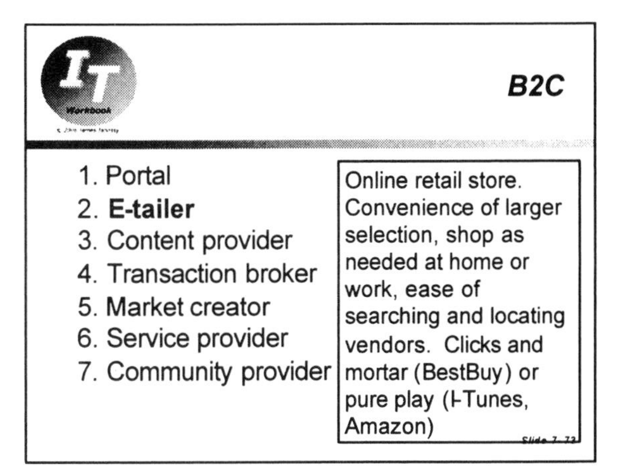
B2C
1. Portal
2. E-tailer
3. Content provider
4. Transaction broker
5. Market creator
6. Service provider
7. Community provider
Online retail store. Convenience of larger selection, shop as needed at home or work, ease of searching and locating vendors. Clicks and mortar (BestBuy) or pure play (I-Tunes, Amazon)

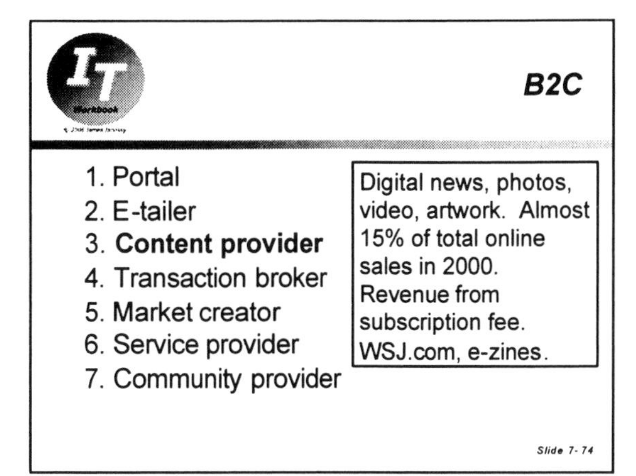
B2C
1. Portal
2. E-tailer
3. Content provider
4. Transaction broker
5. Market creator
6. Service provider
7. Community provider
Digital news, photos, video, artwork. Almost 15% of total online sales in 2000. Revenue from subscription fee. WSJ.com, e-zines.
Slide 7-74

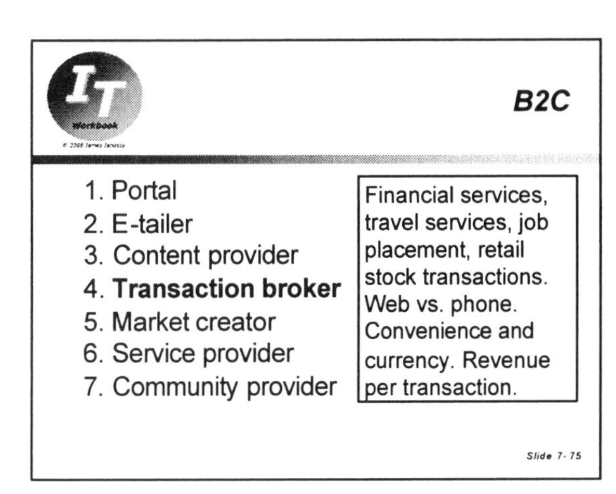
B2C
1. Portal
2. E-tailer
3. Content provider
4. Transaction broker
5. Market creator
6. Service provider
7. Community provider
Financial services, travel services, job placement, retail stock transactions. Web vs. phone. Convenience and currency. Revenue per transaction.
Slide 7-75

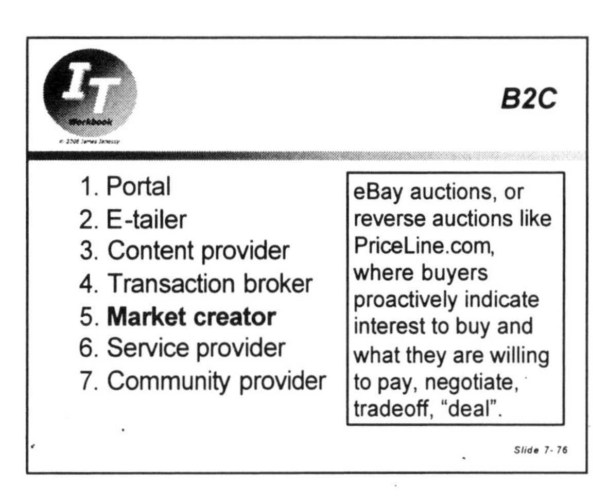
B2C
1. Portal
2. E-tailer
3. Content provider
4. Transaction broker
5. Market creator
6. Service provider
7. Community provider
eBay auctions, or reverse auctions like PriceLine.com, where buyers proactively indicate interest to buy and what they are willing to pay, negotiate, tradeoff, "deal".
Slide 7-76

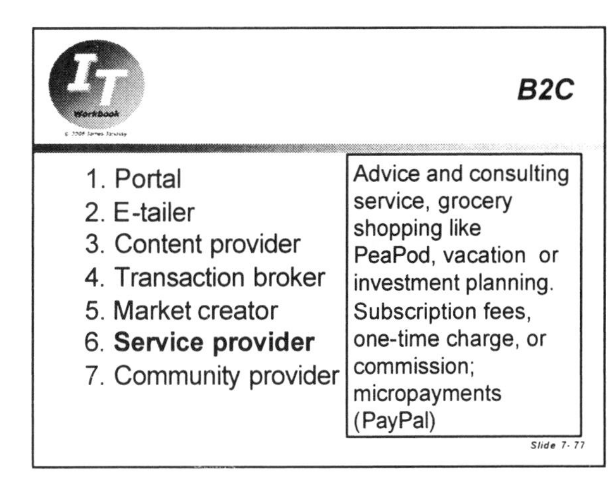
B2C
1. Portal
2. E-tailer
3. Content provider
4. Transaction broker
5. Market creator
6. Service provider
7. Community provider
Advice and consulting service, grocery shopping like PeaPod, vacation or investment planning. Subscription fees, one-time charge, or commission; micropayments (PayPal)
Slide 7-77

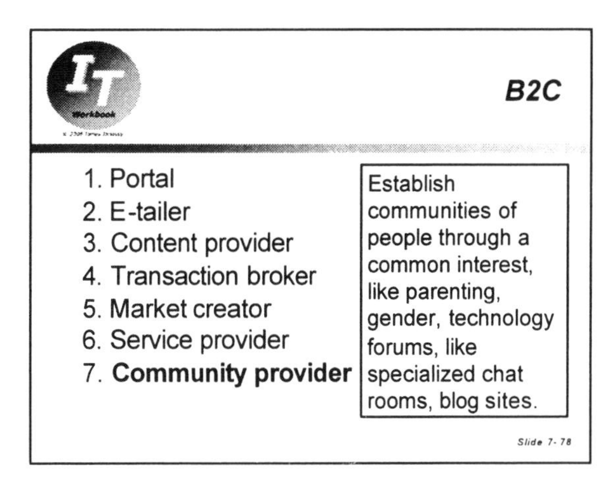
B2C
1. Portal
2. E-tailer
3. Content provider
4. Transaction broker
5. Market creator
6. Service provider
7. Community provider
Establish communities of people through a common interest, like parenting, gender, technology forums, like specialized chat rooms, blog sites.
Slide 7-78

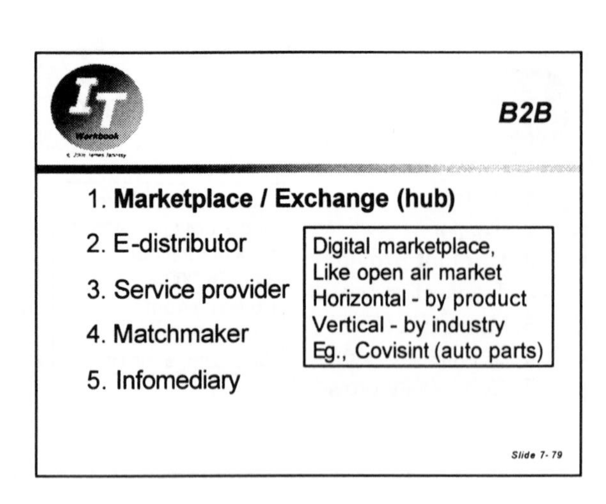
B2B
1. Marketplace / Exchange (hub)
2. E-distributor
3. Service provider
4. Matchmaker
5. Infomediary
Digital marketplace,
Like open air market
Horizontal - by product
Vertical - by industry
Eg., Covisint (auto parts)
Slide 7-79

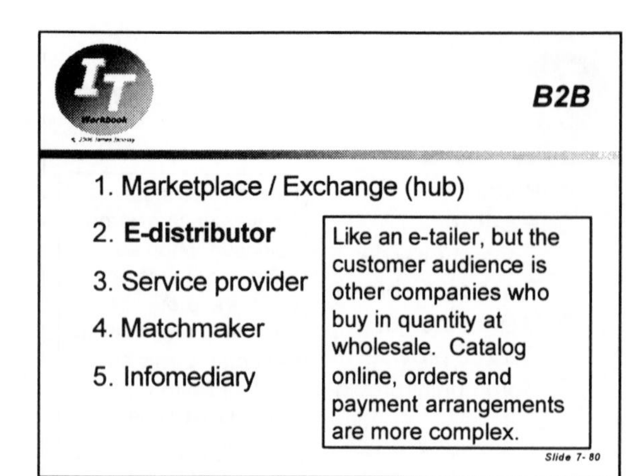
B2B
1. Marketplace / Exchange (hub)
2. E-distributor
3. Service provider
4. Matchmaker
5. Infomediary
Like an e-tailer, but the
customer audience is
other companies who
buy in quantity at
wholesale. Catalog
online, orders and
payment arrangements
are more complex.
Slide 7-80

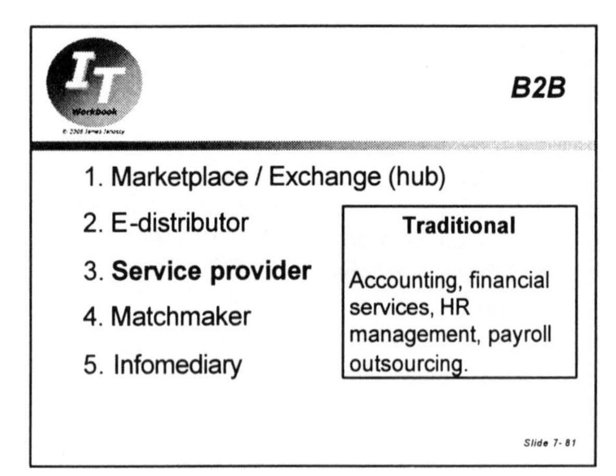
B2B
1. Marketplace / Exchange (hub)
2. E-distributor
3. Service provider
4. Matchmaker
5. Infomediary
Traditional
Accounting, financial
services, HR
management, payroll
outsourcing.
Slide 7-81

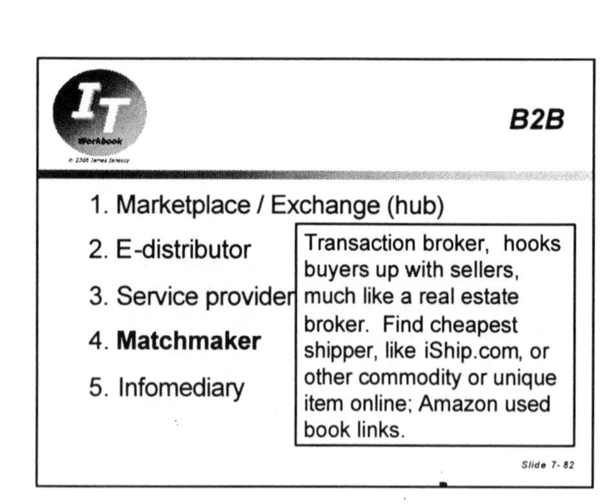
B2B
1. Marketplace / Exchange (hub)
2. E-distributor
3. Service provider
4. Matchmaker
5. Infomediary
Transaction broker, hooks
buyers up with sellers,
much like a real estate
broker. Find cheapest
shipper, like iShip.com, or
other commodity or unique
item online; Amazon used
book links.
Slide 7-82

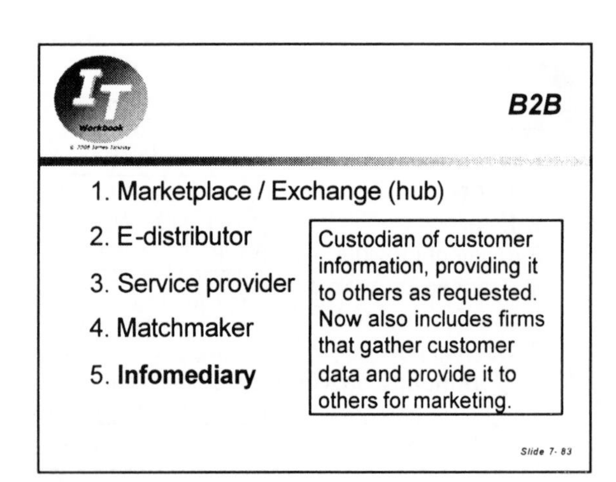
B2B
1. Marketplace / Exchange (hub)
2. E-distributor
3. Service provider
4. Matchmaker
5. Infomediary
Custodian of customer
information, providing it
to others as requested.
Now also includes firms
that gather customer
data and provide it to
others for marketing.
Slide 7-83

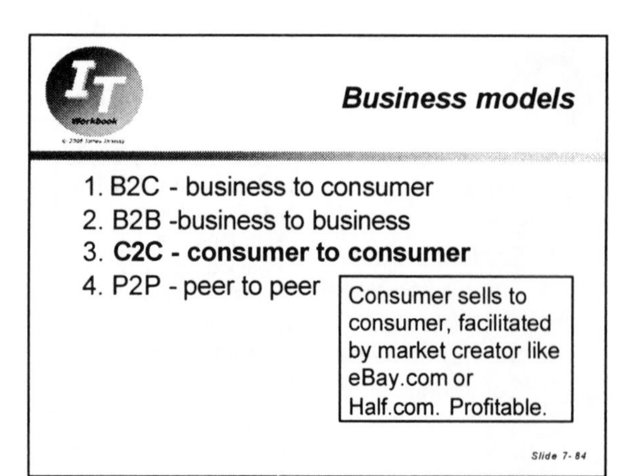
Business models
1. B2C - business to consumer
2. B2B -business to business
3. C2C - consumer to consumer
4. P2P - peer to peer
Consumer sells to
consumer, facilitated
by market creator like
eBay.com or
Half.com. Profitable.
Slide 7-84

Business models

1. B2C - business to consumer
2. B2B -business to business
3. C2C - consumer to consumer
4. **P2P - peer to peer**

Information sharing between consumers. Runs into legal problems when sharing things that don't belong to you!

Slide 7- 85

Internet dangers

- Con artists exploiting the gullible, innocent, or "unaware"
- Real dangers of encouraging naïve users ("newbies") to reveal too much personal information
- "The world's biggest bathroom wall."

Slide 7- 86

Internet dangers

- Blogs, "community" sites easily exploited by criminals
- How is this to be prevented?

Slide 7- 87

Spam, spoofing, phishing

- **Spam:** unsolicited advertising e-mail
- **Spoofing:** manipulating the apparent sender's name to make spam look like
- **Phishing:** spam that spoofs an address of a bank or organization and asks you to "confirm" account details by entering personal information

Slide 7- 88

Opt-in, permission marketing

- **Opt-in:** you agree to receive advertising, sales promotions, e-mails from a preferred vendor
- **Permission marketing:** a synonym for this phenomenon
- Term was coined 1997 by Seth Godin in a book of this name

Slide 7- 89

Opt-out, dangers

- **Opt-out:** offers you a way to stop receiving more of the same
- Believable from a reputable vendor, especially one you have opted-in with
- Dangerous to accept the "opt out" button click on spam, since it just confirms that your address is "live"!

Slide 7- 90

Internet inequalities

- "Digital Divide" is the gap between those who can participate meaningfully in the digitized world and those who are isolated from it
- What are the impacts of lack of access?
- What are the cultural impacts of modernization?

Slide 7- 91

Internet displacements

- State governments and cities derive a portion of their revenues from sales taxes on goods sold; vendors collect and remit to state
- Commerce "hides" from state taxes
- As e-commerce has grown, has negatively affected sales tax revenue

Slide 7- 92

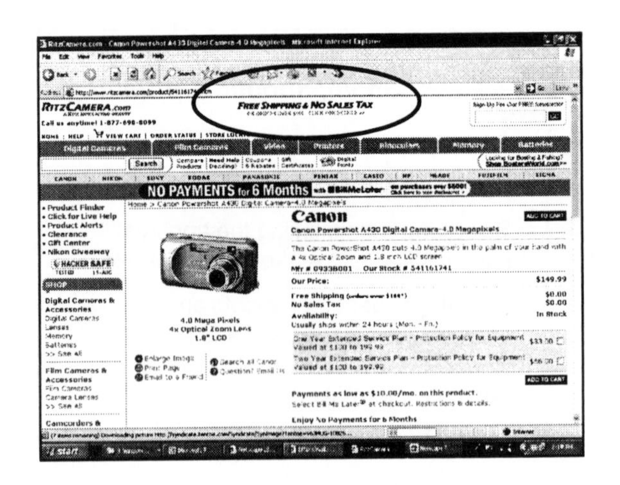

IT Workbook Assignment 7.1

Student name:

Write your answers to the questions below <u>within the box</u>. In each case, please choose your words carefully to answer the specific questions. Avoid simply copying passages from your readings to these answers.

Provide a limited definition of **e-business**, and a broad definition:

1) a limited definition of e-business:

2) a broad definition of e-business:

Provide a definition of the **value chain**, and several things in the value chain of a typical commercial organization:

3) definition of the value chain:

4) one thing in the value chain:

5) another thing in the value chain:

6) another thing in the value chain:

7) another thing in the value chain:

8) another thing in the value chain:

IT Workbook Assignment 7.2

Student name:

Write your answers to the questions below within the box. In each case, please choose your words carefully to answer the specific questions. Avoid simply copying passages from your readings to these answers.

EDI and **EFT** are two early implementation of electronic supports for business. Identify what each of these acronyms stands for, what **EDI documents** contain and why they remain in use even though the Internet has replaced the earlier transmission methods for them:

1) EDI stands for:

2) EFT stands for:

3) what EDI documents contain:

4) why EDI documents remain in use even though the Internet has replaced the earlier transmission methods for them:

Describe the **service** provided by www.PayPal.com (now a subsidiary of Ebay) and why it has flourished (what need it satisfies):

5) service provided:

6) why it has flourished (what need it satisfies):

IT Workbook Assignment 7.3

Student name:

Write your answers to the questions below <u>within the box</u>. In each case, please choose your words carefully to answer the specific questions. Avoid simply copying passages from your readings to these answers.

Identify what "**bricks and mortar**" companies are, what "**bricks and clicks**" companies are, and what "**pure play**" companies are, and provide a brief description of each:

1) bricks and mortar companies are:

2) bricks and clicks companies are:

3) pure play companies are:

Define what **information asymmetry** is, and <u>what effect</u> the World Wide Web has on it:

4) information asymmetry is:

5) the effect of the World Wide Web on information asymmetry:

IT Workbook Assignment 7.4

Student name:

Write your answers to the questions below within the box. In each case, please choose your words carefully to answer the specific questions. Avoid simply copying passages from your readings to these answers.

E-commerce has altered marketing and selling with **seven unique features**. Provide a definition and brief description or example of each of these features:

1) ubiquity:

2) global reach:

3) universal standards:

4) richness:

5) interactivity:

6) information density:

7) personalization and customization:

IT Workbook Assignment 7.5

Student name:

Write your answers to the questions below within the box. In each case, please choose your words carefully to answer the specific questions. Avoid simply copying passages from your readings to these answers.

At digitalenterprise.org/models/models.html Michael Rappa identifies ten e-commerce business models. Identify and provide a brief description of each:

1)

2)

3)

4)

5)

6)

7)

8)

9)

10)

IT Workbook Assignment 7.6

Student name:

Write your answers to the questions below within the box. In each case, please choose your words carefully to answer the specific questions. Avoid simply copying passages from your readings to these answers.

From the point of view of **consumers**, identify and briefly describe **three important advantages** that the digital markets created by the Internet provide:

1)

2)

3)

From the point of view of **sellers**, identify and briefly describe **five important advantages** that the digital markets created by the Internet provide:

4)

5)

6)

7)

8)

IT Workbook Assignment 7.7

Student name:

Write your answers to the questions below within the box. In each case, please choose your words carefully to answer the specific questions. Avoid simply copying passages from your readings to these answers.

Provide the **meaning of the acronym**, and a **brief description** of each of these types of e-commerce:

1) B2C:

2) B2B:

3) C2C

4) P2P

Compare the size (dollar volume) of **B2C** e-commerce to the size of **B2B** e-commerce:

5)

Explain what the term **disintermediation** means in connection with e-commerce:

6)

IT Workbook Assignment 7.8

Student name:

Write your answers to the questions below within the box. In each case, please choose your words carefully to answer the specific questions. Avoid simply copying passages from your readings to these answers.

Describe what is at the core of **public-key encryption**, how it works in making sure that the content of a document has not been altered in transit, how public-key encryption operates in confirming the identity of the sender of a document, and how it operates in keeping the contents of an e-mail secret:

1) what is at the **core** of public-key encryption:

2) how public-key encryption operates in making sure that the content of a document **has not been altered** in transit:

3) how public-key encryption operates in **authenticating the identity of the sender** of a document:

4) how it operates in **keeping the contents of an e-mail secret**:

IT Workbook Assignment 7.9

Student name:

Write your answers to the questions below within the box. In each case, please choose your words carefully to answer the specific questions. Avoid simply copying passages from your readings to these answers.

At www.digitaldivide.org/digitaldivide.html you will find a definition of the term "**digital divide**." Summarize this definition here and explain in your own words how our understanding of this phenomenon has changed in the last several years:

1) definition:

2) how our understanding has changed in the last several years:

Discuss **three reasons** that it is **essential to close the digital divide**, and what your own opinions are in regard to the validity of these reasons:

3)

4)

5)

6) Your **opinions** in regard to the validity of these reasons:

IT Workbook Assignment 7.10

Student name:

Write your answers to the questions below within the box. In each case, please choose your words carefully to answer the specific questions. Avoid simply copying passages from your readings to these answers.

At www.digitaldivide.org/truths.html you will find a discussion of nine "**digital divide truths**." Identify and summarize each of these truths:

1)

2)

3)

4)

5)

6)

7)

8)

9)

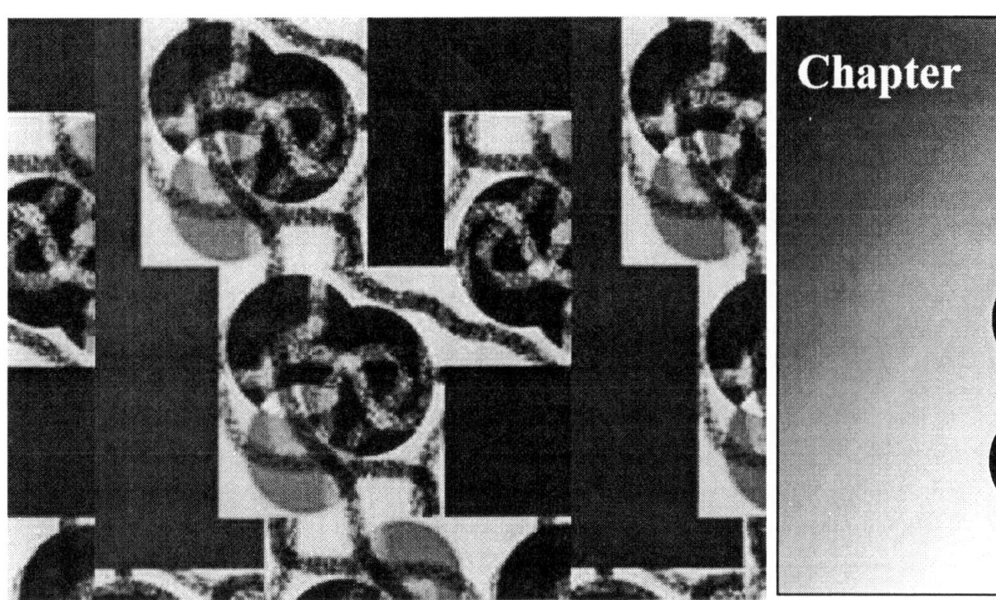

Chapter 8

Modern Information Systems (ERP, CRM, EIS, GIS)

Information systems exist to enable the interaction of an organization with data about its products, services, and customers. In manufacturing industries this means keeping track of supplies of raw materials and the labor and processes to transform them into finished goods. In retail enterprises this means keep track of stock, the wholesale vendors from which it is purchased, and accounting for sales transactions whereby the stock is sold to customers. For educational organizations it means handling all of the myriad tasks associated with recruiting students, organizing, scheduling and conducting classes, recording grades, and issuing transcripts. For all organizations it means recording revenues and expenses in accordance with accounting principles, paying its employees, and maintaining records for tax purposes, and many other similar tasks.

In earlier days many organizations created their own business computer programs. Now organizations often license software for these purposes from vendors. This may be integrated across many business areas as an **enterprise resource planning** (ERP) system. An ERP system may integrate raw materials supply, warehousing, accounting, shipping, marketing, finance, and **customer relationship management** (CRM) in a large common database.

A business system accumulates data as it handles transactions such as orders and payments. Reporting is done from this information for daily operations and to inform organization management of financial status. Reporting is done over longer periods to better understand trends in the business. Trend and status data from internal and external sources is often summarized for management in graphic form by an **Executive Information System** (EIS), and combined with a **Geographic Information System** (GIS) to associate it with geographical areas and topological features.

Learning objectives

Chapter 8
Modern Information Systems (ERP, CRM, EIS, GIS)

This chapter, in combination with the web readings, web links, and podcasts available at www.ambriana.com is focused on providing the knowledge to achieve these learning objectives:

- Understand what an Enterprise Resource Planning (ERP) system is, its origin in the manufacturing environment, and the integration it attempts to achieve across an organization
- Develop an appreciation of the changes that the implementation of an ERP forces in an organization, and its advantages, disadvantages, and hidden costs
- Develop a basic understanding of the role and use of Customer Relationship Management (CRM) systems
- Develop a basic understanding of Executive Information Systems (EIS) and the role they play in decision-making
- Develop a basic understanding Geographic Information Systems (GIS)
- Understand the principle of data ethics that data should be used only for the purpose for which it was collected, the tension that exists in the modern environment for broader use of data for marketing purposes, and the definition of opt-in and opt-out permission marketing arrangements
- Recognize the various careers and career paths that exist in the modern information systems environment.

Podcast Scripts – Chapter 8

These readings of the **Information Technology Workbook** are also accessible in audio form as .mp3 files for free download. Visit the workbook web site, www.ambriana.com. You can listen to this material on your own Ipod or .mp3 player or computer, read it, or both listen and read.

Podcast 56
ERP SYSTEM OVERVIEW

Enterprise resource planning (ERP) describes a broad set of activities, supported by multi-module application software that helps a business manage important parts of its operation. ERP systems typically handle the manufacturing, logistics, distribution, inventory, shipping, invoicing, and accounting operations for a company. ERP software can aid in the control of a number of business activities, including sales, delivery, billing, production, and inventory, quality control, and human resources management. As pundits often note, the most accurate part of the acronym EPR lies in the first letter, which stands for "enterprise." ERP's aim to serve all of the most critical needs of the enterprise as a whole.

ERP's are frequently called **back office systems**, an indication that customers and the general public are not directly involved with them. Contrast this with front office systems such as customer relationship management (CRM) systems that deal directly with the customers, or various e-business systems that deal directly with suppliers.

ERP's are integrated, cross-functional, and enterprise-wide. **Integrated** means that the data needs of all parts of the organization are met with one consolidated database, rather than with separate copies of data for different departments or divisions. ERPs replace the several separate inventory, sales, order-fulfillment, accounting and billing systems that evolved in organizations since computers were commercialized in the 1950's. **Cross-functional** means that the same data is accessed for different purposes by different parts of the organization, and the same software—the many modules of the ERP system—provide the type of access and data visibility required by each part of the organization. **Enterprise-wide** means that when the ERP is implemented, all parts of the organization involved in operations or production share the same understanding of the data, and "information silos" that formerly existed in various areas are replaced by the ERP. This includes manufacturing, warehousing, logistics, information technology, accounting, human resources, marketing, and strategic management.

Because of their wide scope of application within the firm, ERP software systems rely on some of the largest bodies of software ever written. Until the development of the Internet, implementing such a large and complex software system in a company often involved an army of analysts and programmers. The Internet has eased this burden to some extent, because it provides the means for outside software firms to gain access to company computers in order to install standard software updates. The Internet has facilitates flexible communication between workstations—that is, desktop computers—and the ERP, since most ERP's now support a web interface, and any computer running a standard browser can access the ERP.

Some of the bigger players in the ERP market are SAP, Peoplesoft, J. D. Edwards, Oracle, IBM, and Microsoft. Most of these vendors can also supply consulting services to assist firms in preparing for ERP implementation and actually accomplishing it. Independent third-party consulting firms also exist, often specializing in one vendor's line of ERP products, and firms requiring assistance can usually obtain it from such consultants. Such consulting firms often command high rates of pay for their specialized services.

Enterprise resource planning systems are often closely tied to supply chain management and logistics automation systems. Supply chain management software can extend the ERP system to include links with suppliers outside the company, increasing efficiencies in that portion of the operation. With this type of linkage, for example, a manufacturer's ERP may be able to check the inventory levels of various suppliers, and automatically place orders for parts needed to keep an assembly line running smoothly with frequent "just in time" deliveries used in manufacturing as soon as they arrive, minimizing the need for storage.

The payoff for implementing an ERP system is an improvement in the efficiency and competitiveness of a business organization. Once the hurdles involved in an ERP implementation are surmounted, and the organization regains productivity lost to the change in business workflows, the advantages begin to accrue. These include a more rapid ability to shift production to meet shifting demand, efficiencies in volume purchasing and the ability to accomplish just-in-time deliveries of raw materials, and the provision of comprehensive data on organizational operation. A successful ERP implementation takes serious planning and investment. Well handled, the ERP becomes the backbone of the firm. Poorly implemented, an ERP can kill a firm, because it touches and changes so many aspects of its operation at once.

Podcast 57
ERP PRE-INSTALLATION

ERP implementation can be a complex project for large companies, especially those doing business across international boundaries. Consulting companies specializing in ERP implementation can expedite this process. To implement ERP systems, companies often seek the help of an ERP vendor or a third-party consulting company. Consulting in ERP involves two levels, **business consulting** and **technical consulting**.

A **business consultant** studies an organization's current business processes and may assist in evaluating competing ERP systems, to help determine which is the closest match to the organization's needs. This is part of the due diligence that must precede software acquisition. If this vital process is not performed well, the ERP system that is acquired may not suit the organization, and implementing it will be like trying to put a square peg in a round hole. A business consultant may also assist during implementation of the selected ERP system, to configure the settings it provides to best suit the organization's business processes. This consists of loading the appropriate values to system tables to define charts of account, to consolidate and load stock identification codes, tables that supply descriptions for various coded values, and software "switches" (parameters) that tailor the operation of the software. These elements are external to the actual program code of a system, and are intended by the ERP vendor to allow setting up the system to be a close fit to the organization's business practices.

Technical consulting often involves programming to modify vendor-supplied software to suit specific business needs for which the ERP vendor has not provided support. Such programming changes are known as **customizations**. Customizing an ERP package can be expensive and complicated, because many ERP packages are not designed with this in mind. ERP vendors encourage businesses to implement the best industry practices already embedded in the ERP system they have purchased, and to not modify the actual program code of the system. Yet it is almost inevitable that instances will arise where accepting the "standard" way that an ERP handles a situation will result in loss of some advantageous or efficient business practice that distinguishes a firm from its competitors. Therefore, it is almost inevitable that tension will arise between the business units and what they perceive their needs to be, and the implementing organization, often the information technology division. The implementation team strives to implement a "plain vanilla" (no customizations) ERP package, because customizations can seriously complicate the upgrading of the software as the vendor releases new versions. Customizations made once to an ERP must be reinstalled, and tested, every time the vendor releases a new version of the plain vanilla software, seriously increasing the burden of ongoing software maintenance.

Some ERP packages are generic in their reports and inquiries, and customization in that aspect of operation is expected. It's important to recognize that for these software packages it often makes sense to license a reporting software package that interfaces well with a particular ERP, rather than to reinvent such a reporting software suite. A data warehouse implementation is often an appealing solution to reporting from an ERP, since the data warehouse is designed to accumulate historical data. An ERP is essentially a transaction processing system that is tuned for rapid online processing. The data structures used to support an ERP are often not well suited to analytical reporting. The method by which the reporting needs of the organization will be met needs to be considered as a part of implementation planning and budgeting.

To recap, pre-installation tasks for an ERP system include business consulting and technical consulting. Business consulting helps understand the organization's business processes and select and adapt an ERP to suit them. Technical consulting deals with changing the software of the ERP itself, and is usually held to a minimum in the interest of avoiding long term software maintenance issues.

Podcast 58
ERP PITFALLS

Three major pitfalls exist in ERP system implementation. These are inappropriate fit to the organization's business practices, insufficient corporate policies protecting the integrity of the data, and inadequate attention to user training.

ERP's are often seen as too rigid, and difficult to adapt to the specific workflow and business processes of an organization. This is, in fact, cited as one of the main causes of their failure. Very often this complaint means that the problem lies with **inadequate analysis in advance of ERP software selection**, or the assumption by non-technical management that an ERP can easily be reprogrammed to meet the existing practices of the business. And sometimes, ERP vendor salespeople may give the impression that the flexibility of the software is great enough to accommodate any variation in business practices, when in fact this is not the case. ERP's have been developed by analysis of common business practices. If the ERP is not a good fit for the organization, and this is discovered after implementation, it means due diligence was not performed. Making such a package fit will be painful and expensive, and may actually damage the productivity of the firm for a long time.

Problems with **data integrity** after ERP installation can often be due to a rushed implementation, and mistaking setup configuration changes for customization. ERP systems have been designed to allow an organization to make choices in many types of data validations to tailor these in leniency or stringency to the way the organization operates. For example, an ERP function that adds new suppliers to a procurement system table usually performs a scan of existing suppliers to see if the supplier being entered already exists on the system. The scan usually involves "fuzzy" matching of name, address, and other identifiers so that variations in spelling and common errors do not prevent locating possible matches. The ERP system can be set to simply warn if a potential match exists, and permit entry of the new supplier, or block entry due to the apparent attempt to make a duplicate entry. If the ERP as it comes from the vendor sets the validation to simply warn of a potential duplicate entry, but to allow entry, in the interest of making the entry process faster, data integrity issues can later arise and plague the system. A setup choice to make the validation block duplicate entries is incorrectly regarded as a customization, which is frowned upon, because it complicates later maintenance. But if this is simply a configuration setting, it is provided to allow the organization to make the system work as it needs it to. In their initial zeal to remain "plain vanilla" the technical implementers in some organizations, and even at times, inadequately experienced implementation consultants, do not make appropriate configuration settings, and doom the ERP to accumulate data integrity programs that later have an adverse effect.

Many of the problems organizations have with ERP systems are due to the inadequate investment in appropriate and specific training for all personnel involved, including those implementing and testing changes that flow from the vendor, which are known as "patches." The success of ERP systems depends a great deal on the skill and experience of the entire workforce. Unfortunately, as implementation costs rise, some companies cut costs by cutting training budgets. In addition, some small, privately owned enterprises are undercapitalized, meaning their ERP system may be operated by personnel with inadequate education in ERP in general, and in the particular ERP vendor package being used. Appropriate training is often not the type of "canned" training provided by consultants who do not know the business of the organization. Users need to be provided with training tailored to their business function, not the nature of ERP systems in general, or how every button, bell, and whistle works.

To recap, the three major pitfalls in ERP system implementation are:

1. Inappropriate fit to the organization's business practices, due to a rushed or inadequate analysis and selection process
2. Insufficient corporate policies or ERP configuration settings to protect the integrity of data
3. Inadequate or inappropriate user training.

Even with all the challenges an ERP system may present, ERP's are still widely used, and highly praised by those companies who make them work well. But making them work well does not happen accidentally, and it is not usually inexpensive and painless.

Podcast 59
CRM SYSTEMS

CRM stands for Customer Relationship Management, a collection of techniques and processes by which an organization seeks to develop and maintain a consistent and accessible understanding of its interactions with customers. Ultimately, CRM is a strategic process, rather than a technology alone, by which an organization can better understand its customers needs and how to meet them. CRM requires gathering information on every contact with a customer, as represented by data that may exist in several separate computer systems maintained by the organizations, and making it readily accessible by quick search.

CRM is facilitated by the use of an ERP (Enterprise Resource Management) system. ERP's are designed to integrate the operation and data needed by many functional areas of an organization. If an organization uses an ERP that is well designed and carefully implemented, the organization is already a long way towards building a CRM capability. The ERP will already store much of the organization's customer contact data, in the form of transactions.

CRM can help increase business revenues by helping to tailor products and service offerings to customer's needs, offering more responsive and timely customer support, and retaining existing customers. For example, Amazon.com uses a simple form of CRM when it occasionally suggests additional books of possible interest, based on the genre of books a customer is currently ordering. A variation on this is sometimes used, making the customer aware that "other people who have ordered xxx (the book you are ordering) also found these books of interest: aaa, bbb, ccc". These types of attention getters are based on analysis Amazon makes across its product line and sales activity. It's almost like a dynamically configured shelf adjacent to the checkout line in a supermarket—this is where clever marketers place items appealing for impulse buys. But the supermarket shelf is static and all customers are exposed to the same products while they await the checker's attention. Amazon's impulse-buy prompters are tailored specifically to each individual customer: CRM in action.

In addition, by making salespeople more knowledgeable about customers, their prior purchases and experiences with the firm, CRM can make it possible for sales to be closed more readily. In a sense, well-developed CRM aims at making every salesperson or customer service representative more able to act like an especially attentive family insurance agent, who makes it his business to keep track of the first name of each member of the policy holder's family, their birthday, and several personal facts such as hobbies, health, and favorite family vacation spots. A successful insurance agent maintains a lasting relationship with customers. The personal touches help in gaining additional customers by recommendation and word of mouth. A successful salesperson or help desk representative for a manufacturing, distribution, or service-providing organization is enabled by a CRM system to mimic this type of in-depth knowledge of each customer in whatever ways is appropriate to the business in which the firm operates.

CRM systems not only gather together transaction data related to specific customers, they often provide a transaction-based component for recording situations in which a customer contacts the organization with a problem. Help desks use this component to record the contact, access data about the customer and his or her purchases to speak knowledgeably about the situation, and to refer the problem it to an appropriate party within the firm on the customer's behalf if it needs specialized resolution. These CRM components usually provide support for the ongoing entry of notes and comments documenting the status of problem resolution, and include a timing feature to internally follow up on the status of resolution. The record of problems and resolution actions thus forms a collection of data that can be analyzed to proactively identify trends and to serve as an accumulating knowledge database of resolution actions that can speed the resolution of similar problems for other customers.

Podcast 60
ETHICAL DATA MANAGEMENT

ERP and CRM systems acquire and store large amounts of data about customers. Other types of systems, developed for billing purposes or Internet usage measuring purposes also amass huge amounts of data for specific purposes. But there often exists a tension in business organizations between the information technology organization, under the leadership of the Chief Information Officer (CIO) and the marketing area of the organization about the potential uses of this data.

Data that an organization accumulates about its customers and their purchases, and even their Internet search patterns, is an appealing resource to marketing professionals. But as the head of the information technology part of the enterprise, the CIO faces a tough decision. At what point does the use of data represent "function creep," the escalation of the use of data for purposes beyond that for which it was collected? As Scott Thompson, head of Visa's technology division points out, many marketer's would like to mine Visa's 35 billion transactions per year. "There are lots of creative people coming up with ideas" for loyalty programs, target marketing, or partnerships with retailers.[1] But in Thompson's case, Visa has an established data management policy that prohibit such use of customer transaction data. Visa considers that such use of the data would "pollute" the Visa brand and be perceived as a violation of customer's privacy, probably causing a great number of Visa card holders to switch to a competitor such as MasterCard.

But from one point of view, constructing a marketing pitch based on purchase data isn't much different from the kind of coupons some grocery store chains print at the checkout stand,

[1] See http://www.cio.com/archive/070102/pledge.html

specifically for a customer based on the items in the current purchase. If the customer includes cat food in the grocery basket, a coupon for a discount on cat treats or kitty litter might print. If the customer bought hot dogs, a coupon might print offering a discount on a gourmet mustard. This type of immediate use of data is not offensive to many people, probably because it seems just like a human store clerk noticing what you bought and suggesting something that might go well with it.

But customers seem to be much more sensitive to the use of databases of purchase transactions well after the fact, especially when it results in unsolicited marketing pitches in the form of e-mail. Amassing data for billing purposes, then mining it for patterns of purchases, smacks too much of snooping and an invasion of privacy. It links together activity across periods of time, making it seem as if someone is constantly watching what you are doing. Most of us would object to someone following us around and making detailed notes of everything we did. Analyzing our transaction activity at the individual level tends to anger people and produce ill will toward the firm marketing this way. And in some cases, Federal law prohibits using specific data, such as credit reporting information, for any purpose other than offering credit.

Marketing personnel at a major telephone company have suggested using records of where and what times customers call to design customized calling packages for specific individuals. Looking at the past year's of your calls, this would, for example, detect that you often call numbers in a specific area code in another state on weeknights. Some people might appreciate a special "deal" being offered to them by their cell phone provider based on this type of marketing. But many others would feel offended that someone is examining their every calling action and linking together their activity. And if the cell service provider sold the calling pattern data to other firms without their consent, such as to airlines offering service to the area called, most customers would seriously question the integrity of the cell provider, because it is revealing a pattern of your behavior to a third party. For this "function creep" with their calling data, it seems most reasonable that customers should be asked first if they wish to "opt-in" to this type of marketing. If asked, less than 15% of customers usually opt-in, strongly indicating that this marketing activity is regarded as undesirable. As Jack Cranmer, CIO of the Mayo Clinic, has stated, "If you're targeting customers based on data you've collected for some other use, there's where you should start thinking about ethics."[2]

The use, safeguarding, retention, and ultimately the destruction of data accumulated by an organization must be a part of an established **data management policy** of a firm, not something decided on an ad hoc basis by technicians alone. CIO Magazine recently conducted a survey of information technology executives on the issues of data, it use, and protection. The result of this survey indicates that many CIO's ascribe to these six "commandments" of ethical data management:

1. Data is a valuable corporate asset and should be managed as such, like cash, facilities or any other corporate asset.
2. The CIO is **steward** of corporate data and is responsible for managing it over its life cycle—from its generation to its appropriate destruction.
3. The CIO is responsible for controlling access to and use of data, as determined by governmental regulation and corporate policy.
4. The CIO is responsible for preventing the inappropriate destruction of data.
5. The CIO is responsible for bringing technological knowledge to the development of data management practices and policies.
6. The CIO should partner with executive peers to develop and execute the organization's data management policies.

Personnel working the information technology industry must be sensitive to these issues. Since data is very easy to record and store, manipulate,

[2] See http://www.cio.com/archive/070102/pledge.html

and access, it's very easy to confuse the issue of data access with the issue of data ownership. The information technology organization is the steward of data, but doesn't own the data. It belongs to the people identified by it, and it is a mistake to use it in ways beyond the purpose for which it was provided.

Podcast 61
EIS and GIS SYSTEMS

Executive Information Systems (EIS) and Geographic Information Systems (GIS) are two special types of supports that may be used in combination with ERP's or non-ERP systems. Both of these perform specialized services not ordinarily associated with transaction processing systems.

Executive Information Systems sit at the top of the information processing hierarchy. They provide support to the highest level, strategic managers of an organization. These personnel are far removed from the day-to-day transaction processing and business processes of the organization. Typically the personnel at this level are limited to the executive officers—president, chief executive officer (CEO), and perhaps highly placed vice presidents. Their time perspective is multi-year, and their orientation is toward the strategies the firm will pursue to maintain a competitive edge, exploit emerging business opportunities, and respond to changing business and regulatory climates.

Only some of the "information inputs" needed by strategic managers can come from the information systems that run the organization. Analytical reporting from these systems, such as the firm's ERP system, can reveal how the firm itself is performing against its past performance, and against budgeted expense and revenue projections, and performance predictions. But the scope of concern for strategic managers must be broader than this. It must include information about how the firm's competition is doing, financially and in terms of performance and new market development and penetration. A combination of external sources may provide this information, such as business journals, hired research firms, stock market analysis, research and development divisions of the organization, legislative and regulatory analysis, and business conferences. The function of an EIS is to gather this external information, or provide a place for personnel to assemble and consolidate it, for high level presentation and assessment. An EIS may provide queries constructed to provide trend analysis, and readily manipulated analysis tools to provide selective views, graphical presentations, and flexible comparisons and analyses based on performance indicators preprogrammed or implemented by executives themselves.

The role of a **Geographic Information System** is to associate data of interest to the organization to units of geography such as countries, regions, states, counties, cities, zip codes, and smaller division such as census tracts or even city blocks. Rather than portraying data on sales, events, or other occurrences in tabular form as ordinary reporting program do, or in chart form as graphics display tools do, a GIS associates data to places in mapped form.

In order to associate data to mapped places, a GIS must typically provide some method to associate addressed data to geographical areas. This demands access to geographic reference files, which also include digitized mapping information. For example, let's suppose the law enforcement agency of a major city wants to better understand where burglaries are being committed. Police reports carrying addresses of crime scenes and be matched to city blocks or census tracts using a GIS, and then maps produced differentially shading city blocks according to the number of burglaries committed in each during a certain period of time. The density pattern of these events in themselves may be informative. If the incident data is now associated with demographic characteristics of the blocks, obtainable from published United States Census Bureau information, additional insights may be gained into commonalities in the victims of these crimes.

Market researchers are among the heaviest GIS users, both for the formation of marketing campaigns and for the siting of new facilities. Both of these rely on characteristics such as Census Bureau socioeconomic data, reported at the zip code or census tract level, and other measure such as population density and traffic circulation patterns. A GIS can aid in locating areas with concentrations of people having characteristics associated with specific product popularity. An astute firm will use the output of a GIS to guide its placement of print and media advertising and the location of new outlets to maximize the potential for success.

A GIS can also be instrumental in routing vehicles that must traverse the geography. In order to accomplish this, the GIS needs periodic updating of highway locations, intersections, and road conditions. This represents a combination of traditional information processing tasks such as database updating with the interpretation of digitized geographical information about the highway network. The source of much of this geographical data is the United States Census Bureau, which began to create and publish computerized geographic files in the 1980's and has since refined and extended the process. The current implementation of this data is the TIGER/Line files. TIGER is an acronym that stands for "Topologically Integrated Geographic Encoding and Referencing." The TIGER/Line files maintained by the Census Bureau store boundaries, shorelines, the highway network, railroads, and other topographic features in digitized form and are used to support the census conducted in the United States every ten years.

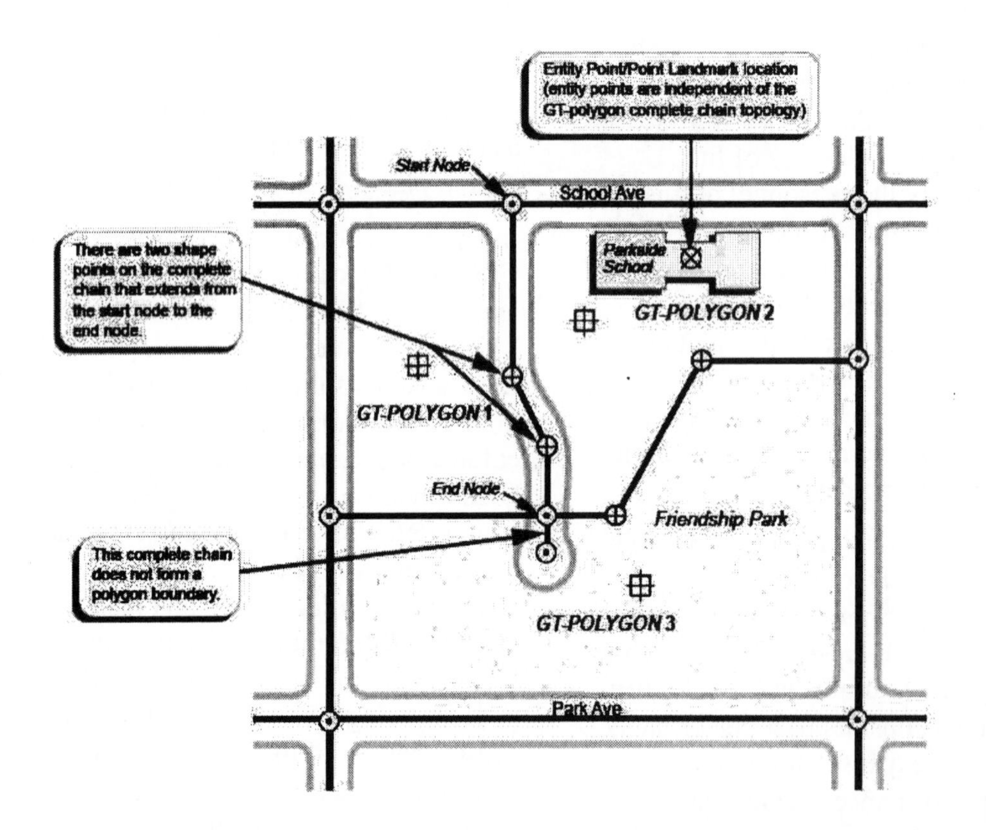

An illustration of the depiction of geographical features as encoded in the TIGER/Line™ files produced and published by the United States Census Bureau. Information about the content of these files is provided at http://www.census.gov/geo/www/tiger/tiger2k/tiger2k.pdf from which this example is drawn.

Lecture Slides 8

Introduction to Information Technology

Enterprise Resource Planning (ERP) systems

(C) 2006 Jim Janossy and Laura McFall, Information Technology Workbook

Course resources

PowerPoint presentations, podcasts, and web links for readings are available at

www.ambriana.com > IT Workbook

Print slides at 6 slides per page

Homework, quizzes and final exam are based on slides, lectures, readings and podcasts

Slide 8-2

Main topics

- Organizations and their info needs
- Non-integrated information systems
- Organizational units and info "silos"
- **Enterprise** Resource Planning system
- Business mergers and acquisitions
- GIS and EIS
- Careers

Slide 8-3

Review of IS systems in business organizations

- **Transaction processing:** interactively processing a purchase, deposit, withdrawal, registration, reservation
- **Batch processing:** retrieve, summarize, select, calculate new information from stored data, or apply bulk updates; "reporting"

Slide 8-4

Data warehousing: a recent phenomenon

- Eliminates batch reporting from the same database as transaction processing systems use
- Covers **longer time periods** for trend analysis than data is maintained for in transaction processing systems
- Makes data **easier** to access for reporting

Slide 8-5

ERPs

- Enterprise Resource Planning systems
- Origin in manufacturing industries, the value chain
- Integrating formerly disparate business support functions and databases
- Real time support

© 2006 J. Janossy Slide 8-6

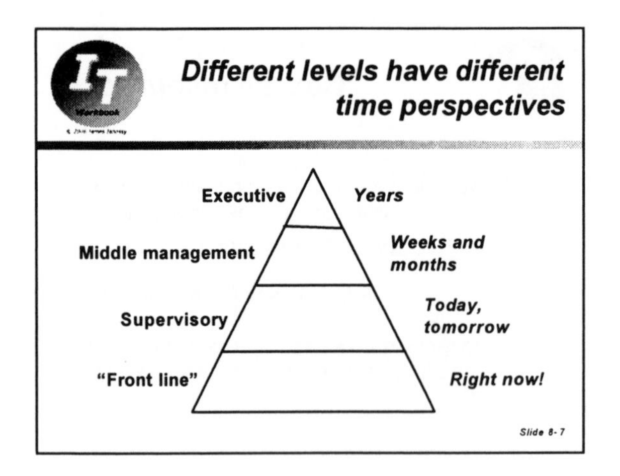
Different levels have different time perspectives
Executive
Years
Middle management
Weeks and months
Supervisory
Today, tomorrow
"Front line"
Right now!
Slide 8-7

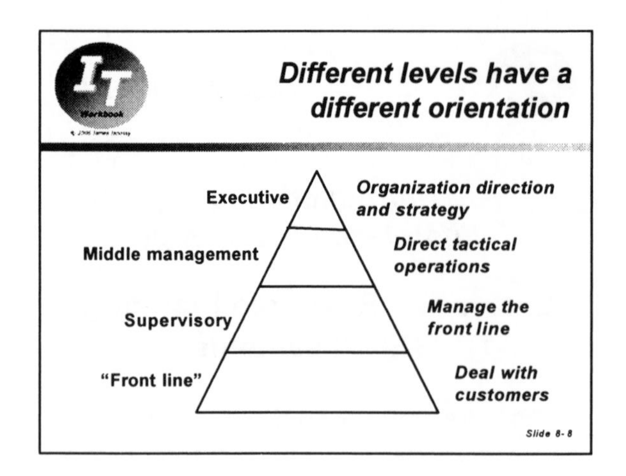
Different levels have a different orientation
Executive
Organization direction and strategy
Middle management
Direct tactical operations
Supervisory
Manage the front line
"Front line"
Deal with customers
Slide 8-8

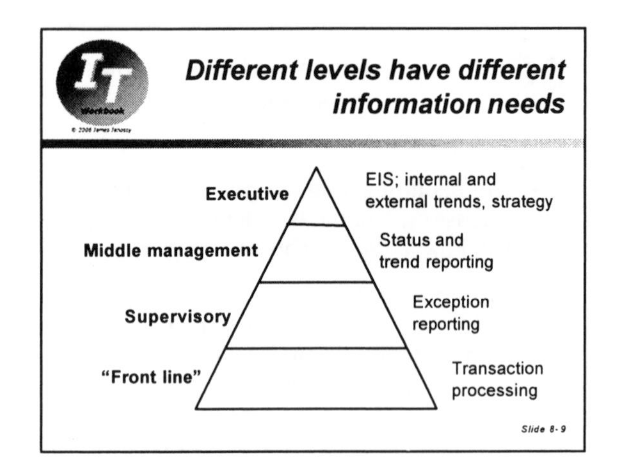
Different levels have different information needs
Executive
EIS; internal and external trends, strategy
Middle management
Status and trend reporting
Supervisory
Exception reporting
"Front line"
Transaction processing
Slide 8-9

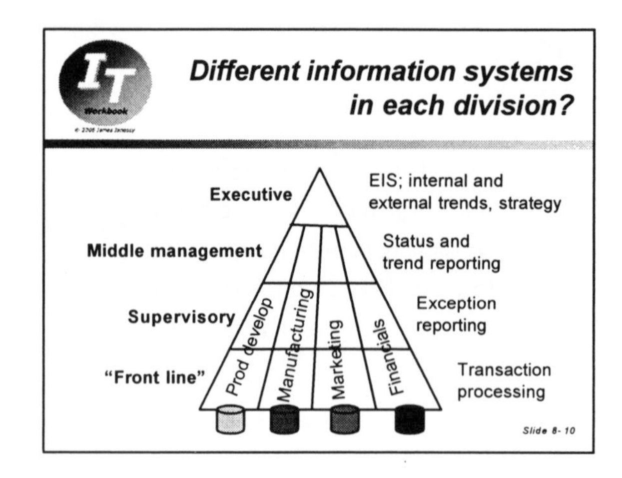
Different information systems in each division?
Executive
EIS; internal and external trends, strategy
Middle management
Status and trend reporting
Supervisory
Exception reporting
"Front line"
Transaction processing
Prod develop
Manufacturing
Marketing
Financials
Slide 8-10

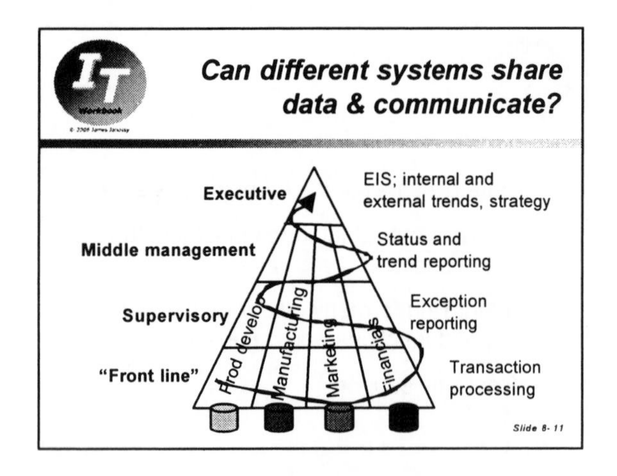
Can different systems share data & communicate?
Executive
EIS; internal and external trends, strategy
Middle management
Status and trend reporting
Supervisory
Exception reporting
"Front line"
Transaction processing
Prod develop
Manufacturing
Marketing
Financials
Slide 8-11

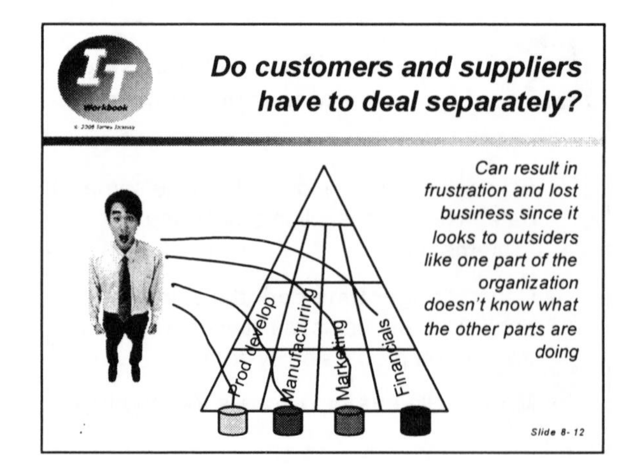
Do customers and suppliers have to deal separately?
Can result in frustration and lost business since it looks to outsiders like one part of the organization doesn't know what the other parts are doing
Prod develop
Manufacturing
Marketing
Financials
Slide 8-12

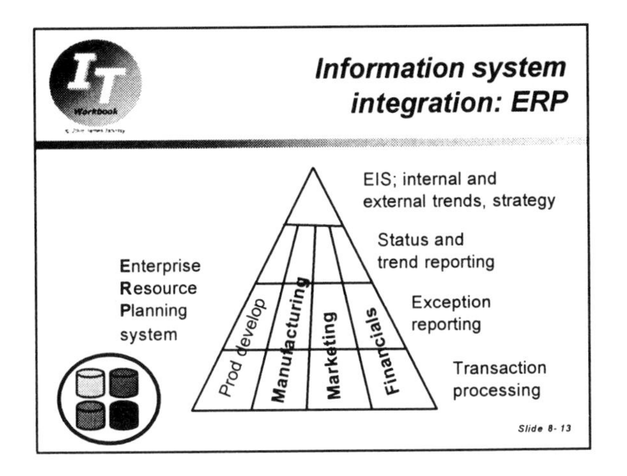
Information system integration: ERP
EIS; internal and external trends, strategy
Status and trend reporting
Exception reporting
Transaction processing
Enterprise Resource Planning system
Prod develop
Manufacturing
Marketing
Financials
Slide 8- 13

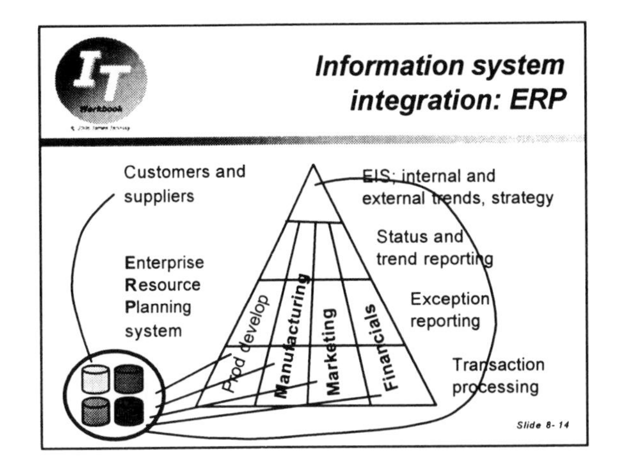
Information system integration: ERP
Customers and suppliers
EIS; internal and external trends, strategy
Status and trend reporting
Exception reporting
Transaction processing
Enterprise Resource Planning system
Prod develop
Manufacturing
Marketing
Financials
Slide 8- 14

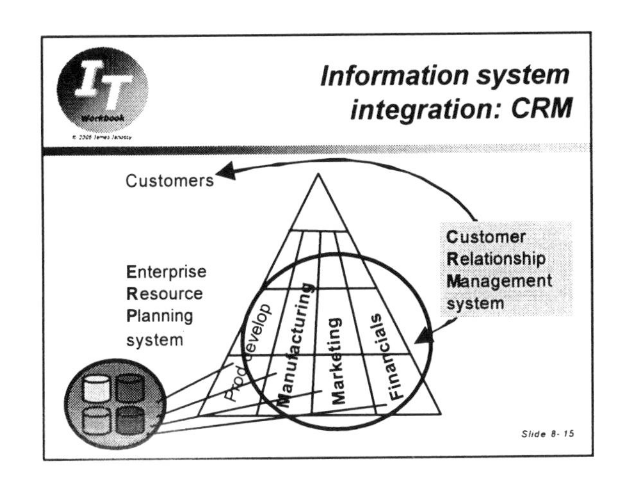
Information system integration: CRM
Customers
Enterprise Resource Planning system
Customer Relationship Management system
Prod develop
Manufacturing
Marketing
Financials
Slide 8- 15

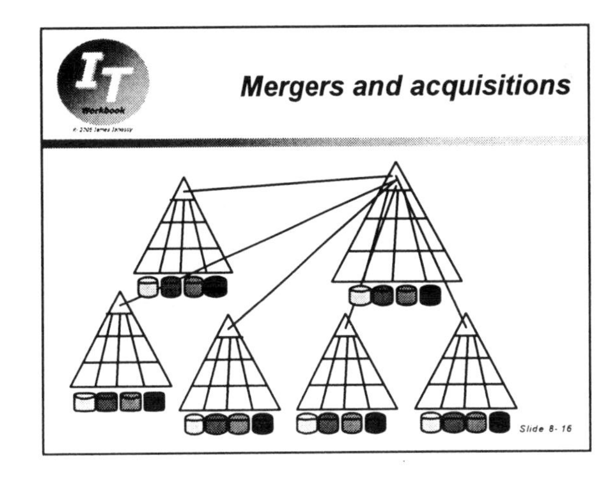
Mergers and acquisitions
Slide 8- 16

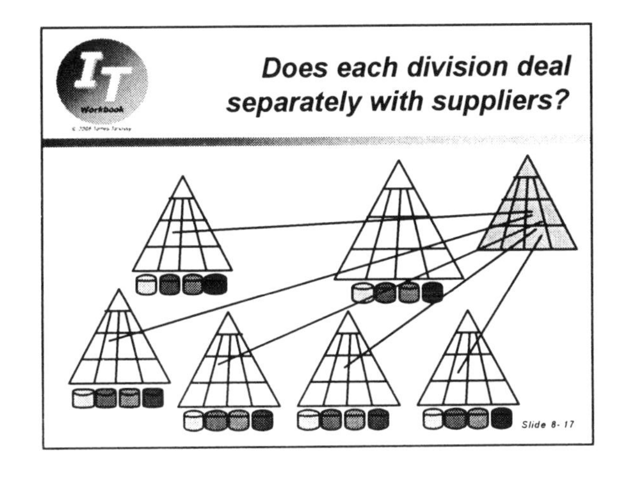
Does each division deal separately with suppliers?
Slide 8- 17

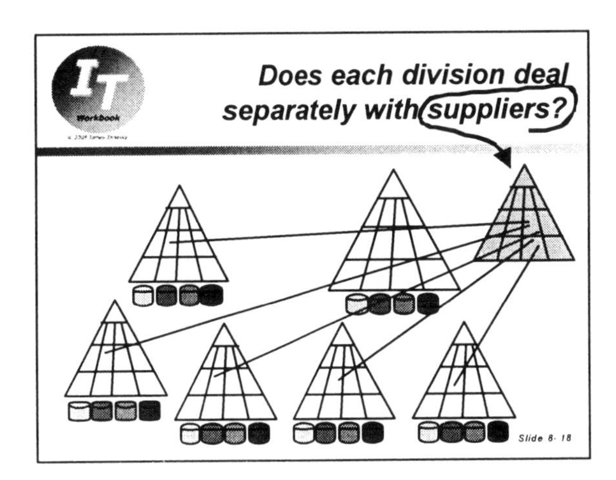
Does each division deal separately with suppliers?
Slide 8- 18

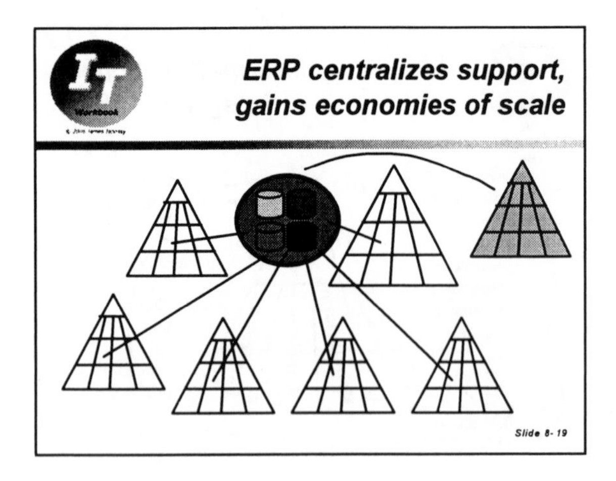
ERP centralizes support, gains economies of scale
Slide 8- 19

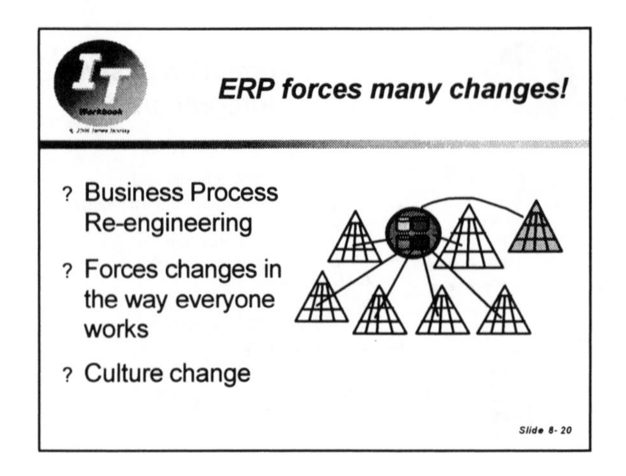
ERP forces many changes!
? Business Process Re-engineering
? Forces changes in the way everyone works
? Culture change
Slide 8- 20

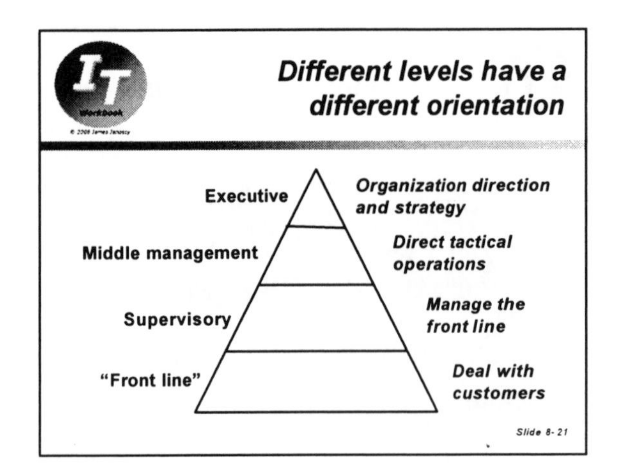
Different levels have a different orientation
Executive
Organization direction and strategy
Middle management
Direct tactical operations
Supervisory
Manage the front line
"Front line"
Deal with customers
Slide 8- 21

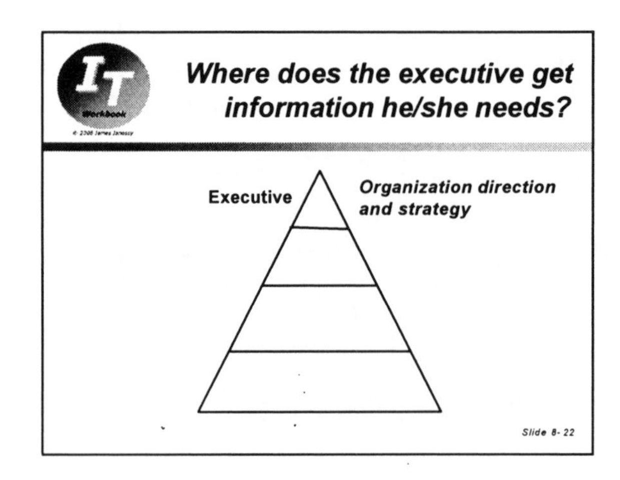
Where does the executive get information he/she needs?
Executive
Organization direction and strategy
Slide 8- 22

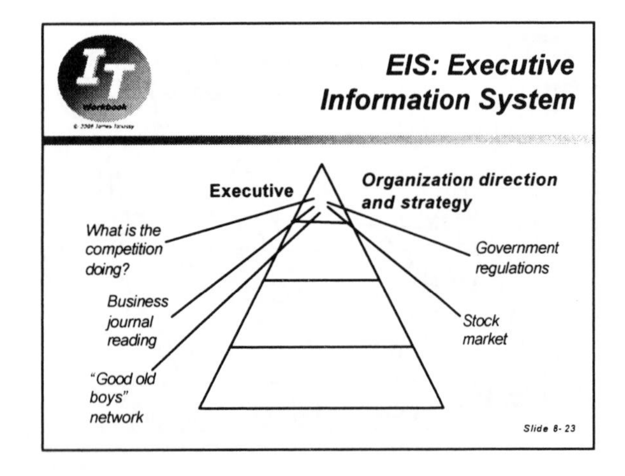
EIS: Executive Information System
Executive
Organization direction and strategy
What is the competition doing?
Business journal reading
"Good old boys" network
Government regulations
Stock market
Slide 8- 23

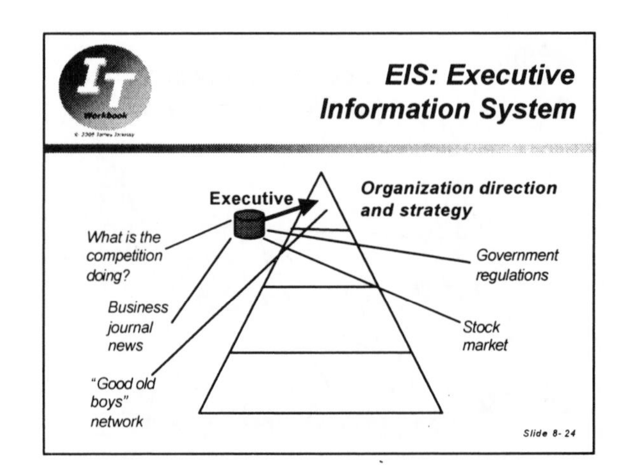
EIS: Executive Information System
Executive
Organization direction and strategy
What is the competition doing?
Business journal news
"Good old boys" network
Government regulations
Stock market
Slide 8- 24

Geographic Information System (GIS)

Customer

Address

What purchased and when

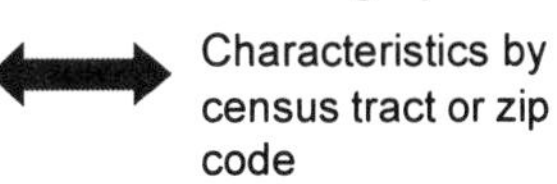

Demographics

Characteristics by census tract or zip code

Associate customer data with demographics of location; "get smarter" about profiles of customers for more efficient targeted marketing

Slide 8- 25

Data Warehousing

Customer

Address

What purchased and when

Demographics

Characteristics by census tract or zip code

Data warehousing using the dimensional model is often employed to associate demographics with purchases for decisionmaking

Slide 8- 26

Ethical use of data: repurposing?

- What can data about customers or clients be used for?
- Collected for one purpose, should not be used for others, like marketing
- **Opt-in:** customer OKs additional uses
- **Opt-out:** customer has to request non-use (not a good route, but often used)

Slide 8- 27

ERP systems

- Ambitious, complex, expensive
- Implementation has had successes and failures
- SAP, PeopleSoft, Oracle, and others
- What kinds of careers do they offer?

Slide 8- 28

Careers

- Database administration
- Network administration
- Development project team (user)
- Development (vendor)
- Training (vendor)
- Web, portal development
- Business analysis, consulting

Slide 8- 29

Database administration

- Which DBMS?
- Oracle is top vendor
- DB2 – IBM mainframe environment
- SQL Server – less expensive
- Sybase, Informix
- Hands-on, certifications, SQL

Slide 8- 30

Network administration

- Telecommunications
- TCP/IP
- Network technician
- Network monitoring
- **Security**
- Internet routers, ISP

Slide 8-31

Development team (user)

- Need business or specific ERP system knowledge + SQL + proprietary language (PeopleCode or ABAP)
- Generally requires hands-on experience with the ERP module (HR, finance, manufacturing, higher education)
- Consulting is common way to gain experience/exposure

Slide 8-32

Development team (vendor)

- Software development? Much is outsourced offshore!
- Generally requires hands-on experience with the specific ERP
- Consulting is common way to gain the experience/exposure
- Requires very good technical skills!

Slide 8-33

Training (vendor)

- Good technical background, but maybe not intensive
- Ability to communicate well, personable
- May require travel (training is sometimes at client site)
- Good way to gain exposure to major ERP users

Slide 8-34

Web development

- Most ERPs have gone to web interface and portals
- Requires good **HTML**, **java**, **.net** skills and experience
- Experience with databases, web interfaces, and SQL is often a necessity

Slide 8-35

Business analysis, consulting

- Often supplied to clients installing ERPs by consulting firms
- Subject matter education is important (business organization, human resources, financials, accounting)
- Good communication skills both oral and written are important
- Often involves travel

Slide 8-36

IT Workbook Assignment 8.1

Student name:

Write your answers to the questions below <u>within the box</u>. In each case, please choose your words carefully to answer the specific questions. Avoid simply copying passages from your readings to these answers.

Identify the **six important parts of a manufacturing business** that an ERP can help manage:

1)

2)

3)

4)

5)

6)

"ERPs are cross-functional and enterprise-wide." Elaborate on this statement and explain what this means, in simpler words that anyone should be able to understand:

7)

IT Workbook Assignment 8.2

Student name:

Write your answers to the questions below within the box. In each case, please choose your words carefully to answer the specific questions. Avoid simply copying passages from your readings to these answers.

Identify and discuss three intended **advantages** of ERP systems:

1)

2)

3)

Identify and discuss three often-experienced **disadvantages** of ERP systems:

4)

5)

6)

IT Workbook Assignment 8.3

Student name:

Write your answers to the questions below <u>within the box</u>. In each case, please choose your words carefully to answer the specific questions. Avoid simply copying passages from your readings to these answers.

Explain what happens to **standalone computer systems** in organizations when an ERP system is implemented:

1)

2)

3)

How can ERP improve a company's business performance? Describe **what needs to happen** in an organization for performance improvement to occur when ERP is brought in:

4)

IT Workbook Assignment 8.4

Student name:

Write your answers to the questions below within the box. In each case, please choose your words carefully to answer the specific questions. Avoid simply copying passages from your readings to these answers.

Identify and briefly discuss the **five major reasons** why companies undertake a conversion to an ERP system:

1)

2)

3)

4)

5)

Identify and briefly discuss the **most common reason** that some companies abandon an ERP implementation on which they may have spent several million dollars:

6)

IT Workbook Assignment 8.5

Student name:

Write your answers to the questions below within the box. In each case, please choose your words carefully to answer the specific questions. Avoid simply copying passages from your readings to these answers.

Identify **ten hidden costs** often associated with ERP system implementation and use:

1)

2)

3)

4)

5)

6)

7)

8)

9)

10)

IT Workbook Assignment 8.6

Student name:

Write your answers to the questions below within the box. In each case, please choose your words carefully to answer the specific questions. Avoid simply copying passages from your readings to these answers.

Discuss the issue of the **training** that is necessary when an ERP system is implemented in an organization, and the various ways that this can be accomplished:

1)

Describe the issue of **customizations** of an ERP system and what they involve, why customizations are sometimes proposed and made, and what the **negative aspects** of making customizations are:

2) customizations of an ERP system are:

3) ERP system customizations involve:

4) customizations are sometimes proposed and made because:

5) the negative aspects of making customizations are:

IT Workbook Assignment 8.7

Student name:

Write your answers to the questions below within the box. In each case, please choose your words carefully to answer the specific questions. Avoid simply copying passages from your readings to these answers.

Describe what an **Executive Information System (EIS)** is, who uses it and for what purpose, and what data it presents:

1) what an EIS is:

2) who uses an EIS and for what purpose:

3) what data an EIS presents:

Describe what a **Geographic Information System (GIS)** is, who uses it and for what purpose, and what data it presents:

4) what a GIS is:

5) who uses a GIS and for what purpose:

6) what data a GIS presents:

IT Workbook Assignment 8.8

Student name:

Write your answers to the questions below within the box. In each case, please choose your words carefully to answer the specific questions. Avoid simply copying passages from your readings to these answers.

Identify the **main reason** that Nestle USA decided to install an ERP at their Glendale, California headquarters and the **five lessons that they learned** from their experience:

1) Main reason for Nestle USA going to an ERP:

2) Lesson #1:

3) Lesson #2:

4) Lesson #3:

5) Lesson #4:

6) Lesson #5:

IT Workbook Assignment 8.9

Student name:

Write your answers to the questions below within the box. In each case, please choose your words carefully to answer the specific questions. Avoid simply copying passages from your readings to these answers.

What are the purpose and goal of **Customer Relationship Management (CRM) systems**, and what are the five keys to successful CRM implementation?

1) **purpose** of CRM systems:

2) **goal** of CRM systems:

3) key #1 to successful implementation:

4) key #2 to successful implementation:

5) key #3 to successful implementation:

6) key #4 to successful implementation:

7) key #5 to successful implementation:

Change management is often mentioned in connection with the implementation of an ERP system. Discuss what this term means and what its primary goal is:

8) what change management is:

9) the primary goal of change management:

IT Workbook Assignment 8.10

Student name:

Write your answers to the questions below within the box. In each case, please choose your words carefully to answer the specific questions. Avoid simply copying passages from your readings to these answers.

In an article in CIO magazine, Scott Berinato stated that *"Data ethics is not just a moral issue; it's the intersection of business and morality..."* (see www.cio.com/archive/070102/pledge.html). Discuss the position in which **data ethics** places Chief Information Officers, how "**function creep**" can lead to unethical uses of data, and what "**opt-in**" and "**opt-out**" mean regard to this.

1) the position in which data ethics places Chief Information Officers:

2) "function creep" can lead to unethical uses of data in this way:

3) "opt-in" means:

4) "opt-out" means:

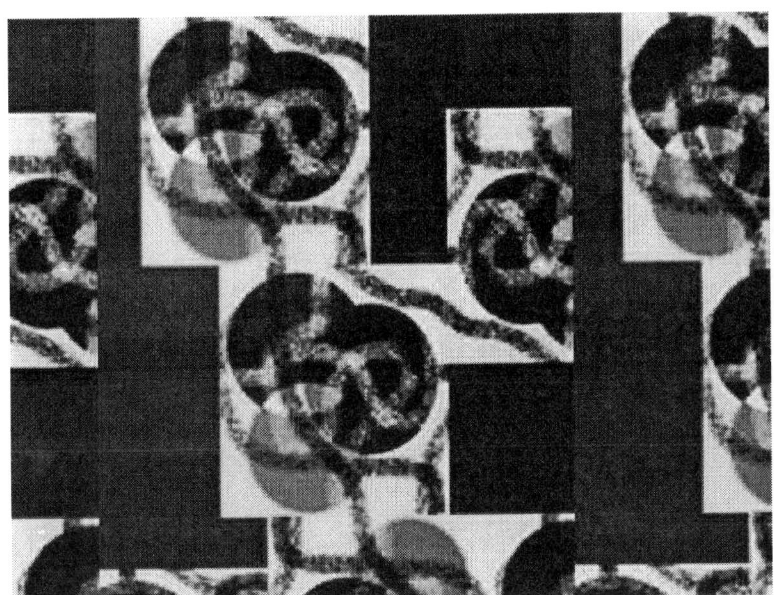

System Development

Two ways exist to obtain the software needed to support a business enterprise: acquire already existing software, or develop it. In this chapter we examine the processes and tools used in the acquisition or development of business software.

Both ways to obtain software begin by developing an understanding of the processes to be automated in an effort known as **systems analysis**. In systems analysis business processes are examined, documented, mapped out, and interfaces between the outside world and the organization identified. Then a choice is made: is there any software available in the marketplace that provides a close enough fit to meet requirements, or must the organization develop its own software to suit it needs? In both cases a **system development methodology** (SDM) is used to structure and document the study.

The **system development life cycle** (SDLC) is a common system development methodology. This "waterfall model" assumes that planning, analysis, design, development (programming), testing, implementation and maintenance can follow linearly in distinct stages. The SDLC employs **data flow diagramming** (DFD) to map information flow, among other tools.

Other SDM's include **prototyping**, **joint application development** (JAD), and **rapid application development** (RAD) with **computer-aided software engineering** (CASE) tools to generate program code and documentation. Each of these was developed to overcome limitations of the waterfall model.

When purchases are planned a **request for proposal** (RFP) is written describing needs and is sent to prospective vendors. Vendors submit proposals and in some cases a **proof of concept** demonstration that may lead to a **contract**.

Learning objectives

Chapter 9
System Development

This chapter, in combination with the web readings, web links, and podcasts available at www.ambriana.com is focused on providing the knowledge to achieve these learning objectives:

- Understand what system development without the use of a methodology ("cowboy coding") is, what it is relevant to, and its disadvantages, and how software engineering is different from it

- Recognize and be able to describe the original intention of W. W. Royce in proposing a "waterfall" model of system developing in 1970, and how his proposal was subsequently misinterpreted and modified by others

- Be able to define and describe criticisms of the waterfall model as commonly understood, which have led to lack of success in some system development efforts that applied it

- Understand and be able to describe what prototyping is in both the hardware and software environments, what "timeboxing" is, what a "proof of concept" is, and what an RFP and contract are

- Be able to recognize and correctly use process, data flow, external entity, and data store symbols on data flow diagrams of simple data handling situations

- Understand and be able to describe the nature of and differences between contemporary system development methodologies, including design-build-test-release, Joint Application Development (JAD), Rapid Application Development (RAD), agile software development, extreme programming, and end-user programming (EUP).

- Be able to describe the purpose of software testing, the common quality attributes associated with it, and the legal liabilities associated with software products that fails in use.

Podcast Scripts – Chapter 9

These readings of the **Information Technology Workbook** are also accessible in audio form as .mp3 files for free download. Visit the workbook web site, www.ambriana.com. You can listen to this material on your own Ipod or .mp3 player or computer, read it, or both listen and read.

Podcast 62
OVERVIEW OF SYSTEM DEVELOPMENT

Two ways exist to acquire the software needed to support a business. Both begin by developing an understanding of the processes to be automated in an effort known as **systems analysis**. In systems analysis, business processes are examined, documented, mapped out, and interfaces between the outside world and the organization identified. Then a choice is made. Is there any software available in the marketplace that provides a close enough fit to meet identified requirements? Or must the organization develop its own software to suit its purposes? In both cases, a system development methodology, abbreviated SDM, is used to structure and document the study.

The **system development life cycle**, often abbreviated **SDLC**, is a common system development methodology. One of the earliest of the system development methodologies, the SDLC is known as the **waterfall model**. The waterfall model assumes that planning, analysis, design, programming, testing, implementation, and software maintenance can follow linearly in stages. The waterfall system development life cycle uses **data flow diagramming**, abbreviated **DFD**, to map information flow.

The waterfall model evolved from a paper written by Dr. Winston Royce in 1970, and is only part of the process suggested by Royce. What has commonly been called the waterfall model was not actually recommended by Royce as a very realistic model for system development, but it has been employed by some, with varying degrees of success or failure, for more than 35 years. In the podcasts associated with chapter 9 of the Information Technology Workbook, we'll discuss Royce's paper, the waterfall model, and why Royce himself suggested alternatives to it.

Other system development methodologies include prototyping, Joint Application Development, and Rapid Application Development. Computer-aided software engineering tools are often combined with these methods to generate program code and documentation. More acronyms abound with these models and tools. Joint Application Development is often called JAD. Rapid Application Development is often called RAD. Computer aided software engineering tools are often called CASE tools. Individual podcasts in this series cover prototyping, JAD, and RAD.

When computer programming, system design, or other automation assistance service purchases are planned, an organization drafts a **request for proposal (RFP)** describing its needs. Composing the RFP is an important process for the organization because it must represent the concurrence of major parts of the organization in the needs that the intended automated system is to meet. The request for proposal is sent to prospective vendors of software products and often to consulting organizations. Vendors and other firms that can meet the requirements of the RFP submit formal proposals, and in some cases provide a **proof of concept** demonstration, that may lead to a contract. In simple legal terms, a **contract** is what results when one party makes an offer to another party, and the second party accepts the offer. In short form, **offer + acceptance = contract**. The RFP is a solicitation of an offer, which says essentially "we have this need: if you're interested, tell us how you might be able to meet it." The response to an RFP, a formal proposal, is an offer. An organization accepts the offer tendered by the proposal by having its authorized officials sign it, creating a contract between the firm in need of software and support, and the vendor whose proposal is accepted. In many cases, some amount of negotiation precedes the drafting and signing of the final contract.

In this chapter of the Information Technology Workbook, we'll introduce you to many of these system development concepts, which make it possible for teams of software engineers to design and develop the programs that harness the power of computers to business purposes.

Podcast 63
COWBOY CODING

When students of computer science first begin to write computer programs, they often do so by simply coding program language statements that seem to make sense in producing an intended action. This type of program coding is sometimes called **cowboy coding**, since it is typified by a wild west mentality in which individuals simply carve out a space for themselves and do what they think is best and most reasonable. In the school environment, it's easy to adopt a programming style based on this type of logic coding and to think that it is the way that actual software is written in the workplace.

Nothing could be further from the truth.

As Wikipedia, the free encyclopedia available on the Internet points out, cowboy coding has a number of disadvantages. For one thing, it often lacks a clear scope and vision. It is similar to what would happen if a contractor started building the foundation of a house, based only on a general notion of the placement, size and shape of the intended house. In this situation, it's easy to envision that much building, tearing down, and rebuilding would probably be necessary since concrete things were being set in place before much thinking about the end result.

Cowboy coding, while it may work for small simple projects, is not suitable for projects with a high degree of complexity or which require the coordinated efforts of multiple software developers. When cowboy coding results in the production of workable software at all, it usually produces software of poor and inconsistent quality, since it lacks processes for software review, testing, and documentation. While cowboy coding is useful for students as a means of trying out various programming language features, it is not appropriate for the professional workplace, which demands a more consistently repeatable process that produces a higher quality product.

Podcast 64
SOFTWARE ENGINEERING

Productive software development follows a method that specifies a structure for the design of a system and its logical components before programming activities are started. In its most highly developed form, software development may be called **software engineering**. It is accurate to call it software engineering, however, only if it is actually conducted in the same way that other forms of engineering, such as mechanical, civil, and electrical engineering are conducted. In order to better understand software engineering, it's informative to examine how these other engineering disciplines function.

An engineering discipline prescribes a minimum level of educational attainment for its practitioners, typically a bachelor of science degree. In achieving such a degree, an engineer learns the basic math, physics, and specialized accumulated knowledge of the field to be able to function capably within it. This preparation makes it possible for an electrical engineer, for example, to apply the correct principles to a problem to accurately describe it, propose a solution, and show that proposed solution to peers, who can judge the solution prior to its implementation. To practice engineering, these steps are followed:

1. The scope of a problem is defined, that is, the requirements that the solution is to meet are stated in such a way that more than one person can understand them.
2. Possible solutions to the requirements are proposed, evaluated as to feasibility and desirability, and a design approach is chosen.
3. Known algorithms, standards, and materials capabilities are applied to the chosen design to produce a circuit diagram, blueprint for a building, or construction documents for a

machine or road. This blueprint specifies what each component of the system is to do and how it is to interface with the other components.

4. Construction is accomplished using accepted practices, and work products are open to review by peers.
5. Planning the work processes includes appropriate steps to test the quality of the products at various stages, first at the component level, and then at the level at which components are combined and integrated. The testing process includes consideration of all of the possible ways in which the engineering solution may fail, and assesses the potential for and tolerance for error in precisely quantifiable terms.
6. Implementation of the new construction is planned and proceeds according to coordinated steps.
7. Documentation is updated as construction work proceeds, so that it is as current as the product constructed.

You can readily see that cowboy coding does not meet these requirements. Cowboy coding excludes all but the program coder from the thought process of evolving a statement of the problem to be solved. It excludes others from the choice of appropriate tools to produce a solution. Cowboy coding vests all knowledge of the intended scope of the solution, and the construction of the product, in just the person attempting it. And cowboy coding does not address the issue of testing the solution or product in any rigorous way.

We don't approach newly constructed buildings with apprehension and demand that they stand empty for a few months before moving in, to make sure the building will not collapse. Buildings are constructed according to an engineering discipline that meets the requirements listed above, and we have confidence in the result. But cowboy coding doesn't scale up, and it's doubtful that a cowboy architect, or a cowboy structural engineer, could create a building in which we would have the same degree of confidence.

Many members of the traditional engineering disciplines question whether computer science is mature enough to support software engineering, with the degree of rigor required by the term engineering. They suggest instead that the term **software development** is more appropriate than software engineering. Given the widespread accessibility of programming languages, and the variety of tasks to which computers can be dedicated, there is no question that much software development has been done under less than rigorous conditions, and in a manner not consistent with engineering professionalism.

For example, computer game development, especially in the early days, has often been accomplished by a single individual, as more of an art form than an engineering activity. It would be incorrect to call such a computer game developer a software engineer. But no general harm results from this software development, since the failure of a computer game is of little consequence to the end user.

But take the case of the software that controls a dialysis machine, to which a patient with kidney failure is connected. The machine extracts blood from the patient, filters toxins out of it, and must feed the blood back into the patient's body at the correct rate, temperature and pressure. Developing this type of software with informal methods is a recipe for disaster. Failure of this type of software can kill the patient. Clearly, the development, programming, and testing of this type of critical software demands a level of structure and discipline more akin to traditional engineering.

Various system development methodologies have been proposed and implemented for software development to organize it into more of an engineering discipline. In the next podcast of the Information Technology Workbook, we'll consider the strengths and weaknesses of the earliest system development methodology, the system development life cycle, or waterfall model, described by Dr. Winston Royce in 1970 as a part of a survey of then-current system development practices.

Podcast 65
ORIGINS OF THE WATERFALL MODEL

Various methodologies have been proposed and implemented for software development to organize it into an engineering discipline. The earliest system development methodology, the **System Development Life Cycle** (the waterfall model), was described in a paper by Dr. Winston W. Royce in 1970. This paper, entitled ***Managing the Development of Large Software Systems***, was presented at a 1970 conference of the Institute of Electrical and Electronic Engineers. The original paper is available on the web and provides a superb basis for understanding the waterfall model, as well as the significant misinterpretations of it that have been made for over 35 years.

In forming his paper, Royce examined several large software system development efforts that had been attempted during the 1950's and 1960's, the first two decades in which computers were commercially available. The first half of Royce's paper described the way many people were attempting to develop software, which mirrored the linear step-by-step process by which buildings are designed and constructed.

It's informative to examine how building design and construction proceeds, in order to better understand how the waterfall model views the software development process. In building construction, an architect develops an idea of what the client wants in terms of building purpose, be it a residence, office building, warehouse, or parking structure. The architect considers the intended use of the new building, its size and capacity, placement on the building site, the desired architectural style, the condition of the soil and the type of foundation needed to support the building.

The architect then develops sketches and proposes alternative general designs. Once the client has committed to the general layout, the architect makes preliminary drawings that delve into the details of the proposed structure, the materials to be used, the internal placement of rooms, and the internal and external traffic circulation patterns. Once the client and local municipal building authority approve the preliminary drawings, the architect composes the working drawings, that is, the blueprints for the new construction.

The blueprints are very specific and highly detailed. The blueprints are explicit manufacturing specifications that convey to a builder, known as a general contractor, exactly what the intended final product will be and how it will be constructed. The blueprints cover everything internal and external to the intended building, including the placement of all supporting members, floors, ceilings, lighting, walls, doors, pipes, wires, ducts, mechanical equipment and surface finishes.

Because final architectural blueprints are a completely thought out and detailed document, they can be, and are, reviewed by structural engineers and the local building regulatory authority prior to the start of construction. When blueprints meet with engineering and local authority approval, a high degree of confidence exists that the actual construction, to be done strictly according to the blueprints, will produce the building that meets the client's purposes.

Architectural blueprints are so thoroughly detailed that contractors bidding on the construction work can make what is called a "materials take off." The materials take off precisely determines the amount of materials and labor required to construct the building, allowing each contractor to propose the dollar amount they would charge to provide the materials and to construct the building. This is possible because the technologies involved, such as foundation digging and forms construction, concrete casting, reinforcing and structural steel work, carpentry, plastering, roofing, plumbing, tiling and wiring are all well established and practiced. The craftsmen in these building trades are well versed in the processes, because the processes have been thoroughly refined through repeated use and application over many years.

When the construction of the new building begins, the contractor chosen for the project works only from the final blueprints. Although we as spectators to a new construction may be left guessing as to what will eventually arise from the foundation, the architect and the contractor already know specifically what the final product will be. In very nearly 100% of cases, the building that is constructed is occupied by the client immediately after completion and satisfies the purposes intended. The process of construction is orderly and well coordinated. The parts of the building fit together perfectly, because the design has considered and addressed all the details.

It was this process of understanding, preliminary design, detailed design, specification writing, construction and implementation that Royce found was being followed by many software developers, having been adapted from the successful work of the building industry. And it was the frequent failure of this adaptation to software development that Royce actually wrote about!

In the first half of his paper, Royce described this adapted building construction process, which assumes that one stage of work can be completed before the next stage begins. The first illustration depicted by Royce shows the stages of defining system requirements, defining software requirements, analysis, program design, program coding, program testing, and system implementation as boxes, with each one flowing sequentially into the next. In this depiction, all of the work in one box was completed before work began in the next box, just as is the case with building design and construction. Since each next box in this first part of Royce's paper was drawn lower on the page than the box preceding it, the arrows between boxes seemed to picture water flowing downward to the right. It was for this reason that the model was called the "waterfall model."

But Royce provided this first depiction not to recommend it, but to point out that this is what some software developers seemed to think was the way to work. In the second half of his paper, Royce pointed out why this process very often did not work well for software development. He documented the reasons why many successful software developers actually did not follow the simple waterfall model! It's ironic that so many people seemed only to read the beginning of his paper and so missed the thrust of his argument.

Podcast 66
WATERFALL MODEL SHORTCOMINGS

The earliest system development methodology, the **System Development Life Cycle** (the waterfall model), was described in a paper by Dr. Winston W. Royce in 1970. This paper, entitled ***Managing the Development of Large Software Systems***, was presented at a 1970 conference of the Institute of Electrical and Electronic Engineers. It is truly ironic that many authors who followed Royce seemed not to have actually read very far into his paper, because they misrepresented what he said and acted as if Royce had recommended simple waterfall model. In fact, Royce did not recommend simple waterfall model for software development.

What Royce actually noted was that many successful software developers had to provide for the fact that as they progressed into a software development effort, things invariably changed. Sometimes what changed was their own understanding of the client's requirements. Many times what changed was the client's set of requirements. And many times what changed was the available technology. Unlike the case with building construction methods, which rely on techniques evolved over thousands of years, computer programming is relatively new, and rapidly evolving. Since the simple waterfall model presumes that requirements can be completely understood and documented, and then not reconsidered, it really was a good fit only for a few types of very stable computerized applications.

One of the earliest applications that the waterfall model fit well was the automation of accounting documents and processes. Computers were first acquired in many organizations for use by the accounting department. Accounting processes are governed by what is known as **GAAP**, that is, **Generally Accepted Accounting Principles**. These principles have evolved over hundreds of years and are published and audited by organizations and government regulatory bodies that oversee the accounting profession. It's easy to see how software could successfully be developed to support accounting processes using the waterfall model, since accounting requirements are so well documented, understood, and relatively standard. However, Royce noted that the waterfall model does not very well support the kind of increase in the knowledge of requirements that occurs in many more innovative software development efforts.

Innovative software development efforts Royce examined included the automation of the airline reservation and ticketing system in the 1960's, accomplished by American Airlines with its online SABRE system. In this type of software development effort, Royce found that developers actually benefited from a form of "spiral" development model. In a spiral model, iterative interactions occur.

Iterative interactions are produced when learning acquired at a later stage, such as program design, makes it necessary to reconsider decisions already made in the prior stage, such as analysis. Royce found that in successful software development efforts, iterative actions not only occurred between successive stages, but sometimes leaped back several stages. For example, discoveries made in software testing might cause reconsideration of decisions made all the way back in the software requirements definition stage! The simple waterfall model provided no ability to handle this type of situation; it is not an appropriate method for innovative software development.

Royce made a number of suggestions for a software development methodology that provides support for iterative interactions. One suggestion was to allow for iteration between stages of the waterfall model and to budget time for reconsidering decisions already made and approved. Royce even went so far as to suggest that some software projects should be planned to be done twice! The first effort would be an early simulation of the final product. The second effort would benefit from the full run through of the first attempt, and could only then be planned, based on this additional knowledge.

Dr. Royce's suggestions led to the development of various alternatives to the simple waterfall model, including prototyping, Joint Application Development, Rapid Application Development, agile programming, and extreme programming. Many of these benefited from a model described by Barry Boehm in 1985, in a paper entitled, ***A Spiral Model of Software Development and Enhancement***. Each of these alternative methodologies attempts to support and actually benefit from the type of iterative interactions that Royce determined were not recognized and supported by the simple waterfall model.

Unfortunately, the notion that the simple waterfall model is the universally correct way to develop all software has plagued the industry for many decades. Many system development methodologies, including those stipulated by some military software development standards, continued to dictate that the simple waterfall model be followed even well after its deficiencies for software development had been identified.

Podcast 67
PROTOTYPING

A **prototype** is a model of an intended creation, put together to quickly illustrate how the final product may look or function. When used in the hardware area, an electronic circuit is designed, then actual components are combined on a "breadboard," a board with fasteners that makes it easy to temporarily connect components and demonstrate the operation of the circuit. The prototype is electrically an implementation of the circuit, but not in the physical form of the product that will finally be manufactured.

When used as a technique of software development, a prototype can help an end user get a better idea of potential automated support than can be provided by a description expressed only in words. Prototyping is one technique recommended by Dr. Winston Royce, the originator of a paper describing the inadequacy of the waterfall model of system development. Royce recommended prototyping as a possible improvement in the system analysis and development stages.

In the analysis stage, information technology personnel discuss automation requirements with end users and attempt to develop an understanding of the inputs, outputs, interfaces, and data transformation processes required. In the development stage, the understanding achieved by information technology personnel is reflected in a description of the form of automation that would best serve the needs of the end users. At both the analysis and development stages, a clear and readily understood way of restating the user's requirements is needed, in order to clarify them and to confirm that information technology personnel correctly comprehend them. A prototype is a vehicle for accomplishing this.

Two types of prototypes exist. A static prototype is composed of static images of inputs, processing screens, and outputs. A simulation prototype is a "throwaway" software creation that can mimic some of the proposed software functionality as a means of exploring the user's requirements.

A **static prototype** does not depend on any computer coding. It can take the form of an inanimate series of "storyboard" images, depicting the contents of computer screens that the user would encounter in using the proposed automated system. This type of prototype is most readily assembled using PowerPoint slides of screen pictures, and, possibly, mockups of documents to be produced by the proposed system. The slides in such a prototype can be arranged in the sequence in which the screens would appear when the end user performs a specific function.

The utility of a storyboard prototype lies in the fact that a picture is almost invariably easier to understand than a word description. The intent of this prototype is to provide a flexible, quickly prepared and easily changed vehicle for end users to evaluate the proposed solution, and to recommend changes to it, to bring it into closer conformity with the needs of their business processes.

In the words of Steve McConnell, a noted Microsoft proponent of static prototyping as a planning and design technique, the cycle of prototype creation, user evaluation, and adjustment continues until the user is excited about the software. Only at this point is the proposed design and operation of the software suitable for further documentation and the actual creation of the software begun in the intended programming language of implementation. Prototyping of this type is an integral part of McConnell's **Design-Build-Test-Release (DBTG)** development method.[1]

Another form of prototype is a software creation that mimics some part of the operation of the intended software. This type of prototype can be created using a software tool designed for the specific purpose of demonstration. Since this type of **simulation prototype** may actually perform some of the proposed functions, it may be even clearer for end users to understand than a static prototype.

However, a software working model prototype presents several potential disadvantages. A simulation prototype may take longer to create than a static prototype, even if a software tool designed for this purpose is employed. The longer time and more work it takes to develop the prototype, the less inclined the developer will be to modify it as the user reacts to it and suggests changes. But this is the whole point of prototyping in the first place!

Even worse is the notion that the actual language of implementation be used to create a prototype. Technical personnel sometimes suggest using the language of implementation to create a simulation prototype thinking that this will save time in the ultimate creation of the program code. But an actual language of implementation, such as C++ or Java, will probably require a

[1] See the ***Software Project Survival Guide***, Steven C. McConnell, Microsoft Press, ISBN 1-57231-621-7.

significant investment in time just to develop a prototype. The delay, and this level of investment in coding, works against regarding the prototype in the appropriate way. At its worst, a simulation prototype in the language of implementation devolves to nothing more than cowboy coding. (This is not necessarily the case in RAD; see podcast 70.)

It is all too common an occurrence that technical personnel who create a prototype in the actual final system implementation language feel compelled to salvage the investment in the prototype by forcing the adoption of this first attempted solution as the final software solution. At the very least, a simulation prototype in the language of implementation often causes technical personnel to force end users to justify changes, rather than regarding the software created for the prototype as the throw away tool that a prototype is intended to be.

Software simulation prototypes can pose even more problems. Once non-technical users see a prototype that provides some semblance of functionality, it may be difficult for them to understand why the actual software implementation should take any significant further amount of time. This can lead to the imposition of unreasonable and potentially unworkable demands that delivery time for the actual software system be accelerated. And finally, experience has shown that with the provision of a software working model prototype, end users are often magnetically drawn to focus on human interface design. It is very easy for user involvement to then center on quibbling concerning screen labels, drop-down box design and screen graphics, to the detriment of fleshing out an understanding of essential business processes.

Podcast 68
PROOF OF CONCEPT < > PROTOTYPE

Closely related to prototypes are "proof of concept" software developments. But these are different techniques, intended for different purposes. Since it is easy to confuse them, this can lead to misunderstandings.

A proof of concept is not a prototype. A prototype is a tool used by information technology personnel to explore and better understand the client's needs. On the other hand, a **proof of concept** is like the test drive you might take in a car you were thinking about purchasing. In making the test drive you would want to put the car through whatever paces you felt would show that it could, or could not, perform as you desired.

To some degree, a proof of concept is part of the "due diligence" that good business practice dictates a client performs before committing to acquire business support or relying on a new technique. While proof of concept developments can resemble simulation prototypes, they are, as the name implies, more oriented to demonstrating the viability of a technique or existing product than to evolving the understanding needed to construct a software system.

For example, a proof of concept is often employed in connection with the production of digital cinema using a new technique. A digital cinema production unit might create a short film using a new animation technique to prove to prospective clients that the technique will meet required purposes, before they can be convinced to invest in a full length production dependent on the technique.

Or a prospective software package vendor, responding to a client's need for a new automated system, might quickly create a demonstration of its packaged software product, set up with parameters closely matching some aspect of the client's business. The demonstration might "prove" to the client that the software package is an adequate fit to some part of the client's needs.

A prototype is an evolving, exploratory development tool, while a proof of concept is a way to confirm the practicality of a specific technique or applicability of a specific solution.

Podcast 69
JOINT APPLICATION DEVELOPMENT

Joint Application Development (JAD) is one response to the inability of the simple waterfall model SDLC that Winston Royce documented as being inadequate to many software development projects. JAD attempts to more or less follow the linear stages described by the simple waterfall model, but to eliminate communication delays between people to be affected by the automated solution. JAD is an intensive fact-finding technique used to identify what users really need in a software product and then to evolve an understanding of processes that can meet those needs, based on four ideas:

1. People who actually do a job best understand the job.
2. People trained in information technology have the best understanding of the possibilities of information technology.
3. Information systems and business processes overlap business areas.
4. The best information systems result from close collaboration between business areas and information technology professionals.

Highly focused workshops are used to implement these simple ideas and foster the communication necessary to allow JAD to work. Properly scheduled, attended, and conducted, these workshops can lead to a shortening of the time needed to complete a project and can improve the quality of the resulting system. The workshops concentrate intensive effort on the front end (analysis and design) processes, rather than on the later stages where software is constructed and tested. By achieving this concentration, costs and delays are minimized since changes are easiest and least expensive to make in the front end work. Changes made in program coding and testing are more difficult, more expensive, and more likely to cause problems.

JAD tries to cut through inefficiencies that crop up in the simple waterfall model due to communication delays. JAD forces knowledgeable representation of all related parts of an organization in the analysis and design processes, instead of having a less structured systems analysis effort. By having sufficiently high level managers and working supervisors from all affected areas present in a workshop, the concerns and interrelationships of work processes can be identified much more quickly than through a process of documentation of individual work areas and later review by members of other business areas.

JAD workshops are not simply "business meetings, " they are organized along rigid lines. Typically they are conducted in a specific environment equipped with discussion support tools. Each session is led by a facilitator, and a separate person is dedicated to serve as records keeper to take notes. Also in attendance are up to five business end users, who are managers or supervisors with direct knowledge and experience with the business processes involved. Developers from the information technology area are also present, as well as subject matter experts whose experience includes both the business processes and information technology. This complement of people is unchanging; each area does not simply send a person to "take notes" and occupy a seat, but to become a continuing and contributing member to the JAD discussions. It's the responsibility of the project sponsor to make sure that participants attend and carry through on their responsibilities. In order for JAD to work, the sponsor must be at a high enough level in the organization to be able to do this.

The business end users present in JAD sessions must span the gamut of actual day-to-day responsibilities. Some must be high enough level to understand the "big picture" of the organization's operation and business strategy. Others must be supervisors who actually participate in the daily work of the business processes involved. As the staff of a major JAD project at the University of Texas point out, if only high level business managers attend the JAD sessions, the results will resemble a theoretical model lacking practical details of operation. If only lower level day-to-day business process participants are involved, the result will be a design that works for today's processes but probably won't be general enough to cope with inevitable changes in the business environment.

At an initial JAD kickoff meeting, the project sponsor, project leader, record keeper, and clients are introduced. Each member of the JAD team is made aware of JAD methods, meeting guidelines, and their responsibilities. The scope and goal of the project is defined and agreed to by all parties. Then JAD meetings are conducted on a regular basis, following this pattern:

- review the current business process
- identify problems in the process
- brainstorm solutions, including considering those used by other organizations
- survey customers for inputs and suggestions
- evaluate the list of generated ideas
- develop course of action, next steps, tasks to be accomplished
- develop the **timebox** for each task [2]
- present design conclusions to project sponsor.

This process repeats until the JAD team has considered all of the problems with the current business process and supports, evolves and refines an intended solution, and oversees its implementation.

[2] **Timebox** means that a definite deadline is set for completion of a task, but the scope of completion is left open. The intent is to complete essential aspects of the task by the deadline specified, but for the person or team performing the task to use reasonable judgment to identify and trade off elements of task scope that can be deferred. For example, suppose your mother-in-law is coming for an extended visit and arriving on a given date. Your task of making your home ready for the guest is inherently time-boxed. Making your spare room into a habitable bedroom demands that certain things be done (cleaning out the junk now stored there, installing a bed and dresser, buying extra linens, and so forth). The list of "things to do" may include some desirable things like repainting the room, having the carpet cleaned, buying and installing new curtains, and installing an air conditioner in the window. If time and perhaps budget are constrained, you quickly identify what you absolutely must accomplish to accommodate your mother-in-law, and what additional things you'll try to do, but if not done would not prevent her stay.

Podcast 70
RAPID APPLICATION DEVELOPMENT

Despite the similarity of its name to JAD, **Rapid Application Development (RAD)** is an entirely different approach to overcoming the inadequacy of the simple waterfall model for software system development. RAD was developed in the 1980's by James Martin, author of several books on business data processing in the mainframe era, and later formalized by IBM in publications in the 1990s. RAD is based on these elements:

- simulation prototyping in the language of implementation, with the intent that the prototype evolves into the actual software product
- iterative development, in which the prototype is enhanced with additional functionality
- time boxing, aimed at deferring some parts of added features to later versions, in the interest of completing iterative cycles quickly
- small teams of experienced and versatile developers and business users
- highly involved management in both the business and technical areas
- software tools and computer-aided design and development techniques.

In RAD, time boxing is intended as an antidote to scope creep, the gradual accretion of functionality that often bogs down an implementation by bits and pieces. However, time boxing must be managed carefully. If poorly handled, it may paint development into a corner, as generality required later is unwittingly forsaken.

RAD projects are conducted by placing a small number information technology developers and knowledgeable business users into a dedicated team, providing unfettered access to technology resources, and using short (typically one or two week) development cycles monitored by high level management. This builds a unity of purpose and vision in the team. It has the added advantage of forcing early discovery of major and potentially fatal obstacles that might otherwise not be recognized until much later in a more traditional development cycle where theory is not reduced to concrete form until later stages.

Computer Aided Software Engineering (CASE) tools were initially a major thrust of RAD methods, when mainframe computers were the primary vehicles of automation in large organizations. In that era, the implementation languages were procedural and included COBOL and PL/1 as supplemented by IBM's pseudoconversational teleprocessing monitor, CICS, and non-relational database management systems. These environments were characterized by programmer productivity of only a handful of lines of code per day. CASE tools were developed to assist in developing data storage arrangements, screen formats, and logic, often allowing programmers to represent requirements in compact form, or even graphically, and using these to generate procedural code and CICS screen maps. With the popularization of graphical user interfaces on PCs with Windows, the implementation of relational databases supporting SQL across all levels of computers, and the proliferation of graphical PC-based tools for design, screen building, database schema generation, and SQL logic implementation, "CASE" should perhaps now be regarded as standing for "common automated software evolution" tools. Given that RAD is not aimed at developing a "big picture" but rather at focused functionality, a RAD team will now typically use whatever productivity tools an organization already has in place to make evolution of functionality as productive as possible, using the often-proprietary toolset of whatever ERP is in local use.

A RAD application starts as a prototype and grows into an application judged to be adequate for the intended purpose. It has the advantage of involving users early in the process, capturing and reflecting some of their system needs in operable software, and making it possible to "grow" the application even as business needs continue to be identified or themselves evolve. It is appealing in contrast to attempting to involve users in a lengthy design process before any functionality is placed in their hands. Properly handled, it can deliver higher quality software.

RAD has led to further refinement and definition of iterative programming methods, including agile programming and extreme programming.

Podcast 71
END-USER PROGRAMMING

The modern office environment includes products such as Microsoft Word, Excel, Access, and similar software products from other vendors. While these are commonly employed by most people for simple tasks such as viewing a spreadsheet or printing mailing labels, they include capable and powerful programming elements. **Macros** can be used to capture sequences of keystrokes to automate repetitive operations and automatically reformat documents. Even more powerfully, **Visual Basic** has been made a part of these tools. Using it, it's possible to create automated support systems combining the functionality of several different members of the Office suite. These can create solutions for complex information handlings tasks such as extracting information from a desktop database, combining it with standard document text, and generating multiple documents and summaries from the same information. It's also possible to regard the spreadsheet grid of Excel as simply an output form and use the Visual Basic in Excel to create a program such as a machine language simulator that uses the spreadsheet as a dynamic visual output device. What makes this especially interesting is that these are end user programming tools, within the reach and understanding of people who are not primarily information technology developers.

When personnel outside of the information technology area use commonly available desktop tools to build local applications, it is often called **end-user programming (EUP)**. End user programming is both beneficial and potentially dangerous. If an end user develops software that makes him or her more productive, it's hard to argue that this is bad. However, one danger in this phenomenon is that the person evolving the software may begin to spend so much time doing it that the work they were actually hired to accomplish suffers. Assuming the software development effort succeeds, the organization may come to depend on the software developed in this way and then become vulnerable to upset when the end user leaves. In many cases, no documentation will exist for the software, and when it needs to be modified, or fails in use, it may be difficult or impossible to maintain. At points such as this, the organization's information technology division is typically called in, and may receive criticism if it can't quickly allocate resources to remedy the situation, since user expectations will have been heightened by having enjoyed the use of the software for a period of time. For all of these reasons, many information technology organizations publicize policies that attempt to distance themselves from the support of end user developed software applications.

Podcast 72
SOFTWARE TESTING

It's humorous to ask software developers and end user programmers what software testing is. Almost invariably, the response will be that "you test software to make sure that it works." In fact, this is not correct.

Testing is a process that is aimed at finding flaws. In some organizations that develop software for especially critical applications, software testers are specialists other than the programmers who create software. It's easy to see why this situation is psychologically different from the common situation of a programmer testing their own code. If a person is employed to test software, they focus on proving that it doesn't work, that it fails in some way, or that it produces incorrect answers. As a tester/specialist, this is what they are paid to do. And, since they didn't write the code, they have no vested interest in "making sure it works"; they relish finding flaws in the software.

Most of the time, the originator of software tests it. This person's mentality is anything but that of the independent software tester, as betrayed by the common answer the question of what testing is. If your goal is to confirm that your software creation works, your creativity is blunted by the desire to see it work. You will not, without serious mental prodding, come up with tests that prove your work is flawed. And, especially if time is tight, as it often is nearing the end of a software development effort, you will be inclined to optimistically interpret correct operation for some cases as sufficient "proof" that the software works.

To overcome our natural bias to be too kind to our code when we test it ourselves, different strategies have developed to orchestrate the testing process. Two types of tests can be implemented.

Glass box testing regards the software as transparent and all of its inner workings visible. This type of testing examines all parts of the code to make sure that each is executed at some time, that loops are not endless, and that all logic branches are exercised. By itself, glass box testing is not sufficient, but the test cases it leads to forming can detect certain types of errors.

Black box testing regards the software as opaque, with its inner workings not visible. In this type of testing, the performance requirements for the software, usually expressed in design documents, are used to form test cases that will exercise all possible combinations of inputs. In each case we check to see whether the software actually produces the intended output for the input.

For both glass box and black box testing, test cases for inputs to the software are formed. In order to be complete, each test case must not only document input data, but also a prediction of what correctly functioning software will produce as output in response. Testing then consists of applying the input, and comparing to see if the software actually does produce the predicted output.

Testing interactive software is more complex than testing batch software, since load testing is required in addition of input/output testing. Load testing investigates whether the software can support the intended number of concurrent users.

Lecture Slides 9

Introduction to Information Technology

Managing System Development

(C) 2006 Jim Janossy and Laura McFall, Information Technology Workbook

Slide 1- 1

Course resources

PowerPoint presentations, podcasts, and web links for readings are available at www.ambriana.com > IT Workbook

Print slides at 6 slides per page

Homework, quizzes and final exam are based on slides, lectures, readings and podcasts

Slide 1- 2

Main topics

- Understand what system development is
- Waterfall model and its inadequacies
- Joint Application Development (JAD)
- Rapid Application Development (RAD)
- Design-build-test-release method (DBTG)
- Software testing (and what it <u>is not</u>)

Slide 1- 3

System development

- Identifying what business users need in automated support functionality
- Documenting the needs
- Designing software to meet those needs
- Constructing and testing the software
- Documenting the software and its operation and use
- Installing the software

Slide 1- 4

Cowboy coding

- Everyone does their own thing
- Write programs just for fun, or to try out a great new idea
- Perhaps results in a "prototype"
- Not scalable (in terms of the product or the team)
- Uncertainty of results...

Slide 1- 5

Software engineering?

- Opposite of cowboy coding
- "Engineering" is a discipline but is software development there yet? It was not in the early years, some debate now
- Scope defined > solutions proposed and evaluated > design > review by peers > construction > documentation

Slide 1 6

A method is needed

- So user requirements can be understood and met
- So feasibility can be assessed
- So project cost and time can be estimated
- So an approved project can be budgeted, and staffed
- So results can be more certain

Slide 1- 7

Ideas and a perspective

- Software projects succeed or fail on how carefully they are planned and how deliberately they are executed
- Most projects can be successful
- A medium sized project has 3 to 25 team members, takes 3 to 18 months, makes 20,000-250,000 lines of code

Slide 1- 8

A successful project is...

- A project that meets its cost, schedule, and quality goals

- Schedule
- Cost
- Quality

} Anyone can maximize 2. But you must know how to **optimize by compromise** to do OK on all 3 !

Slide 1- 9

Method is important

- Well-defined method allows people to spend most of time on productive work
- Poor or no method, people spend a lot of their time “thrashing” (redoing things)
- Effective methods improve productivity by eliminating reinvention of common procedures, help insure consistency

Slide 1- 10

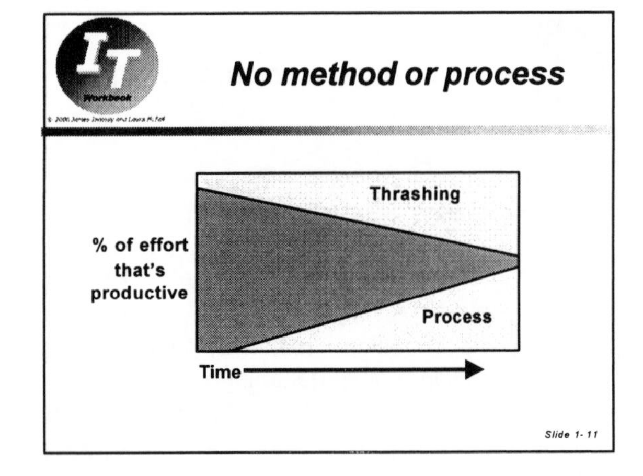

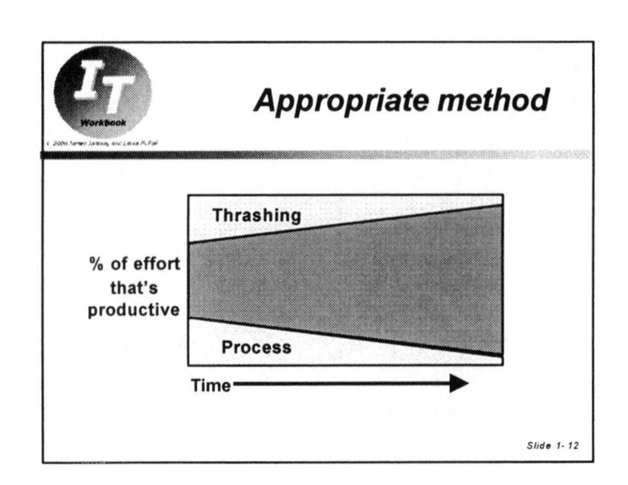

Common software development methods

- **Waterfall – System Development Life Cycle (SDLC)** and data flow diagramming: documented shortcomings!
- **Joint Application Development (JAD)**
- **Rapid Application Development (RAD)**
- **Design-Build-Test-Release (DBTG)**
- Agile programming, extreme programming

Slide 1- 13

Dr. W. W. Royce, 1970

- ***Managing the Development of Large Software Systems*** (a conference paper)
- Studies several major software projects of the 1950's and 1960's, successes and failures to produce acceptable software
- Developed a model from his survey
- Identified what seemed to work

Slide 1- 14

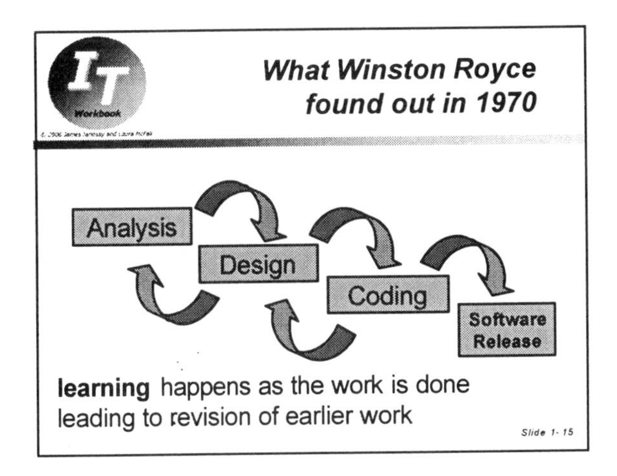

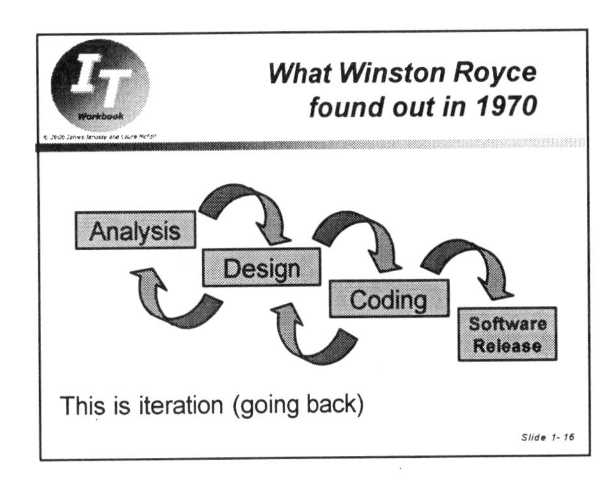

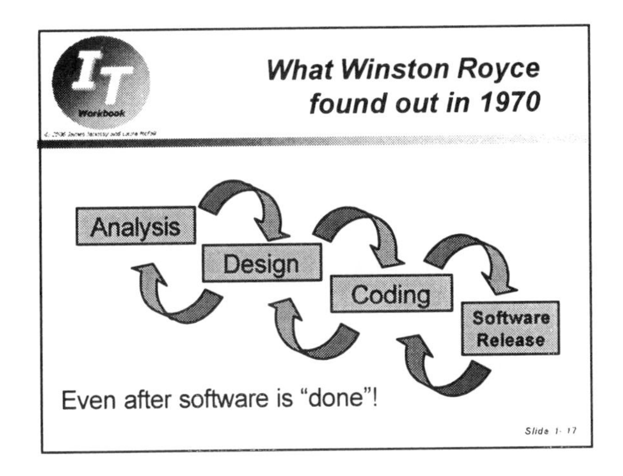

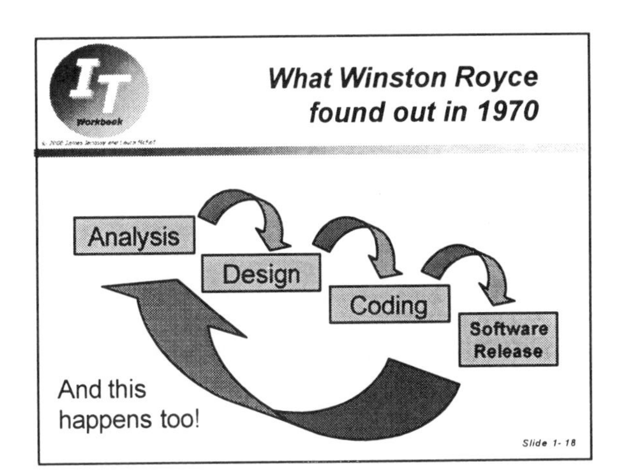

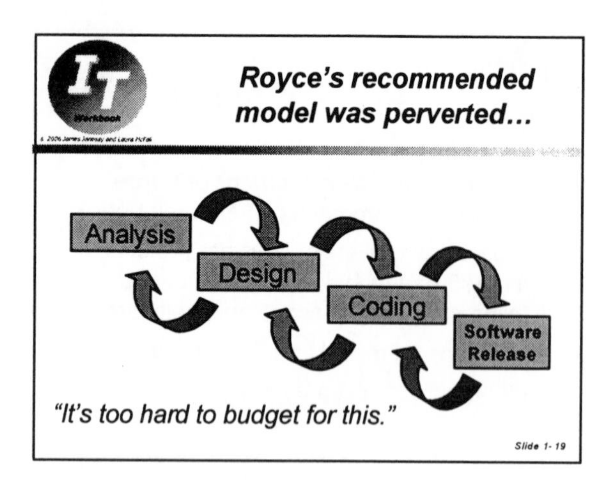
Royce's recommended model was perverted...
Analysis
Design
Coding
Software Release
"It's too hard to budget for this."
Slide 1-19

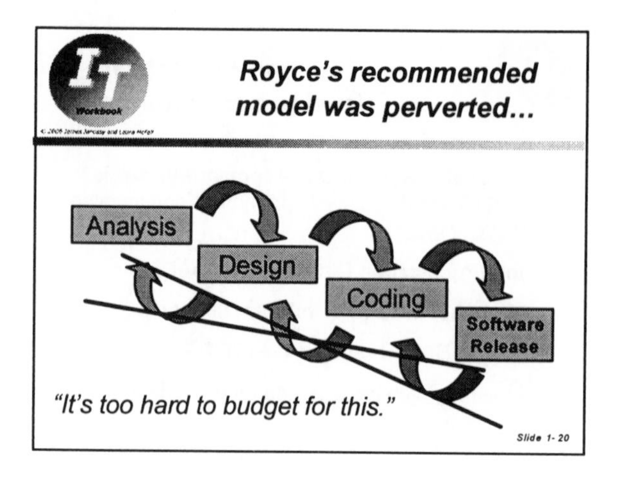
Royce's recommended model was perverted...
Analysis
Design
Coding
Software Release
"It's too hard to budget for this."
Slide 1-20

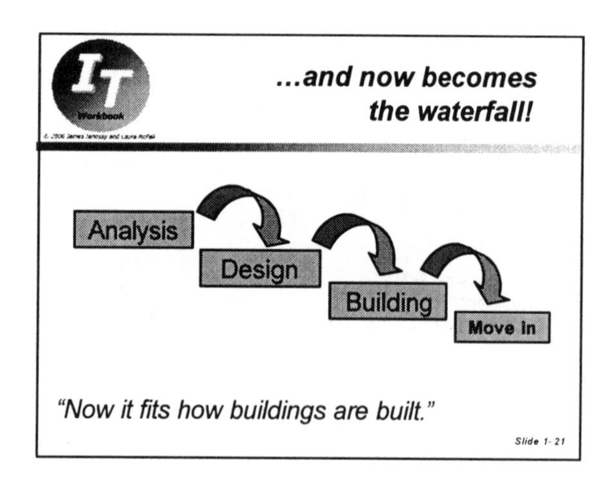
...and now becomes the waterfall!
Analysis
Design
Building
Move in
"Now it fits how buildings are built."
Slide 1-21

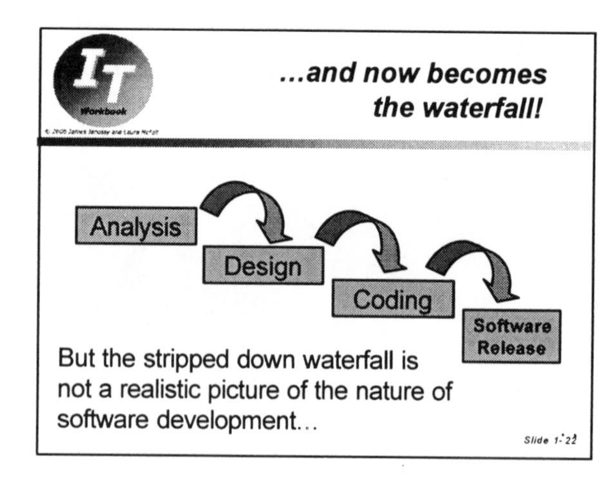
...and now becomes the waterfall!
Analysis
Design
Coding
Software Release
But the stripped down waterfall is not a realistic picture of the nature of software development...
Slide 1-22

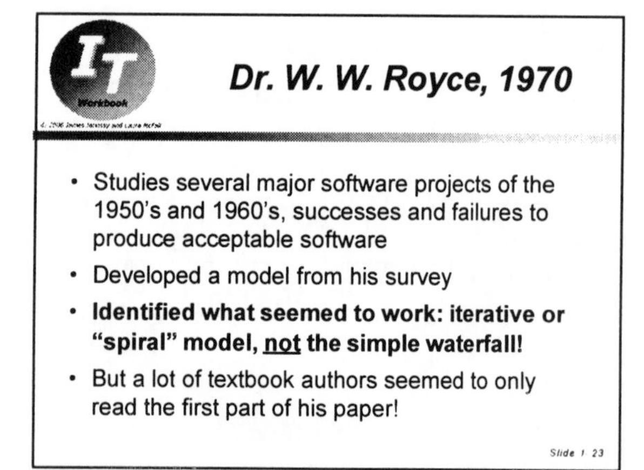
Dr. W. W. Royce, 1970
Studies several major software projects of the 1950's and 1960's, successes and failures to produce acceptable software
Developed a model from his survey
Identified what seemed to work: iterative or "spiral" model, not the simple waterfall!
But a lot of textbook authors seemed to only read the first part of his paper!
Slide 1-23

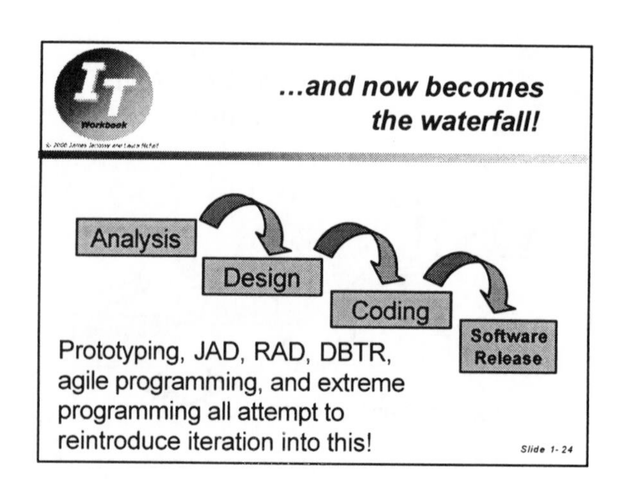
...and now becomes the waterfall!
Analysis
Design
Coding
Software Release
Prototyping, JAD, RAD, DBTR, agile programming, and extreme programming all attempt to reintroduce iteration into this!
Slide 1-24

Common software development methods

- **Waterfall – System Development Life Cycle**

All these try to overcome waterfall shortcomings:

- **Joint Application Development (JAD)**
- **Rapid Application Development (RAD)**
- **Design-Build-Test-Release (DBTR)**
- Agile programming, extreme programming

Slide 1-25

Prototyping

- Possible with hardware: breadboarded circuits after design to try out
- **Two types** of software prototypes
- **Static prototype:** screens, outputs, "storyboard" illustrates operation using sketches, PowerPoint
- **Simulation prototype:** throwaway software providing some functionality

Slide 1-26

JAD – Joint Application Development

- People who do a job understand it best
- Bring high and low level users together with IT in intensive JAD sessions
- Faciltator, records keeper, discussion aids; continuity of participants
- High-level executive support to insure commitment
- Output: an approved design for programming

Slide 1-27

RAD – Rapid Application Development

- Small team of experienced business users and IT software developers
- Simulation prototyping in the intended language of implementation
- Iterations of 1 or 2 weeks length
- Each iteration adds functionality
- Tasks are "timeboxed"

Slide 1-28

Time boxing

- A task is given a definite deadline
- Scope of the task is left open, to be decided by team using tradeoffs
- Goal is to accomplish the heart of the task, defer some functionality to a later stage
- Focuses on minimum acceptable functionality to meet deadline

Slide 1-29

Design-Build-Test-Release

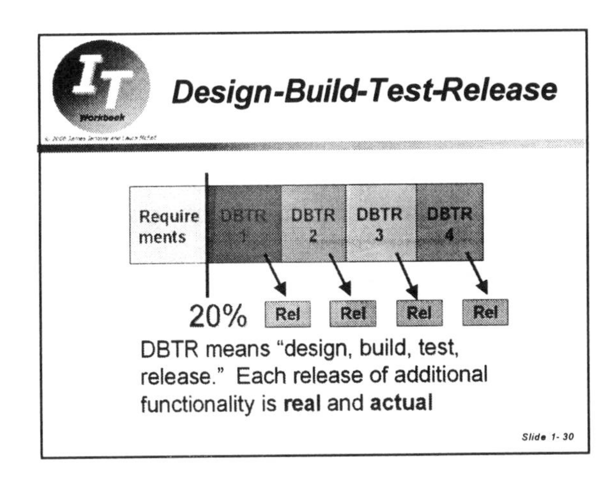

DBTR means "design, build, test, release." Each release of additional functionality is **real** and **actual**

Slide 1-30

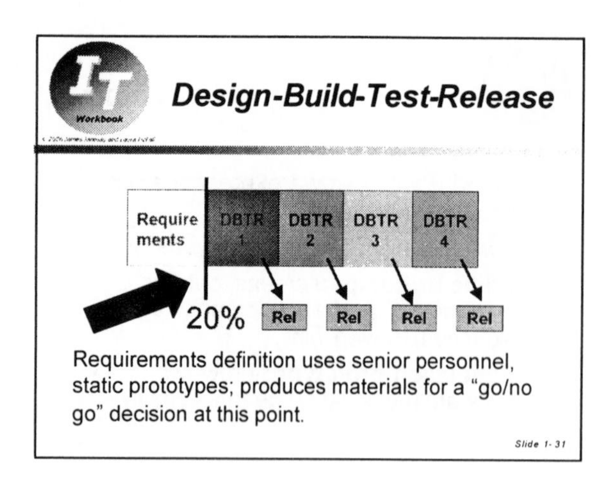

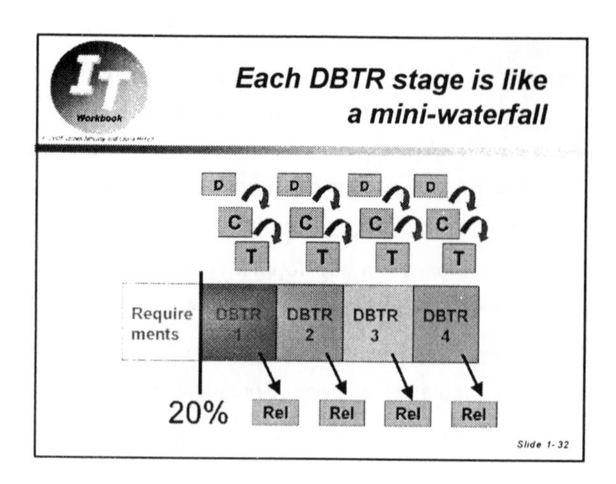

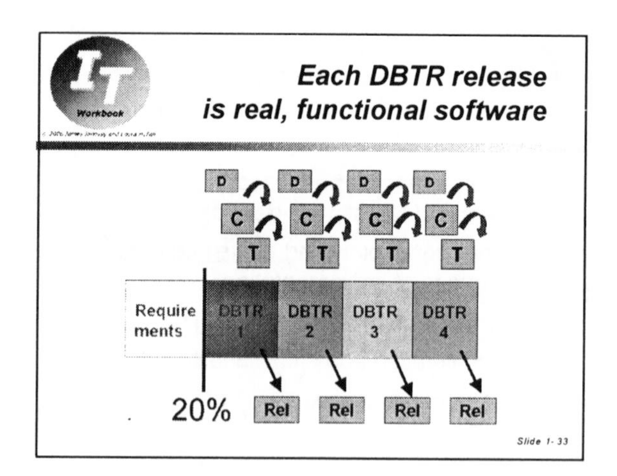

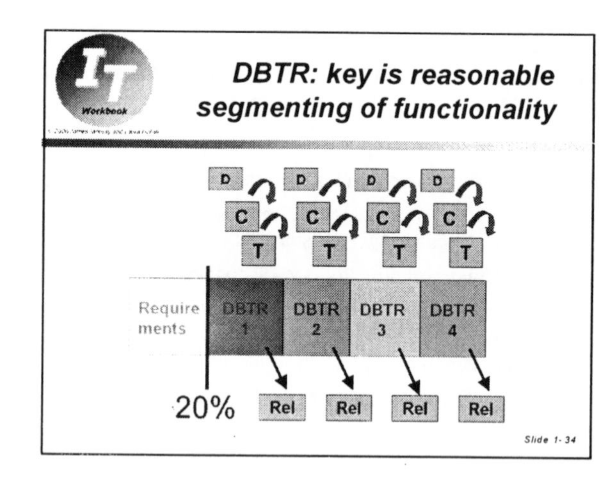

End User Programming

- Non-IT person constructs his/her own local solution using desktop software like Microsoft Office
- Can boost productivity
- Dangers: organization depends on it, originator leaves, no documentation or support!

Slide 1-35

Software testing

- Is not "proving that the software works"!
- It is trying to prove that the software doesn't work!
- **Glass box:** is all code executed?
- **Black box:** do inputs produce intended outputs?
- Test case includes prediction of output

Slide 1-36

IT Workbook Assignment 9.1

Student name:

Write your answers to the questions below <u>within the box</u>. In each case, please choose your words carefully to answer the specific questions. Avoid simply copying passages from your readings to these answers.

Explain what "**cowboy coding**" is and list **three advantages** and **three disadvantages** of it:

1) what cowboy coding is:

2) advantage:

3) advantage:

4) advantage:

5) disadvantage:

6) disadvantage:

7) disadvantage:

Describe what **software engineering** is, and how it differs from cowboy coding:

8)

IT Workbook Assignment 9.2

Student name:

Write your answers to the questions below within the box. In each case, please choose your words carefully to answer the specific questions. Avoid simply copying passages from your readings to these answers.

Describe and discuss what W. W. Royce actually had in mind when he wrote a paper in 1970 that included a description of the **waterfall model** of software development:

1)

Describe how Royce's 1970 proposal for a model for system development has been misinterpreted and misrepresented as the **waterfall model** widely documented in many texts:

2)

IT Workbook Assignment 9.3

Student name:

Write your answers to the questions below within the box. In each case, please choose your words carefully to answer the specific questions. Avoid simply copying passages from your readings to these answers.

Identify and briefly discuss **four criticisms of the commonly-understood waterfall model*** including its relationship to "incremental builds:"

1)

2)

3)

4)

* Not the model as proposed by Dr. Winston Royce, but the linear model universally known as the waterfall model.

Discuss what the technique of **prototyping** is, illustrating your discussion with a **hardware example** and a **software example**:

5) what prototyping is:

6) a hardware example of prototyping:

7) a software example of prototyping:

IT Workbook Assignment 9.4

Student name:

Write your answers to the questions below within the box. In each case, please choose your words carefully to answer the specific questions. Avoid simply copying passages from your readings to these answers.

Identify the data flow diagramming (DFD) symbols for a **process**, a **data flow**, an **external entity**, and a **data store**, including sample labeling:

1) the DFD process symbol, and typical labeling:

2) the DFD data flow symbol, and typical labeling:

3) the DFD external entity symbol, and typical labeling:

4) the DFD data store symbol, and typical labeling:

Use the data flow symbols to depict a simple situation in which a library looks up who has not returned a borrowed video by the due date, and prepares and sends a postcard reminding each borrower to return it, keeping track of and indicating to each borrower how many reminder notices it has already sent about the same video. Make sure you label each symbol appropriately:

5)

IT Workbook Assignment 9.5

Student name:

Write your answers to the questions below within the box. In each case, please choose your words carefully to answer the specific questions. Avoid simply copying passages from your readings to these answers.

Provide an overview of the **design-build-test-release (DBTG)** software development methodology and how it addresses the assessment of **project feasibility**, the **development, testing, and release of software**, and the development of the **user manual**, including the sequence in which these are accomplished:

1) overview of the DBTG software development methodology:

2) how and when feasibility is assessed in the DBTG model:

3) how and when software is constructed, tested, and released in the DBTG model:

4) how and when the user manual is developed in the DBTG model:

Identify the biggest difference between the DBTG software development model and the waterfall model of system development*:

5)

* Not the model as proposed by Dr. Winston Royce, but the linear model universally known as the waterfall model.

# IT Workbook Assignment 9.6	*Student name:*

Write your answers to the questions below within the box. In each case, please choose your words carefully to answer the specific questions. Avoid simply copying passages from your readings to these answers.

Discuss what is meant by the term **timeboxing**:

1) timeboxing is:

Identify and discuss what **agile software development** is, what it **emphasizes** in terms of the software development effort, and what it stresses as the **primary measure of progress**:

5) agile software development is:

6) agile software development emphasizes:

7) agile software development measures progress in terms of:

Provide a brief overview of what **Extreme Programming** is, and **five controversial aspects** of it that make it radically different from the waterfall model or the DBTG model:

8) brief overview of Extreme Programming:

9) controversial aspect #1:

10) controversial aspect #2:

11) controversial aspect #3:

12) controversial aspect #4:

13) controversial aspect #5:

IT Workbook Assignment 9.7

Student name:

Write your answers to the questions below *<u>within the box</u>*. *In each case, please choose your words carefully to answer the specific questions. Avoid simply copying passages from your readings to these answers.*

Briefly explain what **JAD** stands for and involves, and describe its **four basic principles**:

1) what JAD is:

2) principle:

3) principle:

4) principle:

5) principle:

Identify **seven risks** of the JAD approach that can produce an unsuccessful result:

6)

7)

8)

9)

10)

11)

12)

IT Workbook Assignment 9.8

Student name:

Write your answers to the questions below <u>within the box</u>. In each case, please choose your words carefully to answer the specific questions. Avoid simply copying passages from your readings to these answers.

Explain what **RAD** stands for, describe in general its nature, and discuss **two advantages** claimed for it as a system development methodology:

1) what RAD stands for:

2) the general nature of the RAD methodology:

3) claimed advantage #1:

4) claimed advantage #2:

Identify and briefly describe the **six core elements** of RAD:

5)

6)

7)

8)

9)

10)

IT Workbook Assignment 9.9

Student name:

Write your answers to the questions below <u>within the box</u>. In each case, please choose your words carefully to answer the specific questions. Avoid simply copying passages from your readings to these answers.

Discuss what a "**proof of concept**" is, illustrating your discussion with a **software example**:

1) what a proof of concept is:

2) a software example of a proof of concept:

Briefly identify and explain what **outsourcing**, **offshoring**, an **SLA**, and an **RFP** are, and describe <u>the relationship that often exists between these things</u>:

3) outsourcing is:

4) offshoring is:

5) an SLA is:

6) an RFP is:

7) the relationship that often exists between these things:

IT Workbook Assignment 9.10

Student name:

Write your answers to the questions below <u>within the box</u>. In each case, please choose your words carefully to answer the specific questions. Avoid simply copying passages from your readings to these answers.

What is the purpose of **software testing**?

1) this purpose of software testing is:

Identify and briefly discuss the five **common quality attributes** associated with software testing:

2)

3)

4)

5)

6)

In ***Software Negligence and Testing Coverage*** (www.badsoftware.com/coverage.htm) Cem Kaner describes how a firm can be sued for a software malfunction. Under "Software Malpractice" he lists **four claims** that might be the basis of a lawsuit. Identify and briefly describe these claims:

7)

8)

9)

10)

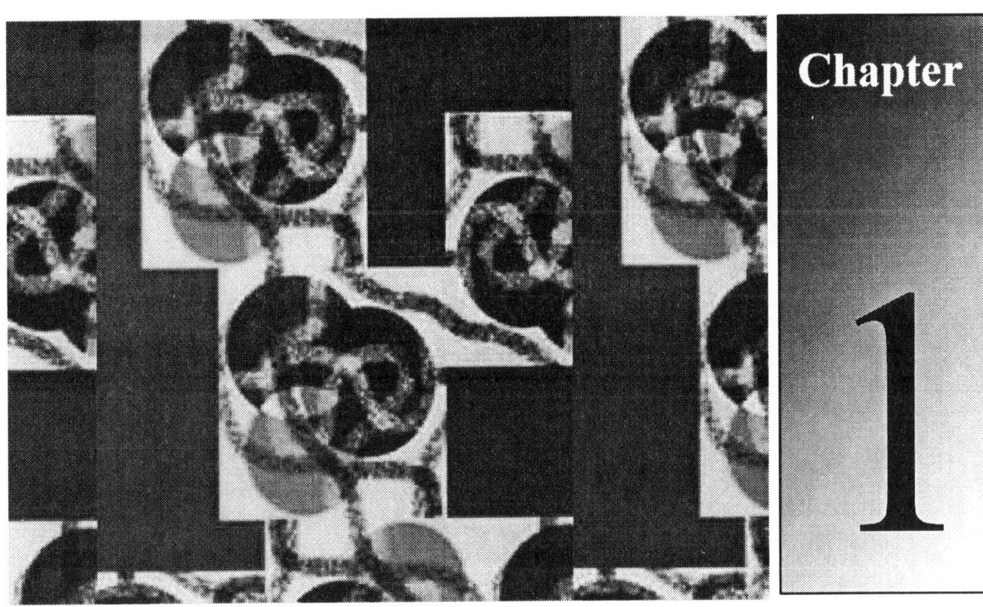

Managing Information Systems

Businesses generate earnings by buying raw materials, adding value to them, and selling them to customers at costs that cover acquisition, labor, and overhead plus an extra margin, called profit. Properly implemented and managed information systems can reduce costs and increase business competitiveness and profitability.

Value chain analysis helps identify the processes within a business that may be improved with information technology. Information technology can be used to distinguish the way a business deals with customers, providing a unique experience. In this way IT can become part of a **differentiation strategy**.

An effort known as **business process reengineering** (BPR) seeks to redesign business processes to reduce cost and improve quality, service and speed. The implementation of automated **knowledge management systems** (KMS) can aid in BPR by capturing and making accessible the knowledge gained by skilled employees. In this chapter, we examine how value chain analysis, business process reengineering, and knowledge management systems are often used in combination in the modern business enterprise.

Software maintenance is an ongoing information technology management responsibility. Complex modern systems mixing various hardware and software platforms require almost constant attention to upgrades and revisions due to vendor and business changes. IT maintenance consumes as much as 50% of the skilled resources available.

Business continuity (disaster recovery) measures are also a critical information technology responsibility. These include backup arrangements, the preservation of critical information in secure offsite facilities, and arrangements to resume business operation following a disaster that affects a primary computer facility.

Learning objectives

Chapter 10
Managing Information Systems

This chapter, in combination with the web readings, web links, and podcasts available at www.ambriana.com is focused on providing the knowledge to achieve these learning objectives:

- Be able to describe and discuss the major parts of the information technology organization presided over by a chief information officer in a large business organization
- Understand what maintenance responsibilities exist in the modern environment, of both the software maintenance and license maintenance varieties
- Recognize and be able to distinguish and describe four forms of software maintenance
- Understand an be able to describe three types of strategies business organizations use to differentiate themselves from their competitors
- Be able to define and describe Business Process Reengineering (BPR), its key principles, and the major difficulty encountered in apply it
- Understand what Knowledge Management is and the factors that drive it
- Understand and be able to discuss the place of end user programming in the modern information technology environment
- Be able to describe the role that Business Continuity Planning (BCP) plays in information technology organizations, and what a BCP manual contains
- Understand the I.T. organization's responsibility to create and maintain secure offsite backups of data assets, and how a storage network array (SAN) and high speed data links can be used to automate this process.

Podcast Scripts – Chapter 10

These readings of the **Information Technology Workbook** are also accessible in audio form as .mp3 files for free download. Visit the workbook web site, www.ambriana.com. You can listen to this material on your own Ipod or .mp3 player or computer, read it, or both listen and read.

Podcast 73
THE I.T. ORGANIZATION

How and when companies and organizations use technology is critical to gaining and holding a competitive edge. Executives in the information technology area play a vital role in the technological direction of their organizations. They do everything from constructing the business plan to overseeing network security to directing Internet operations.

IT executives plan, coordinate, and direct research and facilitate the computer-related activities of firms. They help determine both technical and business goals in consultation with top management, and make detailed plans for accomplishing these goals. For example, they may develop the overall concepts and requirements of a new product or service, or identify how an organization's computing capabilities can effectively aid project management.

IT executives direct the work of **systems analysts**, **computer programmers**, **support specialists**, and other computer-related workers. They plan and coordinate activities such as installation and upgrading of hardware and software, programming and systems design, development of computer networks, and implementation of Internet and intranet sites. They are increasingly involved with the upkeep, maintenance, and security of networks, and analyze the computer and information needs of their organizations from an operational and strategic perspective, determining immediate and long-range personnel and equipment requirements. They assign and review the work of their subordinates and stay abreast of the latest technology to ensure the organization does not lag behind competitors.

The duties of IT executives vary with their specific titles. The **Chief Information Officer (CIO)** oversees the entire information technology area of a firm, and generally reports to the highest executive level. The **Chief Technology Officer (CTO)** reports to the CIO, and manages and plans technical standards and may evaluate the newest and most innovative technologies and determine how these can help their organizations. Because of the rapid pace of technological change, chief technology officers are constantly on the lookout for developments that could benefit their organizations. They are responsible for demonstrating how information technology can be used as a competitive tool that cuts costs, increases revenue, and maintains or increases competitive advantage.

Management information systems (MIS) directors manage information systems and computing resources for their organizations. They may also work under the chief information officer and plan and direct the work of subordinate information technology employees. These managers oversee a variety of user services such as help desks, which employees can call with questions or problems. MIS directors may also make hardware and software recommendations based on their experience with an organization's technology.

Project managers develop requirements, budgets, and schedules for their firms' information technology projects. They coordinate such projects from development through implementation, working with internal and external clients, vendors, consultants, and computer specialists.

Network managers provide a variety of services, from design to administration of the local area network, which connects staff within an organization. These managers direct the staff supporting the organization's connection to the Internet, and often are responsible for network security.

Computer and information systems managers spend most of their time in an office. Most work at least 40 hours a week and may have to work evenings and weekends to meet deadlines or solve unexpected problems. Some computer and information systems managers may experience considerable pressure in meeting technical goals within short timeframes or tight budgets. As networks continue to expand and more work is done remotely, computer and information systems managers have to communicate with and oversee offsite employees using modems, laptops, e-mail, and the Internet.

Advanced technical knowledge is essential for computer and information systems managers, who must understand and guide the work of their subordinates, yet also explain the work in non-technical terms to senior managers and potential customers.

Computer and information systems executives need a broad range of skills, including experience with the specific software or technology used on the job, as well as a background in either consulting or business management. The expansion of electronic commerce has elevated the importance of business insight. Many computer and information systems executives are called on to make important business decisions, and need a keen understanding of people, management processes, and customers' needs.

Because IT organizations are not usually income producers, their budgets, which include funding for computer equipment, software, support, projects, and the manpower to handle all of these, are typically provided by other departments within their organizations that do produce income, such as the sales department. Each year IT executives must defend their spending, and request funds to continue their operations in the upcoming year. This can be a difficult task, especially if business revenue has not met projections for growth or other costs, such as energy, are rising. Department heads tend to want to keep their budgets for their own use rather than giving it to other departments to manage. Organizations recognize the importance of their IT department in maintaining competitiveness and fund IT adequately to maintain the technical competence vital to their survival in our technology-based economy.

Podcast 74
SOFTWARE MAINTENANCE

Software maintenance is the process of enhancing and optimizing deployed software, that is, software that has already been loaded onto computers throughout an organization, as well as correcting defects in software once it has been deployed.

Software maintenance is one of the phases in the system development process, and follows deployment of the software into the field. The software maintenance phase involves changes to the software in order to correct defects and deficiencies found during field usage as well as the addition of new functionality to improve the software's usability and applicability.

In a formal software development environment the developing organization will have mechanisms in place to document and track defects and deficiencies. Software, just like most other products, is typically released with a known set of defects and deficiencies. The software is released with such issues because the development organization decides the utility and value of the software at a particular level of quality outweighs the impact of the known defects and deficiencies.

The known software defect issues are normally documented in release notes so that the users of the software will be able to work around the known issues and will know when the use of the software would be inappropriate for particular tasks. With the release of the software, the users of the software will discover other undocumented defects and deficiencies. As these issues are reported into the development organization they will be entered into a defect tracking system. The personnel involved in the software maintenance phase are expected to work on these known issues, address them, and prepare for a new release of the software which will address the documented issues.

Many software-producing organizations such as Microsoft, provide free software corrections, called **patches**, on their web sites. Microsoft also releases weekly bulletins regarding software updates and concerns. From surveys of real world software development organizations, it would appear that the majority of software maintenance cost is expended for non-corrective actions, that is, evolution of the software to enhance it. About 20 percent of maintenance is, however, **corrective maintenance**, which is performed reactively to correct discovered problems, coding errors, design flaws, and overcome misunderstood requirements.

Several additional categories of software maintenance exist:

Adaptive modifications keep software usable in a changing environment, allowing it to interface with other systems and converting it to use other hardware such as migrating legacy systems to newer platforms. This allows organizations to retire systems such systems gracefully, without a loss of productivity.

Perfective modifications improve performance or maintainability, and include making human interface enhancements, improving the efficiency of the design, and streamlining the business process.

Preventive modifications detect and correct latent faults, preventing system performance from degrading to unacceptable levels.

Some issues to address in performing software maintenance include maintaining control over system modifications, sustaining the system's day-to-day operations during maintenance, and measuring the effort expended on software maintenance.

The word "maintenance" is also used in another way in connection with software. Software license maintenance is an annual cost for continuing support from the vendor, including access to upgrades. Licensing for software can be expensive. For most enterprise software, companies pay 15 to 20 percent of the original licensing cost annually for software updates and technical support. This charge hits their budget year after year, and cost increase clauses may also apply.

Some companies charge maintenance fees based on the list price of the software they've sold, no matter how good a deal was negotiated up front. Companies should be especially wary of this. In a difficult economy, software firms often offer huge discounts to keep income rolling in. To them, software license maintenance fees are an annuity!

It isn't uncommon for a software company to charge a maintenance fee of 20 percent of the list cost of their software. A discount on the original licensing fee can be smothered by the continuing cost of maintenance. For example, let's say the full licensing fee for software is $100,000 but the firm acquiring it negotiates a 70% discount and obtains it for only $30,000. That's a great deal! But the annual maintenance cost will be 20% of the $100,000 list price, which is $20,000 a year. After a few years, the amount the company pays in maintenance fees will far exceed the original license cost. This is the main reason that firms periodically examine their software portfolio (a fancy word that simply means the software they have loaded on their computers) and remove unused software. It makes no sense to maintain software that is not used, by spending money for upgrades and installing them.

Who actually performs software maintenance, and applies the upgrades and corrective maintenance patches that annual software license maintenance pays for? In most organizations, software maintenance on locally developed application software is performed by members of the group that developed the software and its functionality. Vendor-supplied patches, however, are usually applied by a special staff, with the higher levels of security authorization to deal with the software libraries containing "production" program code. A database administration staff usually handles patches to the database management system software, and an operating system or ERP administration team handles patches from the ERP vendor.

Podcast 75
PORTER AND CARR ON DIFFERENTIATION

According to Professor **Michael E. Porter**, who leads the Institute for Strategy and Competitiveness at the Harvard Business School, a three-part category scheme applies the types of strategies commonly used by businesses in marketing their products to competitive advantage. These three strategies are **cost leadership, market segmentation**, and **differentiation**. Porter feels that cost leadership and differentiation are the most important of the three strategies.

Cost leadership emphasizes efficiency. By producing high volumes of standardized products, the firm hopes to take advantage of economies of scale. The more of a product a firm makes, the less each unit of the product costs. Experience curve effects often cause this: as a firm gets more experienced at a task, the more efficient it becomes in performing that task. The resulting product is often a basic, no frills product made at relatively low cost and made available to a very large customer base. Consumable products that get "used up" are often of this variety, such as paper, food products, metal fasteners, and building products.

Market segmentation focuses on a small but profitable market niche. An example of this would be those market segments that can afford to buy Hummers or Corvettes. The number of people who can purchase these vehicles is small, but they represent a very high profit margin for General Motors, and upscale purchasers typically have no financing difficulties and may even pay cash for purchases.

Differentiation involves creating a product that is perceived as unique in the market place. The unique features or benefits of the product should provide superior value for the customer if this strategy is to be successful. Because customers see the product as unrivaled and unequaled, customers tend to remain loyal to the brand even if the price for the product rises. This can provide considerable insulation from competition. A differentiator cannot completely ignore the effect of the cost to buyers, however. Even the people who buy Hummers if the price rose considerably, because other currently more expensive luxury and status cars also exist.

To maintain a differentiation strategy, a firm must have strong research and development, product engineering, and marketing capabilities, good cooperation with distribution channels, and even offer incentives to its customers. The firm must be able to communicate the importance of the differentiating product characteristics, be able to stress continuous improvement and innovation. This is why firms that succeed in exploiting the differentiation model advertise heavily and often maintain a brand association with celebrities who they feel their customers admire.

Nicholas Carr refers to the differentiation strategy in discussing information technology but diminishes the value of IT as a differentiator. In his view, IT was a differentiator between firms in the 1970's era, when it was a leading edge factor in delivering innovative service to customers. However, Carr points out that now, every firm must have an IT capability in order to be in business at all. In his view, excess expenditures on IT are not warranted, and in fact might be harmful if they lead a firm to explore unproven technologies simply for their own sake. Perhaps Carr's advice needs to be applied selectively, with the context of an individual firm in mind. A firm that can gain cost or differentiation advantage by adopting new technology would be foolish to not invest in it. But a firm in which additional IT investment would not gain any advantage, and impose costs in acquisition and an unproductive learning curve, would be foolish to adopt it. Would Carr have advised Nestle not to undertake it's 200 million dollar investment in an ERP in 2000? Nestle claims to have saved much more than this amount in a few years in efficiencies by adopting this leading edge technology. On the other hand, an organization that did not need an ERP to remain competitive, such as a state college system, perhaps has no real need to acquire one, so long as its existing student support system can adequately enroll students, collect tuition, record grades and produce transcripts and other necessary documents. Some things do not have to be leading edge, they just have to provide adequate support.

Podcast 76
VALUE CHAIN

To better understand the activities through which a company develops a competitive advantage and creates shareholder value, it is helpful to separate the business system as a whole into a series of value-generating activities referred to as the value chain. In his 1985 book ***Competitive Advantage,*** Michael Porter introduced a generic value chain model comprised of a sequence of activities common to a wide range of companies. Porter identified the primary and support activities of a firm as follows.

Primary value-adding activities include

- **inbound logistics**, receiving and warehousing raw materials, and their distribution to manufacturing processes as they are needed
- **production and operations**, the processes of transforming inputs into finished products and services
- **outbound logistics**, the warehousing and distribution of finished goods
- **sales and marketing**, including advertising, sales representation, and order taking
- **maintenance and service**, the support of customers after products and services are sold to them.

Activities that support these primary functions include organizational structure, control systems, and company culture, human resources management, which includes recruiting, hiring, training, development, and compensation; research and development, and procurement, or purchasing inputs such as materials, supplies, and equipment.

The firm's profit depends on its ability to efficiently perform these activities so the amount the customer is willing to pay for products is more than the cost to the firm of the activities in the value chain. What the customer is willing to pay must cover the cost of production with enough left over for the firm to make a profit on activities performed. It is in these activities that a firm has the chance to generate superior value. A competitive advantage may be achieved by reconfiguring the value chain to provide lower cost to customers or better differentiation.

The value chain model is a useful analytical tool for defining a firm's core competencies and the activities in which it can pursue a competitive advantage. Core competencies are those things a firm can do well that provide customer benefits, that are hard for competitors to imitate, and that can be leveraged widely to as many products and markets as possible.

There are two ways a firm can pursue a competitive advantage. The first, cost advantage, can be achieved by better understanding costs and squeezing them out of the value-adding activities in the value chain. The second, an advantage in differentiation, can be achieved by focusing on those activities associated with core competencies and capabilities in order to perform them better than the firm's competitors.

Once the value chain is defined, a firm may create a cost advantage by reducing the cost of individual value chain activities, or by reconfiguring the value chain itself. A cost analysis can be performed by assigning costs to value chain activities. In this way, the value chain is similar to systems analysis, but applied in a very specific way.

Porter identified ten cost drivers related to these activities. These drivers are economies of scale, learning, capacity utilization, linkages among activities, interrelationships among business units, the degree of vertical integration, timing of market entry, the firm's policy of cost or differentiation, geographic location, and institutional factors such as regulation and union activity. A firm develops a cost advantage by controlling these drivers better than its competitors.

A firm can also recognize a cost advantage by reconfiguring the value chain, making structural changes such as a new production process, setting up new distribution channels, or developing a different sales approach. For example, FedEx structurally redefined express freight service by acquiring its own planes and implementing a hub and spoke system.

Differentiation stems from uniqueness, and can come from any part of the value chain. A differentiation advantage may be achieved by changing individual value chain activities to increase uniqueness in the final product, or by reconfiguring the value chain itself.

Porter identified a number of drivers of uniqueness, including policies and decisions, linkages among activities, timing, location, interrelationships, learning, integration, better service as a result of larger scale, and institutional factors. Many of these also serve as cost drivers. Differentiation often incurs greater costs, resulting in tradeoffs between cost and differentiation. Sometimes, however, a firm may be able to reduce cost in one activity while enjoying a cost reduction in another, such as when a design change simultaneously reduces manufacturing costs and improves reliability, and service costs are also reduced.

A firm can reconfigure its value chain in several ways to create uniqueness. It can "forward integrate" in order to perform functions once performed by its customers. It can "backward integrate" in order to have more control over inputs. It may implement new process technologies or utilize new distribution channels. Ultimately, the firm may need to be creative to develop a novel value chain configuration that increases product differentiation.

Because technology is employed to some degree in every value creating activity, changes in technology can impact competitive advantage by incrementally changing the activities themselves or by making new configurations of the value chain possible. Various technologies are used in both primary and support value activities. These can include transportation, material handling and storage, communications, testing, and information systems. Many of these technologies are used across the value chain. For example, information systems are seen in every activity. To the extent that these technologies affect cost drivers or uniqueness, they can lead to a competitive advantage. Even Nicholas Carr would have to admit that information technology, applied selectively in making value chain and activity improvements, could affect the cost or differentiation capacity of a firm. The key words are "applied selectively." New technology for technology's sake is not a wise investment.

A firm may specialize in one or more value chain activities and outsource the rest. **Outsourcing** means clearly specifying what the work is, and buying it—goods or services—from other firms. Outsourcing is clearly advantageous when it lowers the cost of an item below that which it would cost the firm to produce it itself, such as low-volume printing or desktop maintenance. The extent to which a firm performs upstream and downstream activities is described by its degree of vertical integration. In outsourcing, managers may also consider whether the activity is one of the firm's core competencies from which a cost advantage or product differentiation is realized, what the risk of outsourcing is great, and whether the outsourcing of an activity can result in business process improvements such as reduced lead time or higher flexibility. A firm may wish to outsource to more than one supplier in order to reduce risk.

A firm's value chain is part of a larger system that includes the value chains of upstream suppliers and downstream distribution channels and customers. Porter calls this series of value chains the **value system**. Linkages exist not only in a firm's value chain, but also between value chains. While a firm exhibiting a high degree of vertical integration is poised to better coordinate upstream and downstream activities, a firm having a lesser degree of vertical integration nonetheless can forge agreements with suppliers and channel partners to achieve better coordination. For example, an auto manufacturer may have its suppliers set up facilities in close proximity in order to minimize transportation costs, and reduce parts inventories and the risk of delivery delays. Clearly, a firm's success in developing and sustaining a competitive advantage depends not only on its own value chain, but also on its ability to manage the value system of which it is a part.

Podcast 77
BUSINESS PROCESS RE-ENGINEERING

Business Process Re-engineering is the fundamental rethinking and radical redesign of business processes to achieve dramatic improvements in critical, contemporary measures of performance, such as cost, quality, service, and speed. Understanding the fundamental operations of a business is the first step prior to re-engineering. Business people must ask the most basic questions about their companies and how they operate.

Why do we do what we do?

Why do we do it the way we do?

Three kinds of companies undertake re-engineering. Some companies find themselves already in deep trouble and have no choice but to examine the activities in their value chain and change them, or go out of business. Other companies that foresee themselves in trouble, because of the changing economic environment, and begin examining the activities in their value chains. An finally, some astute companies that are in peak condition see re-engineering as a chance to further their lead over their competitors.

Radical redesign means disregarding all existing structures and procedures, and inventing completely new ways of accomplishing work. Re-engineering is about business reinvention, and begins with no assumptions, and takes nothing for granted. Re-engineering is not about making marginal improvements or modifications to tune up existing processes, but about achieving dramatic improvements in performance.

In traditional business structures, organizations are divided into departments and process is separated into the simplest of tasks, distributed across departments. This specialization is based on the assembly-line organization of work, which was revolutionary in about 1910. This type of task-based thinking needs to shift to process-based thinking, in order to gain efficiency. The following example illustrates the characteristics of re-engineering.

IBM Credit Corporation is in the business of financing the computers, software, and services that IBM Corporation sells. IBM Credit's operation was comprised of five steps, as follows:

1. When an IBM field sales representative called in with a request for financing, one of the operators in the central office wrote down the request on a piece of paper.
2. The request was then dispatched to the credit department, where a specialist checked the potential borrower's creditworthiness, wrote the result on the piece of paper, and dispatched to the next link in the chain, which was the business practices department.
3. The business practices department was in charge of modifying the standard loan covenant in response to a customer request. Any special terms would be attached to the request as needed.
4. The request went to the price department where a pricer determined the appropriate interest rate to charge the customer.
5. Finally, the administration department turned all this information into a quote letter that could be delivered to the field sales representative.

The original process took an average of six days. From the sales representative's point of view, this turnaround was too long. The customer could be seduced by another computer vendor in the interim and decide to buy competing equipment. To improve this process, IBM Credit tried several minor changes, none of which were successful. Eventually, two senior managers at IBM Credit took a customer request and walked themselves through all five steps in the process. They found that performing the actual work took just ninety minutes. Clearly, the problem did not lie in the tasks and the people performing them, but in the process itself. IBM Credit replaced the specialists and the old process. The credit checkers, pricers, and so on, were replaced with generalists. Now, a generalist processes the entire request from beginning to end with no handoffs.

How could one generalist replace four specialists? The old process design was, in fact, founded on a deeply held but deeply hidden assumption. The assumption was that every bid request was unique, thereby requiring the intervention of four highly trained specialists. In fact, this assumption was false. Most requests were simple and straightforward. Processing consisted of finding a credit rating in a database, plugging numbers into a standard model, and pulling standard clauses from a file and assembling them into a document. These tasks fell well within the capability of a single individual when he or she was supported by an easy-to-use computer system. IBM Credit developed a sophisticated computer database to support their generalists. In most situations, the system provides guidance and data to generalists. In really tough situations, the generalists can get help from a small pool of actual specialists who are can work with the processing team as needed.

The new turnaround became four hours instead of six days. The company achieved a dramatic performance breakthrough by making a radical change to the process. That is the definition of re-engineering. IBM Credit did not ask, "How do we improve the calculation of a financing quote?" They did not ask "How do we enhance credit checking?" Instead, the question asked was, "How do we improve the entire credit issuance process?"

The preceding example appears simple and attractive. However, the closer we look at it, the more questions about the benefits of re-engineering arise. For example, what are the underlying costs for the implementation of the radical change? People need intensive training for their new skills and their styles. The new skills affect the ways in which they think, and behave, and their attitudes. It is critical to appreciate, that what they believe, is important about their work. What are the implications of the radical change to the organization, especially the human issues? Organizations are communities of people who cannot be treated as machines. People may resist change and fear losing their jobs. Re-engineering requires that people take on more responsibility. It requires that employees learn and change constantly. This may cause discomfort for people who seek stability in their lives. Retraining must be taken into account and provided.

Information technology plays a crucial role in business re-engineering. IT is essentially an enabler. However, many people look at technology through the lens of their existing tasks. That is, they think only about applying computerization to what already exists. In this example, IBM Credit might have tried to digitize the request application and send it to different departments via a computer network. Such computerization would have accelerated the movement of pieces of paper from one department to another but it would have then just increased the time requests waited in an electronic in-box in each department, doing nothing to improve the overall process.

Redesigning processes should not be constrained by information technology. Rather, after the business process is redesigned, we should then seek the best technology to implement the revised process. Similarly, past investments in information technology should not be allowed to constrain the redesign.

Podcast 78
BUSINESS CONTINUITY PLANNING

Business continuity planning (BCP) is a methodology used to create a plan that defines how an organization will resume critical functions, within a predetermined time period, after a catastrophic disruption. BCP is part of organizational efforts to reduce operational risk. In many organizations, BCP was formerly called **disaster recovery**.

Business continuity planning is not a new concept. Plans for dealing with adverse conditions are evident from the beginning of human history. Stockpiling water and food in a fortress in case of attack was a prudent measure in ancient times. Installing life boats on ships is a common practice in case of a sinking. In the years prior to January 1, 2000 governments anticipated computer failures, called the "Year 2000" or "Y2K" problem, in important parts of the nation's infrastructure such as power transmission, telecommunications, and the health and financial industries. Regulatory agencies required those industries to formalize business continuity planning manuals to protect the public.

Regulatory and business focus on BCP waned due to the problem-free Y2K rollover. But this lack of interest came to an abrupt end on September 11, 2001 when simultaneous terrorist attacks devastated the twin towers of the World Trade Center in New York City, caused serious damage to the Pentagon, and caused the crash of a commercial flight in Pennsylvania. In watching the businesses affected by the destruction of the World Trade Center it became evident that every business needs to have a business continuity plan in order to recover from a natural or man-made disaster. Any type of organization may create a BCP manual. Every organization should have one, in order to insure survival of the organization's information assets and its ability to carry on.

Proof that firms may not invest enough time and resources in BCP preparations are evident in disaster survival statistics. Fires permanently close 44% of the businesses they affect. In the earlier 1993 World Trade Center bombing, 150 businesses out of the 350 affected by the relatively limited damage of the explosion failed to survive the event. Conversely, the firms with well developed and tested BCP manuals which were affected by the September 11 attacks were back in operation within days.

The following six items summarize what any business continuity plan must address:

- alternative methods for business process handling
- IT systems back up and recovery
- premises recovery
- customer service recovery
- administrative and operations recovery
- insurance coverage and claim filing procedures.

A BCP manual for a small organization may be as simple as a printed manual stored safely away from the primary work location. It would contain the names, addresses, and phone numbers of staff and vendors, along with the location of the off-site data back up storage media, copies of insurance contracts, and other critical materials necessary for organizational survival. At the other extreme, for a large organization, a BCP manual may outline a secondary work site, technical requirements and readiness, regulatory reporting requirements, work recovery measures, the means to re-establish physical records, the means to establish a new supply chain, and the means to establish new production centers.

The development of a BCP manual has five phases:

- analysis
- solution design
- implementation
- exercising the plan
- maintaining the plan.

The **analysis phase** in the development of a BCP manual consists of a threat analysis, an impact analysis, and impact scenarios with resulting documentation. An impact analysis differentiates between the critical and non-critical functions within the organization. The impact analysis also results in an understanding and documentation of the recovery requirements for each critical function.

The goal of **solution design** is to identify the most cost effective disaster recovery solution that meets two main requirements from the impact analysis stage: the minimum application software and application data recovery requirements, and the time frame in which this must be available. The solution design phase determines the crisis management command structure. It also defines the location of a secondary work site, if necessary, the telecommunication architecture between primary and secondary work sites, and data replication methodology between primary and secondary work sites to keep the second site ready.

The **implementation phase** is the execution of the design elements identified in the solution design phase. This consists of negotiating contracts with suppliers of off-site data storage, second site vendors, telecommunications arrangements, and any other services or premises required.

The purpose of **exercising the plan** is to confirm that the solution satisfies requirements for the organization's recovery. Plans may fail to meet expectations due to insufficient or inaccurate recovery requirements, solution design flaws, or solution implementation errors. At a minimum, exercising the plan is generally conducted at least every two years. Problems identified in the initial plan exercise phase should be rolled up into the maintenance phase and retested in the next exercise.

Maintenance of a BCP manual is broken down into three periodic activities: confirmation of information in the manual, the exercise and verification of technical solutions, and the exercise and verification of documented recovery procedures. As a part of ongoing maintenance, the question must be answered: does the replacement equipment and software work as intended?

All organizations change over time. A BCP manual must therefore be updated to remain relevant. Changes that should be identified and updated include staffing changes, address and telephone numbers of staff, clients and vendors, internal reorganizations, and whether any systems used in the execution of critical functions have been modified.

Podcast 79
KNOWLEDGE MANAGEMENT

Knowledge Management (KM) refers to the systematic collection, preservation, and provision of accessibility to information held by key individuals in an organization. KM is particularly concerned with developing ways to preserve information for decision-making and problem solving that resides in the experience of knowledgeable employees, since turnover or retirement would remove this knowledge if it is not captured. KM deals with identifying and organizing knowledge leading to the reuse and leveraging of existing knowledge. Many elements of knowledge, however, are not readily reduced to digital form, such as judgment, leadership, persuasiveness, aesthetics and humor.

Knowledge Management has benefited from developments in three related field: information management, quality improvement efforts, and the human capital movement. These are thrusts that developed in the last quarter of the twentieth century.

Information management focuses on how, independent of technologies, information is actually managed and manipulated. Information management deals with data, structured messages, and documents. This field has been spurred in major ways by the development of capable Internet search engines such as Google, which can be adapted to search private text databases within organization intranets.

Quality improvement efforts have resulted in the attempt to organize and document practices and procedures through efforts such as ISO-9001. Knowledge about organizational functions not otherwise captured are being reduced to documented form under these efforts.

The **human capital approach** recognizes the benefit to organizations from providing training to workers and regarding employees as an investment. To conduct this training on organization-specific processes and assets, the accumulated knowledge of existing skilled workers is essential.

Lecture Slides 10

Introduction to Information Technology

Managing Information Systems

Course resources

PowerPoint presentations, podcasts, and web links for readings are available at www.ambriana.com > IT Workbook

Print slides at 6 slides per page

Homework, quizzes and final exam are based on slides, lectures, readings and podcasts

Slide 10- 2

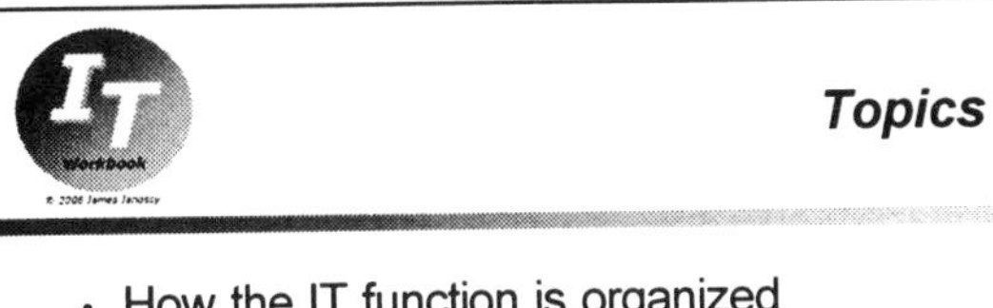

Topics

- How the IT function is organized
- Software maintenance
- Differentiation and the value chain
- Business Process Re-engineering (BRP)
- Business Continuity Planning (BCP)
- Knowledge Management (KM)

Slide 10- 3

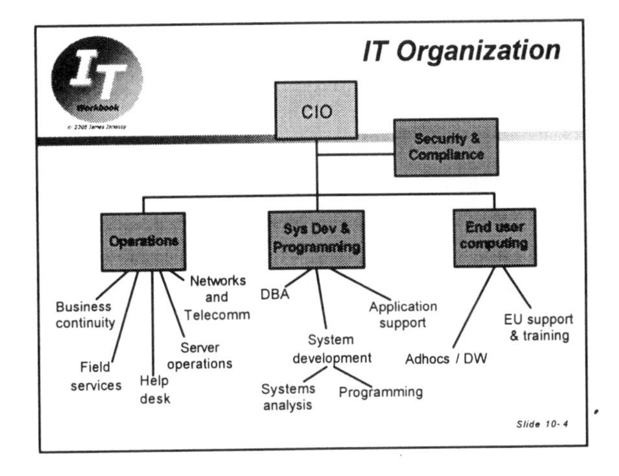

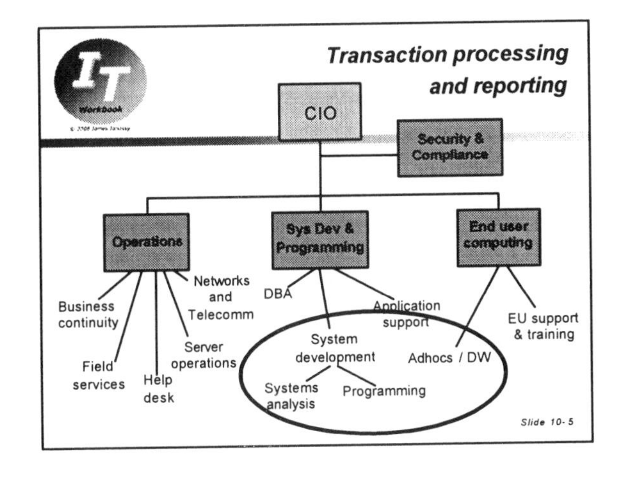

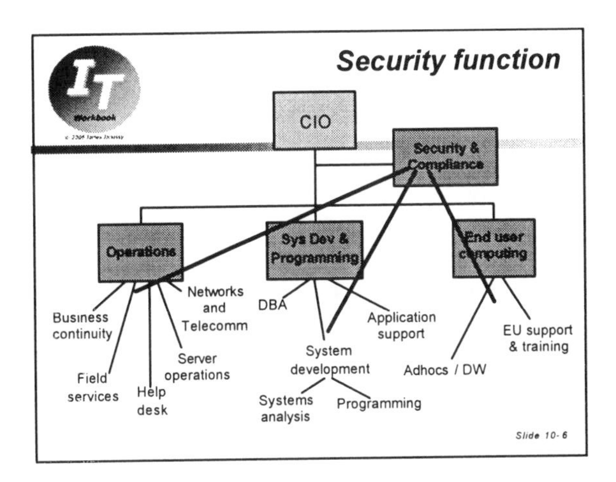

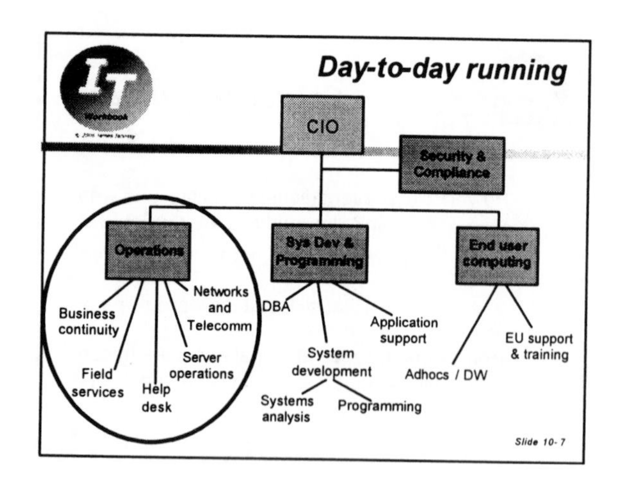

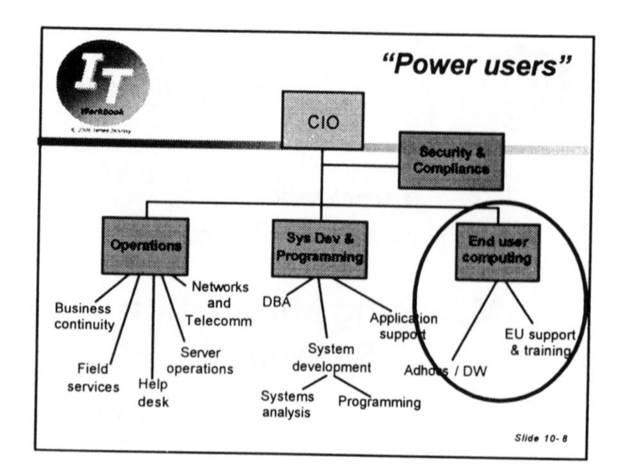

Budgeting: how is IT funded?

- Organizations work with annual budgets (expenditure allotments)
- Top and middle executives decide strategy, place funding to implement it
- Does IT "charge back" internal customers or not?

Slide 10-9

Software maintenance

- Does software "wear out?"
- Four types:
  - Corrective
  - Adaptive
  - Perfective
  - Preventive
- **How big an IT expense?**

Slide 10-10

Software maintenance

- Does software "wear out?"
- Four types:
  - **Corrective**
  - Adaptive
  - Perfective
  - Preventive
- **How big an IT expense?**

Fix errors detected after the software has been placed into production; perform "patching"

Slide 10-11

Software maintenance

- Does software "wear out?"
- Four types:
  - Corrective
  - **Adaptive**
  - Perfective
  - Preventive
- **How big an IT expense?**

Modify the software to meet changing business requirements

Slide 10-12

Software maintenance

- Does software "wear out?"
- Four types:
  - Corrective
  - Adaptive
  - **Perfective**
  - Preventive
- **How big an IT expense?**

Enhance the software ("perfect it") to give it greater functionality and utility; some patching qualifies here

Slide 10-13

Software maintenance

- Does software "wear out?"
- Four types:
  - Corrective
  - Adaptive
  - Perfective
  - **Preventive**
- **How big an IT expense?**

Change the software to keep it running as the machine/software environment changes; upgrade it; patching

Slide 10-14

Software license maintenance

- Different from code-change software maintenance
- An every-year cash flow to the vendor of a software package like an ERP
- Nominally supposed to pay for "support"
- May be as high as 20% of full list license cost for a software package!

Slide 10-15

Differentiation of a business from its competition

- Value chain helps understand business process and cost of each step
- Cost of products to customers is one differentiator
- Service level is another differentiator
- Uniqueness of product and branding are other valuable differentiators

Slide 10-16

Does I.T. Matter Anymore?

"By now, the core functions of IT have become available and affordable to all."

"Worrying about what might go wrong may not be as glamorous a job as speculating about the future, but it is a more essential job right now."

—Nicholas G. Carr

Read Nicholas Carr's article at:

http://workingknowledge.hbs.edu/archive/3520.html

Slide 10-17

Does having IT differentiate businesses anymore?

- Service level is a differentiator
- Carr argues no. Why?
- Because every business is now expected to have competent IT.
- Carr is addressing executive management. What does he recommend?

Slide 10-18

Differentiation of a business from its competition

- Porter's three generic strategies, along two dimensions
- Market scope strategy (focus)
- Uniqueness strategy (differentiation)
- Cost strategy (cost)

Slide 10-19

BPR: Business Process Re-engineering

Re-engineering is the fundamental rethinking and radical redesign of business processes…

to achieve dramatic improvements…

in critical, contemporary measures of performance, such as…

cost, quality, and speed.

Slide 10-20

BPR: Business Process Re-engineering

"Don't automate, obliterate."

"If it ain't broke, break it."

Michael Hammer, 1990

Slide 10-21

BPR: Business Process Re-engineering

- IBM case study, processing of proposals for equipment financing
- Needed to reduce time to prepare for customer
- Speed up existing processes with higher speed communication, or change the process?

Slide 10-22

BCP: Business Continuity Planning

- Planning actions that will be taken should physical plant be damaged or destroyed
- Backing up data offsite (daily)
- Arrangements for alternative facilities in advance
- Periodically trying out (exercising) recovery actions

Slide 10-23

KM: Knowledge Management

- Recording, organizing, and sharing the information held in the heads of employees
- Involves databases of problems encountered, solutions, other business knowledge; **"leverage" existing expertise**
- Helps maintain business efficiency in the face of turnover

Slide 10-24

IT Workbook Assignment 10.1

Student name:

Write your answers to the questions below within the box. In each case, please choose your words carefully to answer the specific questions. Avoid simply copying passages from your readings to these answers.

Identify and describe the roles of the **major organizational units** that a chief information officer (CIO) in a large business organization typically oversees:

1)

2)

3)

4)

Describe the types of activities and functions that **end user programming** in a large business organization involves, and what type of personnel engage in end user programming:

5)

IT Workbook Assignment 10.2

Student name:

Write your answers to the questions below within the box. In each case, please choose your words carefully to answer the specific questions. Avoid simply copying passages from your readings to these answers.

Software maintenance and **software license maintenance** sound very similar but they are very different activities. Both are conducted by the information technology area. Briefly define each:

1) software maintenance is:

2) software license maintenance is:

Four types of **software maintenance** exist. Identify these four types, and provide a brief description of each that distinguishes each from the other types:

3)

4)

5)

6)

IT Workbook Assignment 10.3

Student name:

Write your answers to the questions below *__within the box__*. *In each case, please choose your words carefully to answer the specific questions. Avoid simply copying passages from your readings to these answers.*

Given how frequently changes occur in the environment of personal computers, it's easy to get the idea that older languages and software have largely faded away. However, as Jussi Koskinen points out in Software Maintenance Costs (see www.cs.jyu.fi/~koskinen/smcosts.htm) this is not true. Answer these questions about this legacy software:

1) This many lines of source code were being maintained in 2000:

2) An average Fortune 100 company maintains this many lines of source code:

3) The amount of code maintained doubles in this number of years:

4) As of 2004, more that 70% of the code maintained as active business applications is written in this language:

5) The annual cost of software maintenance in the United States in 2003 was:

Software license maintenance charges for database management systems and ERP systems are often based on the list price of software, even if a given organization negotiated a discounted purchase price. Suppose the list price of a license to use a particular ERP is $800,000 but your firm negotiated a huge discount, of 80%, so you only paid $160,000 to obtain it. But suppose the annual maintenance charge for the license is 16%, and it is based on the license list price.

6) How much would your firm be expected to pay every year for "maintenance"?

7) What would your firm receive for paying the annual maintenance fee?

8) What would happen to your firm's ability to use the software if your firm did not pay the annual maintenance fee?

IT Workbook Assignment 10.4

Student name:

Write your answers to the questions below <u>within the box</u>. In each case, please choose your words carefully to answer the specific questions. Avoid simply copying passages from your readings to these answers.

Nicholas Carr of the Harvard Business School wrote the article ***Why I.T. Doesn't Matter Anymore*** in 2003. After reading it, answer these questions about it:

1) Carr's main argument is:

2) What percentage of its gross revenue does Carr claim the typical company spends on I.T?

Carr's three "new rules for I.T. management" are:

3)

4)

5)

Given Carr's arguments (see above), **what specific actions *(plural!)* would you recommend** to an organization that was operating profitably, enjoyed customer good will, and was meeting the growth targets of its business plan, and was considering installing an ERP system to replace its locally-developed manufacturing and sales software, which was currently meeting its needs?

IT Workbook Assignment 10.5

Student name:

Write your answers to the questions below within the box. In each case, please choose your words carefully to answer the specific questions. Avoid simply copying passages from your readings to these answers.

What are Michael Porter's **five primary value chain activities**, and for which of these are information systems a relevant technology?

1) activity #1:

2) activity #2:

3) activity #3:

4) activity #4:

5) activity #5:

6) information systems are a relevant technology for these activities:

Michael Porter has described **three general types of strategies** commonly used by businesses. Identify these strategies and for each, describe a specific way that I.T. might support that strategy:

7) strategy #1:

8) how I.T. might support strategy #1:

9) strategy #2:

10) how I.T. might support strategy #2:

11) strategy #3:

12) how I.T. might support strategy #3:

IT Workbook Assignment 10.6

Student name:

Write your answers to the questions below within the box. In each case, please choose your words carefully to answer the specific questions. Avoid simply copying passages from your readings to these answers.

What is the **definition of Business Processing Reengineering** (BPR) articulated by Michael Hammer and James Champy in 1990?

1)

As Kevin Lam notes, the definition of BPR by Hammer and Champy contains **four important keywords**. Identify these keywords, and discuss the meaning of the keyword in the context of this definition:

2)

3)

4)

5)

IT Workbook Assignment 10.7

Student name:

Write your answers to the questions below within the box. In each case, please choose your words carefully to answer the specific questions. Avoid simply copying passages from your readings to these answers.

In the example case of IBM Credit Corporation finance request processing cited by Kevin Lam in ***A Study of Business Process Reengineering***, what was the essence of the problem, what was the first attempt to resolve problems and how did it make the situation worse, and what was the final solution that reduced turnaround time from six days to four hours?

1) the essence of the problem was:

2) the first attempt to resolve problems was:

3) the first attempt to resolve problems made them worse by:

4) the final solution that reduced turnaround time from six days to four hours was:

What is the **most difficult part of the organization to** change in BPR?

IT Workbook Assignment 10.8

Student name:

Write your answers to the questions below within the box. In each case, please choose your words carefully to answer the specific questions. Avoid simply copying passages from your readings to these answers.

Knowledge Management (KM) is an activity broader than information technology, but supported by it. I.T. organizations also benefit from KM. Identify and briefly discuss the **three principal aims** of KM:

1)

2)

3)

Interest in Knowledge Management as a field is driven by **six factors**. Identify these factors and briefly describe each of them:

4)

5)

6)

7)

8)

9)

IT Workbook Assignment 10.9

Student name:

Write your answers to the questions below within the box. In each case, please choose your words carefully to answer the specific questions. Avoid simply copying passages from your readings to these answers.

Describe what role **Business Continuity Planning (BCP)** plays in the management of an information technology organization:

1)

Identify and describe ten things that the a **BCP manual** for a large organization will contain:

2)

3)

4)

5)

6)

7)

8)

9)

10)

11)

IT Workbook Assignment 10.10

Student name:

Write your answers to the questions below within the box. In each case, please choose your words carefully to answer the specific questions. Avoid simply copying passages from your readings to these answers.

When an organization or an individual relies on information technology and stores data in machine-readable form, making and preserving regular data backups is vital. Since some of the data preserved in this way may be especially sensitive, ethical and legal obligations exist to safeguard it when it leaves secured premises. Describe the way that M.E. Kabay suggests doing this in his *Security Strategies Newsletter* in the May 18, 2006 issue of ***Network World.***

1)

Some people claim that the measures suggested by M.E. Kabay (see above) are not necessary. Read the article by Jo Maitland in the June 17, 2005 issue of ***Storage Management News*** and **summarize the arguments** of these people:

2)

A **SAN** provides the means to make high volume backups without human intervention or the use of magnetic tapes. Explain what SAN stands for, what a SAN is, describe how it can automate the process of making off-site backups, and describe an important added advantage it can provide:

3) "SAN" stands for:

4) a SAN is:

5) a SAN can automate the production of off-site backups by:

6) an important added advantage of a SAN-based backup is:

Sample final exam

The following pages include a worksheet for you to take notes concerning the logistics of your class final exam, and a complete, full-length sample final exam.

You are encouraged to use the worksheet to record the date, starting time, and what you can and cannot bring and use in your final exam. You should also record what assignments are due no later than the final exam so you can make sure to not lose track of that as you focus on reviewing for the exam!

The sample final exam is typical of those in use at DePaul University where this workbook is used in classes in the undergraduate computer science and liberal arts curriculum. This exam may or may not reflect the type of final that your instructor chooses to use. However, it is helpful for you to see the content of this sample final exam because it provides a good tool for focusing your attention on important concepts and facts documented in these learning materials. Use it as a study guide in combination with the learning objectives at the start of each chapter.

This final is conducted in a two-hour time period as a closed-book exam. If you have mastered the learning objectives of each chapter and are comfortable with the questions in this sample final exam, you should be well prepared for any form of test covering the same subject matter.

Check the workbook web site at **www.ambriana.com** for updates and additional materials that may be made available to aid you in your review for your course final exam.

This page is provided for your use in taking notes about your class final exam

Final exam date and start time:	
Final exam location:	
Materials you can bring to the exam:	
Items you cannot use in the exam:	
Other items due to be turned in at the final exam:	

Additional notes:

Info Tech Workbook Sample Final Exam

Student name:

This exam is closed book, closed notes. Write your answer under each question within the box provided. Your answer must fit in the box. Each numbered question counts for 5 points.

1. Briefly describe a) what **transaction processing** is, b) what **batch processing** is, and c) how they are different:

a) what transaction processing is:

b) what batch processing is:

c) transaction processing and batch processing are different in these ways:

2. Identify and describe four categories of digital computers based on their size, computing power and the general number of concurrent (simultaneous) users that each can support:

Category of computer	Size/power	Number of concurrent users

3. Describe the significant difference(s) between **application software** and **system software**:

4. Provide a concise description of a modern **database management system** (DBMS) including coverage of what it actually does:

5. In each of the following network topologies is each computer an **active participant** in the transmission of network traffic or is it **passive**? Consider each of the network topologies and briefly state in each case the basis for your answer:

a) bus:

b) star:

c) ring:

d) mesh:

6. Describe the difference between the way that the **traditional telephone system** routes and supports communication between end points and the way that the **Internet** does this:

7. From the point of view of **consumers**, identify and briefly describe **three important advantages** that the digital markets created by the Internet provide:

a) Advantage #1:

b) Advantage #2:

c) Advantage #3:

8. Identify and briefly describe **three intended advantages** of ERP systems:

a) ERP advantage #1:

b) ERP advantage #2:

c) ERP advantage #3:

9. Describe a) how the **waterfall model of system development** orchestrates the processes involved in developing new software and b) the **major specific shortcoming(s)** of it:

a) the waterfall model orchestrates system development in this way:

b) the major shortcoming of the waterfall model are:

10. Describe (a) what **office automation** is and (b) name **three common tasks** that are handled by office automation programs:

a) what office automation is:

b) three common tasks handled by office automation systems are:

11. The **stored program concept** for computing machinery was described by John von Neumann in 1945. Describe a) what this concept states and b) explain what it contributed to making computers flexible and powerful.

a) the stored program concept states that:

b) the stored program concept contributes this to making computer flexible and powerful:

12. Identify and describe each of the **first four generations of computer programming languages** and explain how each is different from the generation that preceded it:

a) first:

b) second:

c) third:

d) fourth:

13. Describe a) the concept of **data stewardship** and b) the **guiding rule** for the ethical use of data accumulated by an organization:

a) data stewardship is:

b) the guiding principle for the ethical use of data:

14. Identify a) what purpose **Bluetooth** and **Wi-Fi** serve, and explain how they are b) similar and how they are c) different:

a) purpose:

b) how Bluetooth and Wi-Fi are similar:

c) how Bluetooth and Wi-Fi are different:

15. Explain a) what a **packet** is and what's in it, and b) its approximate **size**:

a) what a packet is and contains:

b) the approximate size of a packet:

16. Explain what **disintermediation** is in terms of the Internet and digital markets:

17. Describe what needs to happen in an organization for performance improvement to occur when **ERP** is brought in:

18. Discuss what W. W. Royce actually had in mind when he wrote a paper in 1970 that included a description of the **waterfall model of software development**:

19. Explain a) what **software license maintenance** is, what b) **software maintenance** is, identifying c) the different types and kinds of support each type of software maintenance provides:

a) what software license maintenance is:

b) what software maintenance is:

c) the different types of software maintenance and the kind of support each provides:

20. In connection with **JAD** and **RAD**, a) state what the acronym JAD means, b) explain concisely how JAD is done, c) state what the acronym RAD means, d) explain concisely how RAD is done, and e) describe how each confronts the major problem(s) of the waterfall model:

a) what JAD is:

b) how JAD is done:

c) what RAD means:

d) how RAD is done:

e) how JAD and RAD confront the major problem(s) of the waterfall model:

